BONE TISSUE ENGINEERING

BONE TISSUE ENGINEERING

Edited by
Jeffrey O. Hollinger, Thomas A. Einhorn,
Bruce A. Doll, and Charles Sfeir

Library of Congress Cataloging-in-Publication Data

Bone tissue engineering / edited by Jeffrey O. Hollinger... [et al.].
p. cm.
Includes bibliographical references and index.
ISBN 0-8493-1621-9 (alk. paper)
1. Bone regeneration. 2. Tissue engineering. 3. Bones--Transplantation. 4. Bones--Cytology. I. Hollinger, Jeffrey O.
[DNLM: 1. Bone and Bones--cytology. 2. Models, Animal. 3. Research Design. 4. Tissue engineering--methods. WE 200 B7134 2004].

RD123.B648 2004
617.4'710592--dc22 2004050333

This book contains information obtained from authentic and highly regarded sources. Reprinted material is quoted with permission, and sources are indicated. A wide variety of references are listed. Reasonable efforts have been made to publish reliable data and information, but the author and the publisher cannot assume responsibility for the validity of all materials or for the consequences of their use.

Neither this book nor any part may be reproduced or transmitted in any form or by any means, electronic or mechanical, including photocopying, microfilming, and recording, or by any information storage or retrieval system, without prior permission in writing from the publisher.

All rights reserved. Authorization to photocopy items for internal or personal use, or the personal or internal use of specific clients, may be granted by CRC Press, provided that $1.50 per page photocopied is paid directly to Copyright Clearance Center, 222 Rosewood Drive, Danvers, MA 01923 USA. The fee code for users of the Transactional Reporting Service is ISBN 0-8493-1621-9/05/$0.00+$1.50. The fee is subject to change without notice. For organizations that have been granted a photocopy license by the CCC, a separate system of payment has been arranged.

The consent of CRC Press does not extend to copying for general distribution, for promotion, for creating new works, or for resale. Specific permission must be obtained in writing from CRC Press for such copying.

Direct all inquiries to CRC Press, 2000 N.W. Corporate Blvd., Boca Raton, Florida 33431.

Trademark Notice: Product or corporate names may be trademarks or registered trademarks, and are used only for identification and explanation, without intent to infringe.

Visit the CRC Press Web site at www.crcpress.com

© 2005 by CRC Press

No claim to original U.S. Government works
International Standard Book Number 0-8493-1621-9
Library of Congress Card Number 2004050333
Printed in the United States of America 1 2 3 4 5 6 7 8 9 0
Printed on acid-free paper

Preface

This book has been designed to emphasize the fundamentals of bone tissue engineering. The editors have identified key topics, thought-leaders, researchers, and clinicians who understand basic elements of bone and the translation of that knowledge to exciting applications for patients.

The book is organized into four sections. The first section edited by Bruce Doll, DDS, PhD contains three chapters with up-to-date information on the development of the skeletal system, cell lineage and progression, extracellular molecules, and the physiology of bone dynamics.

The second section includes four chapters and is edited by Charles Sfeir, DDS, PhD. In two chapters, there is an emphasis on the basic elements of bone, including signaling molecules and pathways, and the regulatory interactions among molecules and pathways, as well as an in-depth, contemporary explanation of the organic and inorganic substrata. The third and fourth chapters in this section build on previous chapters and include scaffold design and development, with two distinguished groups offering different innovative approaches for bone tissue engineering scaffolds.

Section three, coedited by Jeffrey O. Hollinger, DDS, PhD and Thomas Einhorn, MD is the logical sequence from the first two sections. Section three has three chapters that focus on fundamental statistics, animal models, and key outcome techniques in biomechanics and tissue morphology quantitation that will guide the maturation of experimental bone tissue-engineered designs to the clinic. This section will answer important questions: how do I determine how many experimental animals will be necessary to detect statistical significance? What is the best experimental preclinical model for a particular clinical indication?

Thomas Einhorn, MD and Jeffrey O. Hollinger, DDS, PhD are the coeditors of section four, with three chapters discussing opportunities for tissue engineering and bone, as well as thematic opportunities for tissue engineering in the spine, craniofacial, and dental areas.

I want to thank my coeditors and contributors for their highly informative chapters. The comprehensive nature of the chapters with extensive bibliographies will make this book an invaluable resource for students, clinicians, and scientists interested in bone tissue engineering.

Editors

Jeffrey O. Hollinger, DDS, PhD is the Director, Bone Tissue Engineering Center (BTEC), Carnegie Mellon University, and is a tenured Professor of Biomedical Engineering and Biological Sciences.

Dr. Hollinger is a Professor in Orthopedics and Plastic Surgery at the University of Pittsburgh Medical School and has an adjunct appointment at the McGowen Institute for Regenerative Medicine, in addition to being an Associate Director for the Pittsburgh Tissue Engineering Initiative. He serves on the scientific board of several companies.

Dr. Hollinger had been at the Oregon Health Sciences University from 1993 to 2000 as a tenured professor of Surgery, Anatomy, and Developmental Biology at the School of Medicine. He was also the president and director for research for the Northwest Wound Healing Center.

Prior to 1993, Dr. Hollinger was in the United States Army at the Walter Reed Army Institute of Research in Washington, DC, where he was chairman of Physiology and Director for the bone program. He retired as a Colonel after 20 years of active military duty.

Dr. Hollinger's training in dentistry at the University of Maryland School of Dentistry was followed by a Residency Program and postgraduate work in physiology at the University of Maryland resulting in a PhD.

Dr. Hollinger's research focuses on bone tissue engineering and includes polymers, gene therapy, cells, signaling molecules, and surgical models to test bone regenerative therapies. Dr. Hollinger has published over 150 peer-reviewed articles, abstracts, chapters in texts, and texts, including an in-press textbook on Bone Tissue Engineering. He is an editor and reviewer for 15 clinical and scientific publications.

Thomas A. Einhorn, M.D. is Chairman of the Department of Orthopaedic Surgery and Professor of Orthopaedic Surgery and Biochemistry at Boston University School of Medicine. A graduate of Rutgers University and Cornell Medical College, he completed his internship at the Hospital of the University of Pennsylvania, orthopedic residency at St. Luke's-Roosevelt Hospital in New York City, and a fellowship at the Hospital for Special Surgery. His professional interests include research on the repair and regeneration of bone and cartilage, reconstructive surgery of the hip and knee, and the treatment of metabolic bone disease. He has served as Chairman of the Orthopaedics and Musculoskeletal Study Section of the National Institutes of Health, President of the Orthopaedic Research Society, President of the International Society for Fracture Repair, and Chairman of both the Committee on Examinations and the Council on Research and Scientific Affairs of the American Academy of Orthopaedic Surgeons. Currently, he serves on the Board of Trustees of the Orthopaedic Research and Education Foundation and the National Osteoporosis Foundation. His awards include the *American British Canadian Traveling*

Fellowship, Marshall R. Urist Award, and *Kappa Delta Award.* He is Deputy Editor for Current Concepts Reviews for *The Journal of Bone and Joint Surgery*, and serves on the Editorial Boards of *The Journal of Bone and Mineral Research and Bone.* Since 1997, he has been listed by Woodward/White as one of *The Best Doctors in America.* An author of over 100 scientific articles, his future goals are dedicated to exploring the role of molecular medicine in orthopedic surgery.

Bruce A. Doll, DDS, PhD is Chairman of the Department of Periodontics, School of Dental Medicine at the University of Pittsburgh, Pittsburgh, Pennsylvania. In addition, he is currently an Associate Research Professor at Carnegie Mellon University's Bone Tissue Engineering Center. Prior to his appointment at the University of Pittsburgh's School of Dental Medicine, Doll was an Assistant Professor at Oregon Health Sciences University School of Dentistry, where he was director of the graduate program in periodontology.

Dr. Doll received his DDS from the State University of New York at Buffalo. He subsequently specialized in periodontics and trained at the Naval Dental School, Bethesda, MD. He received his PhD from the Pennsylvania State University for studies in site-directed mutagenesis and enzyme kinetics.

Charles Sfeir, DDS, PhD is Assistant Professor in Oral Medicine and Pathology at the School of Dental Medicine at the University of Pittsburgh, Pennsylvania. In addition, he is a faculty member of the McGowan Institute of Regenerative Medicine and also holds adjunct positions in the Biomedical Engineering and Materials Science and Engineering Departments at Carnegie Mellon University. Prior to his appointment at the University of Pittsburgh's School of Dental Medicine, Dr. Sfeir was an Assistant Professor at Oregon Health Sciences University School of Dentistry. His professional interests include research on the regeneration of bone, the role of extracellular matrix protein in mineralized tissue regeneration, and development of a non-viral gene delivery system based on calcium phosphates.

Dr Sfeir received his DDS, periodontal residency, and PhD from Northwestern University in Chicago. His PhD research studies focused on the role of posttranslational modification of noncollagenous proteins of dentin/bone.

Contributors

Scott D. Boden, M.D., Ph.D.
Emory Spine Center
Department of Orthopaedic Surgery
Emory University School of Medicine
Atlanta, Georgia

Adele L. Boskey, Ph.D.
Hospital for Special Surgery
New York, New York

James P. Bradley, M.D.
Department of Plastic and Craniofacial Surgery
University of California at Los Angeles
Los Angeles, California

Phillip Campbell, Ph.D.
Institute for Complex Engineering Systems
Carnegie Mellon University
Pittsburgh, Pennsylvania

Tien Min Chu, Ph.D.
Department of Biomedical Engineering
Indiana University Purdue University
Indianapolis, Indiana

Dennis M. Cullinane
Orthopaedic Research Laboratory
Department of Orthopaedic Surgery
Boston University Medical Center
Boston, Massachusetts

Bruce Doll, D.D.S., Ph.D.
School of Dental Medicine
University of Pittsburgh
Pittsburgh, Pennsylvania

Henry J. Donahue, Ph.D.
Musculoskeletal Research Laboratory, Department of Orthopaedics and Rehabilitation
Department of Bioengineering
Department of Surgery and Materials
Department of Science and Engineering
Center for Biomedical Devices
Functional Tissue Engineer and Materials Research Institute
The Pennsylvania State University
University Park, Pennsylvania

Thomas A. Einhorn, M.D.
Department of Orthopaedic Surgery
Boston University Medical Center
Boston, Massachusetts

Carol V. Gay, Ph.D.
Dept. of Biochemistry and Molecular Biology
The Pennsylvania State University
University Park, Pennsylvania

Jeffrey O. Hollinger, D.D.S., Ph.D.
Bone Tissue Engineering Center
Carnegie Mellon University
Pittsburgh, Pennsylvania

Scott J. Hollister, Ph.D.
Skeletal Engineering Group
Department of Biomedical Engineering
Department of Surgery
Department of Mechanical Engineering
University of Michigan
Ann Arbor, Michigan

Julie Jadlowiec, B.S.
Bone Tissue Engineering Center
Carnegie Mellon University
Pittsburgh, Pennsylvania

Sanjeev Kakar, M.D., M.R.C.S.
Department of Orthopaedic Surgery
Boston University Medical Center
Boston, Massachusetts

Hannjörg Koch, M.D.
Department of Orthopaedic Surgery
University of Greifswald
Greifswald, Germany

Cheng-Yu Lin
Skeletal Engineering Group
Department of Mechanical Engineering
University of Michigan
Ann Arbor, Michigan

Kacey G. Marra, Ph.D.
Department of Surgery
Department of Bioengineering
University of Pittsburgh
Pittsburgh, Pennsylvania

Mark P. Mooney, Ph.D
Department of Oral Medicine and Pathology
Department of Anthropology
Department of Surgery – Division of Plastic and Reconstructive Surgery
Department of Orthodontics Cleft Palate-Craniofacial Center
University of Pittsburgh
Pittsburgh, Pennsylvania

John M. Rhee, M.D.
Emory Spine Center
Department of Orthopaedic Surgery
Emory University School of Medicine
Atlanta, Georgia

Robert T. Rubin, M.D., Ph.D.
Center for Neurosciences Research
Drexel University College of Medicine
Allegheny General Hospital Campus
Pittsburgh, Pennsylvania

Kristy T. Salisbury
Orthopaedic Research Laboratory
Department of Orthopaedic Surgery
Boston University Medical Center
Boston, Massachusetts

Rachel D. Schek
Skeletal Engineering Group
Department of Bioengineering
Department of Oral Medicine, Pathology, and Oncology
School of Dentistry
University of Michigan
Ann Arbor, Michigan

Charles Sfeir, D.D.S., Ph.D.
Department of Oral Medicine and Pathology
University of Pittsburgh
Pittsburgh, Pennsylvania

Christopher A. Siedlecki, Ph.D.
Musculoskeletal Research Laboratory
Department of Orthopaedics and Rehabilitation
Department of Bioengineering
Department of Surgery and Materials
Department of Science and Engineering
Center for Biomedical Devices
Functional Tissue Engineer and Materials Research Institute
The Pennsylvania State University
University Park, Pennsylvania

Michael I. Siegel, Ph.D.
Department of Arthropology
Department of Orthodontics
Cleft Palate-Craniofacial Center
University of Pittsburgh
Pittsburgh, Pennsylvania

Juan M. Taboas, Ph.D.
Skeletal Engineering Group
Department of Bioengineering
Department of Oral Medicine,
Pathology, and Oncology
School of Dentistry
University of Michigan
Ann Arbor, Michigan

Edwin Vogler, Ph.D.
Musculoskeletal Research Laboratory
Department of Orthopaedics and
Rehabilitation
Department of Bioengineering
Department of Surgery and Materials
Department of Science and
Engineering
Center for Biomedical Devices
Functional Tissue Engineer and
Materials Research Institute
The Pennsylvania State University
University Park, Pennsylvania

Table of Contents

SECTION I

Basic Bone Biology and Tissue Engineering

1 Developmental Biology of the Skeletal System

Bruce Doll

Contents

The principal calcified tissue of vertebrates is bone. Other calcified tissues in vertebrates include calcified cartilage, which is present to some extent in most bones and the dental tissues — enamel, cementum, and dentin.

Bone develops by the process of ossification, osteogenesis, as a specialized connective tissue. During ossification, osteoblasts secrete an amorphous material, gradually becoming densely fibrous — osteoid. Calcium phosphate crystals are deposited in the osteoid (i.e., mineralization), thereby becoming bone matrix. Osteoblasts become surrounded during the mineralization process, and the cells become osteocytes. Osteoblast secretion does not become entirely fibrous. The secretion also forms an amorphous adhesion between the fibers.[1]

Chemical and mechanical influences on the nature of ossification and of the absorption of bone comprise the process of osteogenesis. The general mechanisms through which characteristic external form and internal structure are determined, organized, and maintained as integrated elements within the skeletal framework complete the continuous process. Prenatal development of bones, the environment in which ossification begins, the relationship of ossification to general growth patterns, shape and structure of bones during prenatal growth, and the sequence of ossification are the focus of this chapter. Certain bones have been chosen to illustrate these points. Previous aspects of osteogenesis appeared mainly descriptive, however; new

0-8493-1621-9/05/$0.00+$1.50
© 2005 by CRC Press LLC

data amplify a growing appreciation for the molecular mechanisms underlying the sequence of specific skeletal growth patterns.[2–5]

OSTEOGENESIS IN THE FETUS

INTRAMEMBRANOUS OSSIFICATION

The mandible, parts of the clavicle and most bones of the craniofacial complex are formed by intramembranous ossification.[6] The process is preceded by fibrocellular proliferation, which forms during the embryonic period. The flat bones of the skull are defined by cells that differentiated directly into bone-forming osteoblasts.[7] Cell culture with postnatal marrow osteogenic precursors suggested that there are two potential paths of differentiation — bone or cartilage, dependent upon the micro-environment.[8] These cells secrete a matrix of type I collagen and other molecules.[9] The primary center of each bone is first indicated by an increase in cells and fibers. Differentiating cells, osteoblasts, synthesize phosphatase, and concurrent formation of a calcified organic matrix cements the fibers together. The matrix and ground substance contains a complex mix of mucopolysaccharides, in varying degrees of polymerization.[10–14]

Osteoblasts surround the initial bony trabecula. Some of the osteoblasts become enclosed in the matrix being formed around them, thereby becoming osteocytes. Osteoblasts also divide and continue to add bone to the trabecula. Similar changes occur proximal to these events.[7] New trabeculae are formed and the center of ossification expands, connecting trabecula in a radiating pattern. Trabeculae increase in thickness as new bone is added to their surfaces. Osteoid refers to the bone matrix that has not been calcified. (Formation and calcification of bone matrix are simultaneous in rats.[15])

Bony trabecula growth and orientation are unique for each bone. Parietal and frontal bones exhibit primary trabeculae radiating as a network parallel to the surface of the skull. Some are directed at right angles and thus add to the thickness of the center. The primary trabeculae become interconnected by secondary trabeculae. Enclosed spaces contain vascular connective tissue, the forerunner of hematopoietic tissue. The spaces become smaller as trabeculae increase in thickness, thereby increasing the density of bone in the initial sites of ossification. The radiating pattern becomes less apparent. The combination of the denser central plate, with a more peripheral, trabeculated region of advancing ossification, together with free islands and nodules at the peripheral margins, constitute the early morphological character or central plate of bone. Throughout the period of ossification, with the expansion of the central plate, there is a gradual reduction in the amount of open reticulum and marginal zones.[16] When the bones have grown to occupy their definitive regions and come into closer relationships with other bones, a bony border begins to form around peripheral edges as the trabeculae become interconnected. Fibroblast growth factor (FGF) ligands are expressed in the coronal sutures separating the parietal and frontal bones. The expression pattern for FGF-18 and FGF-20 contrast with FGF-9 in the calvarium and FGF-20 and-9 in the limb buds. Temporal and spatially distinct expression indicates specialized nonredundant roles in proliferation and differentiation.[17,18] Subsequent increments in growth decrease sharply, and reorganization becomes prominent.

Membrane bone formation comprises certain activities common to all bones and some unique to the craniofacial region. These have to do with the factors controlling early formation of differentiation, the relation of skull bones to the meninges and the brain, the mechanisms of absorption and reconstruction in a changing radii of curvature, the significance of sutural positions and the problem of sutural growth, the formation of diploe, and the relationship between bones. The pressures of growing and expanding intracranial contents subject the cranial bones to mechanical influences during the prenatal period. These influences may be important in trabecular orientation. Once they have reached their definitive size and relation to other bones, the cranial bones accommodate the growth of the brain primarily by a change in curvature.[17] Deposition of bone on the external surface progressively increases toward the periphery, with absorption of bone from the inner surface, although to a lesser degree since bone increases in thickness. Sutural growth is not emphasized in the prenatal period. Absorption of bone is a minimal influence within the diploic spaces rather than on surfaces or at sutures because the formation of diploe is a postnatal event.[19,20]

Bone formation by intramembranous ossification is primarily fibrous. Collagen fibers in each trabecula are gathered in course bundles, which are arranged in an irregular plexuslike manner with lacunae of variable shape. Reconstruction and formation of lamellar systems, and the resorptive mechanism forming diploe, occur after birth. The connective tissue adjacent to each surface of the bone primarily forms a fibrous periosteum. The inner surface becomes a dura mater. The cranial bones consist of a single plate, featuring a mixture of fine and coarse-fibered bone, compact bone, containing spaces filled with loose tissue and vessels.[19]

Secondary Cartilage

Secondary cartilage is associated with certain membrane bones, especially the clavicle and the mandible.[21] It forms after ossification begins and is characterized by larger cells and less intercellular matrix than in the case of typical hyaline cartilage. Although it is not part of the cartilaginous primordium, the secondary cartilage participates in growth processes, similar to cartilage that precedes ossification in other bones.[22–24]

Endochondral Ossification

Bone formation begins through the adhesion of cells into clusters or condensations. Early in the embryonic period, cells with oval or round nuclei are packed together in mesenchymal condensations; their cytoplasms appear to have a syncytcial arrangement. SOX 9, a transcriptional control factor for chondrocyte differentiation, is expressed in mesenchymal condensations before and during chondrogenesis. A runt-related transcription factor 2 isoform, Runx2-I (also known as Cbfa1, Osf2, and AML3), has an exclusive role in the early commitment stage of intramembranous or endochondral bone-forming processes.[25,26] Colony bone factor beta (Cbfβ, or polymavirus enhancer binding protein 2β (PEBP2B)) is expressed in developing bone and forms a functional interaction with Runx2.[27,28] Their concerted activity contributes to the differentiation of cells into chondroblasts and osteoblasts. An intercellular matrix is formed containing compounds characteristic of cartilage matrix. The intercellular matrix appears first at the center of the mesenchymal condensation.[6] The cells become separated by a matrix (Table 1.1).

This process represents the beginning of chondrification. The proliferation of chondrocytes contributes to the growth of the bone precursor. Hematopoietic stem cells interact with the stroma to establish the main site for hematopoiesis after birth. Chondrocytes secrete a matrix rich in type II collagen and proteoglycans. The matrix contains aggrecan and other proteoglycans (Table 1.1). SOX 9 and other transcription factors constitute a genetically directed cascade culminating in cartilage precursors of each bone.[29] The deposition of the matrix spreads peripherally to the margin of the original condensation.[30] The mesenchymal cells orient to become a perichondrium and contribute to subsequent growth. The cartilage enlarges through chondrocyte proliferation and matrix production. Encapsulated in the matrix, the chondrocytes undergo hypertrophy and subsequently secrete collagen. Centrally located chondrocytes mold the shape, hypertrophy, and synthesize type X collagen. Each skeletal part becomes outlined and resembles the adult in form and arrangement before the embryonic period is over. A fraction of the chondrocytes flattens, forming a columnar orientation that directs the bone lengthening. Cell-to-cell contact appears important to chondrocyte differentiation.[31] Transcription factor, c-Maf, is expressed in hypertrophic chondrocytes during fetal development and is required for normal chondrocyte differentiation during endochondral bone development. The rate of hypertrophic chondrocyte proliferation contributes to bone lengthening.[32]

Growth hormone accelerates the rate of soluble collagen synthesis and the proliferation of cartilage in the epiphyseal plates. Growth hormone is a single polypeptide, 44 kDa (188 amino acids) chemical entity of high activity, originating from the pituitary. Hydrolysis of growth hormone yields smaller molecules displaying the same biological activity as the complete molecule. The stimulation of epiphyseal cartilage under the influence of growth hormone is accompanied by similar responses in the pancreas, intestinal mucosa, adrenal cortex, liver, and adipose tissue.

Table 1.1
Cartilage Matrix Proteins

Proteins derived from procollagen: Chondrocalcin (C-propeptide of type II procollagen) role in cartilage calcification[93–95]

Proteins with γ-carboxyglutamic acid: Matrix gamma-carboxyglutamic acid (Gla) protein (MGP): modulates BMP activity; developmentally regulated inhibitor of cartilage mineralization by controlling amount of mineral formed[63, 96–98]

Proteins with glycosaminolgycan side chains: HAPG1 (biglycan), HAPG2 (decorin) Human cartilage glycoprotein 39 (HC gp-39), aggrecan, lumican, fibromodulin, biglycan (S1) interact with collagen, and TGF beta 1 appears to modulate the synthesis and accumulation of decorin, biglycan, and fibromodulin[93, 99–104]

Phosphorylated glycoproteins: Bone sialoprotein (BSP) is a possible nucleator of hydroxyapatite crystals, working in concert with osteoadherin; Thrombospondins: Cartilage oligomeric matrix protein (COMP) interacts with fibronectin, expression is sensitive to mechanical compression; Thrombospondin 1–4 (TSP 1,2,3 & 4): unique roles implicated in cell migration and development. [105–112]

Growth plate chondrocytes integrate multiple signals during normal development. Along with growth hormone, FGFs, insulin-like growth factor I (IGF-I), parathyroid hormone-related protein (PTHrP), Indian Hedgehog Protein (Ihh), and bone morphogenetic proteins (BMPs) regulate bone growth. FGFs initiate multiple pathways that result in the inhibition of chondrocyte proliferation and the down-regulation of growth-promoting molecules. The activation of FGFR3 inhibits both chondrocyte proliferation and differentiation.[33–36] Serum response factor binds to the FGFR3 gene and is a determinant of chondrocyte gene expression.[37] FGFR3 signaling inhibits Ihh signaling and BMP4 expression in the cartilage and perichondrium. FGFR3 signaling is genetically upstream of Ihh, BMP, and PTHrP signaling pathways, suggesting that FGFR3 signaling through its endogenous ligand may be a global coordinator of bone growth.[38] BMP signaling modulates the Ihh/PTHrP signaling pathway that regulates the rate of chondrocyte differentiation.

Thyroid hormone (T3) enhances the expression of aggrecanase-2/ADAM-TS5 (a disintegrin and a metalloproteinase domain with thrombospondin type I domains) mRNA in mice. Aggrecanase-2/ADAM-TS5 is involved in aggrecan breakdown during endochondral ossification.[39]

During endochondral bone formation, an avascular tissue (cartilage) is converted into one of the most highly vascularized tissues (bone) in the vertebrate body. Ossification begins with invasion of the calcified hypertrophic cartilage by capillaries. Apoptosis of the terminal hypertrophic chondrocytes, degradation of the cartilage matrix, and deposition of bone matrix by osteoblasts accompany neovascularization of the growth plate. Remodeling of the extracellular matrix (ECM) results in a cavity filled with vascular channels containing hematopoietic cells. MMP9, MMP13, and vascular endothelial growth factor are key regulators for the remodeling of the skeletal tissues. They coordinate not only matrix degradation but also the recruitment and differentiation of endothelial cells, osteoclasts, chondroclasts, and osteoprogenitors. Matrix metalloproteinases (MMPs) degrade most components of the extracellular matrix (ECM), as well as many non-ECM molecules. MMPs participate in the following: (1) degradation of ECM to allow cell migration; (2) alteration of the ECM microenvironment resulting in alteration in cellular behavior; (3) modulation of biologically active molecules by direct cleavage or release from ECM stores; (4) regulation of the activity of other proteases; and (5) cell attachment, proliferation, differentiation, and apoptosis.[40–42]

The appearance of a collar between the hypertrophic cartilage cells and the overlying perichondrium or periosteum is the initial step in primary ossification of the cartilaginous anlage of a future bone. Capillary vascular buds penetrate the hypertrophic cartilage. In the elaboration and removal of bone, the capillary cells provide a transudate upon which the progression of the replacement of the anlage is dependent. The time of onset of circulation to the primary and secondary ossification centers varies.

The bone marrow is the natural source of blood cells throughout life. However, the skeleton is the last site to be involved in hematopoiesis during embryonic fetal life. Hematopoiesis begins in the human embryo as early as 4 weeks after conception from elements of the yolk sac as hemangioblastic foci. The last organ in the embryo to produce cells is the bone marrow. Hematopoiesis begins there during the fifth month, coinciding with the onset of placental circulation. Remnant mesenchymal tissue following trabecula formation in bones of intramembranous origin develops into hematopoietic cells and into endothelial elements delimiting the earliest capillaries. In the bone of endochondral origin, the mesenchyme resorbs cartilage and penetrates into the spaces between newly deposited primary bone trabeculae and together with vessels develop into myeloid tissue. The development of the hematopoietic marrow is conditional on the rate of incoming blood supply to the surrounding bone — development being most vigorous when the blood supply to the bone is greatest. Stimulants to hematopoiesis are conveyed to the marrow through the bone venules that empty into the marrow sinusoids. Early in postnatal life, all the skeletal segments are full of hematopoietic red marrow, whereas in adults most of the shafts of the long bones are filled by yellow marrow, some of which can be replaced by red marrow following increased demand.

All the vessels in relation to the blood supply of bone are either afferent or efferent, depending upon whether they enter the bone and provide nourishment or exit it with marrow components. Within the bone, the two systems possess distributing and collecting branches that come together in capillaries. The three primary components of the afferent system of the long bone are the principal nutrient artery, the metaphyseal arteries, and the periosteal arterioles. The principal nutrient artery penetrates the diaphyseal cortex without branching and subsequently divides into ascending and descending medullary arteries. The metaphyseal arteries enter each metaphysis and branch to provide the blood supply of the constituent cancellous bone. They also anastomose with the ascending and descending medullary arteries. The periosteal arterioles enter diaphyseal cortex along heavy fascial attachments and aid in the supply of compact bone close to where they enter, anastomosing locally with arterioles derived from the medullary arteries.[43–45]

The efferent vascular system of a long bone is comprised of the large emissary veins and vena comitans of the principal nutrient artery, the cortical venous channels, and the periosteal capillaries. Emissary veins and nutrient vena comitans arise from the central venous sinus of the medulla and are concerned chiefly with the drainage from the hematopoietic elements of the marrow. Cortical venous channels are numerous and drain most of the compact bone, emptying fascicular venules into the venous system of the surrounding soft tissues. The normal flow of blood through the full thickness of the diaphyseal cortex is functionally centrifugal — from medulla to periosteum. Periosteal capillaries are distributed throughout the diaphyseal cortex. They are in continuity with the capillaries of the external layers of cortex and are terminations of the medulla-derived circulations that nourish these layers.[46]

At specific locations, secondary ossification centers, chondrocytes cease proliferation, hypertrophy, and attract vascular invasion along with osteoblasts. Chondrocytes continue to proliferate between regions of bone of primary and secondary ossification centers, defining an area, a growth plate. A distinct disc of cells between the bone of the secondary ossification center and the primary spongiosa characterizes the growth plate. Hypertrophic chondrocytes undergo an apoptotic process as their surrounding matrix is mineralized and replaced by trabecular bone. Several signaling mechanisms contribute to the generation of the growth plate. Bone morphogenetic protein-2 (BMP-2) regulates growth plate chondrogenesis during development and postnatal bone growth. NF-kappaB regulates BMP-2 gene expression in chondrocytes. Growth plate chondrocytes possess leptin receptors. Leptin enhances chondrocyte proliferation and subsequent cell differentiation.[47,48] Estrogen receptors alpha and beta and androgen receptor are expressed in the growth plate throughout pubertal development.[49] The cartilage is gradually replaced by bone, and the growth plate disappears during adolescence.

When cartilage ossifies, the first phosphatase appears in the perichondrium near the regions where ossification of the matrix will first appear. The cartilage cells show signs of enzymatic activity when they begin to hypertrophy; phosphatase appears at first in the nuclei and then spreads to the cytoplasm and the matrix. Once the matrix slows activity, the cytoplasmic reaction decreases. Calcification appears to occur only in the presence of extracellular phosphatase. Phosphatase has several roles: association with the production of phosphate ions that secure the precipitation of calcium as bone salt, association with the formation of the organic matrix of bone, and minimizing the formation of ester phosphate on bone crystals, thereby permitting continued growth of the crystals. Phosphatase activity is pronounced in periosteal osteoblasts, then, they retain this activity to a varying degree as an osteocyte. After injury to bone, large numbers of phosphatase - positive pre-osteoblasts migrate into the injured area associated with the production of calcified osteoid.[50,51]

At the microscopic level, bone has two forms: woven and lamellar. Woven or primary bone is considered immature bone. Woven bone is characterized by a coarse fiber arrangement with no orientation. It has more cells per unit volume than lamellar bone, the mineral content varies, and the cells are randomly arranged. The comparatively disoriented collage fibers endow the bone with isotropic mechanical properties. Applied forces from any direction are adapted to in a similar fashion. Lamellar bone formation begins approximately 1 month after birth. Woven bone is resorbed by 1 year of age. By age 4, most normal bone is lamellar bone. Lamellar bone is found in several structural and functional systems: trabecular lamellae, outer and inner circumferential lamellae, interstitial lamellae, and osteons with concentric lamellae. The highly organized, stress-oriented collagen of lamellar bone gives it anisotropic properties. Anisotropic properties dictate different mechanical behavior of lamellar bone depending on the orientation of the applied forces. The greatest strength in the bone is exhibited parallel to the long axis of the collagen fibers.[52,53]

Woven and lamellar bone are structurally organized into trabecular (spongy and cancellous) bone and cortical (dense or compact) bone. Cortical bone has 4 times the mass of trabecular bone, but a slower metabolic turnover due to a significantly smaller surface area.[54]

Cortical bone is found as the envelope in cuboid bones, and it composes the diaphysis in long bones. Due to the relatively rapid growth in humans, cortical bone is made up of layers of lamellar bone and woven bone with the vascular channels located mainly in the woven bone — plexiform bone.

The most complex type of cortical bone is haversian bone. Vascular channels are surrounded by lamellar bone. The arrangement of bone around the vascular channel is called the osteon — an irregular, branching, and anastomosing cylinder composed of a more or less centrally placed neurovascular canal surrounded by cell-permeated layers of bone matrix. Osteons are usually oriented in the long axis of the bone and are the major structural units of cortical bone. Cortical bone becomes a complex of many adjacent osteons and their interstitial and circumferential lamellae.

APPENDICULAR SKELETON

The cartilaginous models of bones such as the humerus are sharply defined with well-developed perichondrium. They grow by apposition from perichondrium and from tissue forming interzones of joints, and through multiplication of cartilage cells already formed, together with an increase in intercellular matrix. The most immature cells are at the ends where most growth occurs. Cells are small and lack a specific arrangement. From this initial appearance, cells proliferate and form rows of flattened, closely packed cells. Intercellular matrix increases, the cells enlarge, and their cytoplasm becomes vesicular. Once they have reached their maximum size, they undergo vacuolization and are termed hypertrophic chondrocytes. Hypertrophic chondrocytes are primarily a regulatory cell for bone growth, directing localized mineralization, inducing vascularization by production of vascular endothelial growth factor and other factors, and attract chondroclasts. Chondroclasts digest matrix. These are found in the center or midsection of the shaft. Under the direction of the hypertrophic chondrocytes, adjacent perichondral cells become osteoblasts. The osteoblasts secrete matrix forming a bone collar. The hypertrophic cells subsequently liquefy and disappear. The remaining cartilage matrix is a scaffold upon which osteoblasts migrate concurrent with vessel ingrowth, resulting in the formation of the primary spongiosa. The shaft is narrower here due to both the relatively small number of cells originally present and the process of cartilage growth by the addition of cells at the ends and then by the multiplication of cells further from the immature ends of the bone.[55–57]

Ossification begins at the middle of the cartilaginous model. A thin layer of osteoid is laid down between the perichondrium and that portion of the shaft containing hypertrophied cartilage cells and by extending around the shaft containing hypertrophied cartilage cells and by extending around the shaft forms a ring or collar. The osteoid is quickly calcified, thereby forming a bone collar. The bone collar is directly in contact with the cartilage. Osteoid formation precedes definitive bone. The inner cells of the periosteum differentiate into osteoblasts, and alkaline phosphatase can be found in them. These cells line up in relation to bundles of collagen fibers and trabeculae begin to form, just as in the case of intramembranous ossification described previously. The bone collar gradually becomes multilayered.[57–59]

In the cartilage itself, alkaline phosphatase appears in the hypertrophied cartilage cells and then in the matrix of this zone. Calcification of this zone can occur concurrently or subsequent to the formation of the bone collar.

The time between the formation of the primary bone collar, calcification of cartilage matrix, and the subsequent vascular invasion of the cartilage is variable with each bone.[60,61] Following the formation of the bone collar, small cellular masses derived from the periosteum penetrate the collar at several points. Cellular penetration appears soon after collar formation in the long bones, whereas days or weeks may intervene between the formation of the bone collar and vascular invasion in the metacarpals. In the mammalian growth plate, MGP is expressed by proliferative and late hypertrophic chondrocytes apparently required for chondrocyte differentiation and matrix mineralization.[62,63] Cellular masses invading the cartilage contain chondroclasts and undifferentiated cells but few formed blood vessels. Cartilage cells are rapidly destroyed and then replaced by the proliferating and differentiating cells derived from the invading tissue. Blood vessels and a circulation are quickly established. Some of the cells differentiate into osteoblasts, while others are forerunners of blood-forming cells. This invasive, proliferating process extends toward the ends of the bone. As invasion proceeds, the osteoblasts form bone around the calcified cartilage matrix in the region of original entry and at the advancing zone of cartilage removal.

The rapid advance toward the ends of the bone leaves behind a loose network of endochondral trabeculae, which fuses with the multilayered periosteal shell. This network contains in its meshes a vascular tissue, the forerunner of hematopoietic tissue. Concurrently, the cartilage growth phases remain present as the calcified cartilage matrix becomes thinner and the periosteal ossification extends toward the ends of the bone.

Periosteal ossification precedes the endochondral ossification and maintains a level with the zone of hypertrophied cartilage until the epiphyseal regions are reached. The advance consists of a continuing deposition of bone matrix adjacent to the hypertrophied cartilage. New trabeculae are formed and the shaft increases in thickness. The trabeculae form at right or acute angles to the bone present and surround vessels running longitudinally in the periosteum.

In endochondral ossification, neural cell adhesion molecule, NCAM, is highly expressed in osteogenic buds as seen in the epiphysis and diaphysis of tibia and vertebrae. In intramembranous ossification, NCAM is seen in osteogenic condensation of calvaria and in the periosteum of tibial diaphysis. The expression is transient because NCAM is not expressed in mesenchymal cells before osteogenic condensation, and NCAM expression is lost in osteocytes in later stages.[1]

Many of the endochondral trabeculae are removed rapidly, leading to the formation of a marrow cavity. Endochondral trabeculae removal continues as the zone extends toward the epiphyses so that relatively a few remain. The remaining trabeculae become progressively larger as more bone is laid around them and fuse to form a network. Endochondral ossification reaches the epiphyses by the beginning of the fourth fetal month. Hypertrophic chondrocytes are arranged in longitudinal columns with thin strips of calcified matrix between the cell columns. Vascular components invade and destroy the column ends that point toward the shaft, leaving bars of calcified matrix. Bone is then deposited around these bars. Many of the endochondral trabeculae are removed while others form an irregular network in the interior of this part of the shaft.

As the length increases, the bone maintains its shape by remodeling. The process begins during the fetal period and becomes prominent during the fourth or fifth month. Near the end of the first trimester, other changes occur: ossification has reached the metaphyses; orderly growth zones begin to form; remodeling is under way; and growth in length continues at an orderly, symmetrical, and slower rate. Muscle and ligament attachments are shifted and modified by a continuing process of removal of periosteal bone and formation of new bone.

Remodeling is a series of concurrent activities in the growing bone. The diaphysis widens on approaching the epiphyses. These wider regions, often termed the metaphysic, must be reduced as the bone lengthens to prevent club-shaped ends from being formed. Bone is removed from the metaphyseal external surfaces and laid down internally (endosteum). Endochondral trabeculae that fuse with the internal surface of the primitive periosteal compacta becomes incorporated in this compacta as bone is formed by endosteal ossification. The removal of bone from the external surface is most prominent in the region just below the level of the growth zone, to the extent that the periosteal bone may be completely removed, exposing endochondral trabeculae. The periosteal bone above this level remains encircling the growth zone.

Osteoclasts are numerous where bone is being removed. The diaphysis increases in diameter by surface accretion and endosteal removal. As a result, those endochondral trabeculae that had become incorporated into the inner part of the compacta are removed.

Reduction in metaphyseal width does not take place at each end of every bone, nor does it appear to occur around the entire circumference. The distal end of the humerus narrows in the anteroposterior direction but widens from epicondyle to epicondyle. Periosteal bone thickens above the growth zone, and removal of bone from the surface occurs only in medial and lateral epicondylar regions. Local processes related to development of ridges, grooves, and torsion may also modify general processes.

The epiphyses grow in width primarily by addition of cartilage cells from perichondrium above the bone collar. The bone collar grows upward by addition of new bone to the upper edge; yet, the length is not changed because bone at the lower edge is removed.

At birth, the shaft is composed of thick compacta, in which trabeculae are directed mainly longitudinally, and a short marrow cavity. The inner surface of the compacta is irregular and trabeculated. Lamellar bone and Haversian systems or osteons (secondary ossification) begin to form during the third trimester, but mainly develop postnatally.[64,65]

The vascular supply of bone becomes well established during the prenatal period. Initial vascularization of the shaft and marrow occurs through numerous sites, and a single nutrient artery is not present in the early stages. The epiphyses become vascularized prior to epiphyseal ossification.

Although the general appearance of the growth zone for the diaphysis is similar to that seen after an epiphyseal plate is well established, a discussion of the epiphyseal ossification that leads to the formation of this plate is not within the scope of this section. Yet there are certain phases of this growth that are first indicated in prenatal life.[66]

During the early fetal period, epiphyseal cartilages are invaded by blood vessels that form cartilage canals, preceding ossification by many months or years. These cartilage canals are filled with loose connective tissue, containing one or more arterioles, venules, capillaries, and a few nerve fibers accompanying the blood vessels.

Ossification in the head of the humerus is usually indicated at or before term. Cartilage cells in the region of the future center enlarge and become vesicular. Intercellular matrix calcifies followed by vascular budding from adjacent cartilage canals invading this area, eroding the matrix, and depositing bone around cartilage remnants. Endochondral ossification spreads rapidly in all directions. In most long bones, however, epiphyseal ossification, including the distal end of the humerus, begins postnatally, even though canals appear early in the prenatal period. Generally, those centers that appear first (as in the head of the humerus, distal end of the femur, and proximal end of the tibia) are in those ends from which most growth in length occurs.[58,67,68] Following establishment of an epiphyseal growth center, its growth zone is in the deep part of the articular cartilage.[69] Hence, postnatal growth in length is derived from articular growth zones as well as from epiphyseal plates.

An exception to this pattern in the appendicular skeleton is the clavicle. The clavicle is a long bone that is preceded by a membranous rather than a cartilaginous stage. Once ossification begins, cartilage forms secondarily and thereafter takes part in growth of length.[70]

The humerus is a representative illustration of the embryonic and fetal developmental of the skeletal long bones. The growth plate established during development shares molecular signaling cascades with other bones displaying a growth plate.

The physiological FGF ligand that regulates endochondral ossification is not identified. Gain of function and loss of function mutations in *Fgfr3* and comparison to transgenic mice that overexpress an FGF in the growth plate suggest that an FGF ligand is the rate-limiting signal in the FGF pathway regulating endochondral ossification. During embryonic development, growth plate morphology and proliferation of chondrocytes were not affected by overexpression of wild-type FGFR3 in the growth plate. A broadened growth plate results when FGFR3 is lost, supporting the possibility that chondrocyte proliferation is insensitive to FGFR signaling. The insensitivity may be due to excess FGF ligand saturating FGFR signaling pathways, or the inhibitory effect of FGFR3 expression is limited by the dominant action of other mitogens.[71]

FGF receptors 1 and 3 are both expressed in the epiphyseal growth plate. FGFR1 is expressed in hypertrophic chondrocytes, possibly maintaining the hypertrophic phenotype of these cells, regulating the production of unique extracellular matrix products of hypertrophic chondrocytes, or in signaling their eventual apoptotic death.[71] *Fgfr3* is expressed in proliferation chondrocytes and in the cartilage of the developing embryo, prior to the formation of ossification centers. The expression pattern suggests a direct role for FGFR3 in regulating chondrocyte proliferation, and possibly differentiation[72] FGFR3 is an essential regulator of endochondral bone growth.

Mutations in *Fgfr3* result in mild to severe dwarfism (Table 1.2). All of the mutations identified in FGFRs are autosomal dominant and frequently arise sporadically. Many of these disorders result from point mutations in the coding sequence of the *Fgfr* gene, resulting in a single amino acid substitution.

Bone formation requires the proper coordination of proliferation, differentiation, and movement of multiple cell types. Such coordination requires a number of local signaling pathways involving paracrine factors and cell–cell interactions. Roles for BMPS, transforming growth factor b, FGF, Indian Hedgehog (IHH) IGFs, and PTHrP have been demonstrated through the study of transgenic and gene knockout mice and human genetic disorders.

PTHrP is important in bone growth. Karaplis demonstrated that mouse lacking the PTHrP gene have foreshortened columns of proliferating chondrocytes in bones of endochondral origin.[113] They exhibit remarkable foreshortening of the columns of proliferating chondrocytes. Mice die at birth probably due to the inability to breathe normally owing to small rib cages (Table 1.3). Expression patterns support a role for PTHrP acting upon the PTH/PTHrP receptor on chondrocytes to maintain them in the proliferative pool.

Table 1.2
Mutations in FGF Receptor Gene

Clinical diagnosis	Gene mutation		Phenotype
Hypochondroplasia (HCH)	*Fgfr3* missense N540K	Bellus 1995*Nat. Gen.* 10: 357–359	Mild form of dwarfism
Achondroplasia (ACH)	*Fgfr3* missense, gain of function G346E, G375C, G380R	Shiang 1994 *Cell* 78: 335–342 Naski 1996 *Nat. Gen.* 13: 4977–4988 Li 1997 *Oncogene* 14: 1397	Shortening of proximal and distal long bones, frontal bossing, depressed nasal bridge, increased receptor tyrosine kinase activity
Thanatophoric dysplasia	*Fgfr3* missense, gain of function R248C, S249C, G370C, S371C, Y373C, K650E	Rousseau 1995 *Nat. Gen.* 10: 11–12, 1996 *Hum. Mol. Genet.* 5: 509–512 Li 1997 *Oncogene* 14: 1397	Similar to ACH, most common form of lethal-neonatal skeletal disorder, clinically similar to homozygous ACH, increased receptor tyrosine kinase activity
Syndromic craniosynostosis	*Fgfr2*	Webster and Donoghue 1997 *Trends Gen.* 13: 178–182	Syndactly, dwarfism is not a common feature

PTHrP and IHH may participate in a negative feedback loop that enhances coordination of chondrocyte differentiation. IHH are synthesized by cells as they are leaving the proliferative compartment and turning on the hypertrophic cell program. IHH increases the synthesis of PRHrP, which then acts to delay the movement of chondrocytes from the proliferative to the hypertrophic compartments. PTHrP action delays the appearance of cells that synthesize IHH (Table 1.3).

Vertebrate limbs represent a diversified range of anatomical patterns. These anatomical changes have their basis in embryonic development as the body plan is laid out.[4,73] During normal limb development, the ridge is induced in the apical ectoderm by a signal from underlying mesenchymal cells. The ectoderm responds to that signal by activating expression of genes such as Fgf4 and organizing itself into a pseudostratified, columnar epithelium. Limb position and axial skeletal identity are regulated by Hox gene expression along the primary body axis. Somites that form the vertebrae are regionalized by nested domains of *Hox* gene expression. The lateral plate mesoderm, which forms the limbs and body wall, is regionalized by *Hox* gene expression. Molecular regionalization of the lateral plate is an important step in determining the position at which limbs develop relative to the main body axis. Differential *Hox* gene expression is operating in the axial skeleton and in the lateral plate mesoderm.[74]

Along the neural axis of the vertebrate, there are specific patterns of *Hox* gene expression. A pattern suggests a correlation between *Hox* gene expression boundaries and where the forelimb, flank, and hind limb will develop. However, the mechanism dictating limb-bud position is independent of the initiation of limb-bud formation.

Table 1.3
Mutations of Growth Factor Genes

Growth factor	Mutation/phenotype	Proposed function	Reference
PTHrP	Knockout/Mouse: Foreshortening of the columns of proliferating chondrocytes	Maintain chondrocytes in a proliferative mode, hinder differentiative mode	113
PTH/PTHrP receptor	Knockout/Mouse: Chondrocytes become hypertrophic and a bone collar forms at the anterior rib portions	Maintain chondrocytes in a proliferative mode, hinder differentiative mode	114
PTHrP	Transgene/Mouse overexpressing PTHrP Delay in chondrocyte differentiation	Controls chondrocyte differentiation	115
PTH/PTHrP receptor	Mouse/Constitutively active receptor Delay in chondrocyte migration to hypertrophic phenotype and hypertrophic chondrocyte apoptosis		116
IHH	Overexpression/Chicken: conversion of hypertrophic chondrocytes was delayed, increased expression of PTHrP at the bone ends Knockout/Mouse no limb bones but vertebrae, skull, and scapula are present.	IHH and PTHrP participate in a negative feedback loop in the growth plate	117

In summary, long-bone and epiphyseal development progress through structural stages. These stages include limb-bud formation with a uniform distribution of mesenchymal cells and the formation of an apical ectodermal ridge; mesenchymal condensation; cartilage differentiation; formation of a primary center of ossification; epiphyseal cartilage vascularization with formation of cartilage canals; vascular invasion of the developing secondary ossification center; bone formation and marrow cavitation in the secondary ossification center with formation of hematopoietic marrow; fullest relative extent of secondary-ossification center development in epiphyseal cartilage; thinning of the physis; and resorption of the physis with establishment of continuity between epiphyseal and metaphyseal circulations.

Development of the Mandible

In the maxilla and mandible, ossification is similar but the pattern of growth and orientation of trabeculae are different. Central plates develop more rapidly, borders are smooth and there are few if any bone islands or nodules. The trabeculae extending from the central plate are short. The general trabecular pattern is more complex than that of the parietal or frontal bones, and the reconstructive growth is more complex in view of the necessity of providing for teeth and their migration as well as maintaining the maxilla and mandible in proper relation.

A sequence of transitory stages defines condylar growth by molecules that are synthesized by cells in the condyles. SOX 9 was expressed by cells in the proliferative prechondrocytic layer and by chondrocytes. Type X collagen was expressed only by hypertrophic chondrocytes and precedes the onset of endochondral ossification. Vascular endothelial growth factor (VEGF) is expressed by hypertrophic chondrocytes.[30,75]

The mandible is a membrane bone in which cartilage forms secondarily and is closely associated with Meckel's cartilage. Together with the clavicle, they are the first bones to begin to ossify.

In embryonic chicks, there are three types of mesenchymal cells in the early mandibular arch: myogenic, chondrogenic, and osteogenic. The cells are independent in origin and distinct at the time of appearance. The myogenic cells do not arise in the mandible, but migrate into it from another source. Other experimental work indicates the possibility that the cartilage of the trabeculae and most visceral arches, including Meckel's cartilage, may come from neural crest cells. Parathyroid hormone-related protein, PTHrP, may influence skeletal tissue histogenesis by affecting the differentiation of mandibular mesenchymal cells into chondroblasts and osteoblasts.[76] Osteogenic cells are characteristically associated with a thickened patch of mouth epithelium, which corresponds to the rudiment of the enamel organ in mammals.

Meckel's cartilages begin to form in human embryos at about 6 weeks.[77] Ossification begins on the outer side of the ventral part of the cartilage, in the region of the future mandibular body, in a condensation that extends posteriorly along the lateral side of the cartilage. By 7 weeks, this condensation spreads from about the symphysis in front to about the level of the auriculotemporal nerve behind.[78,79] This form of the temporomandibular joint is also indicated at this time.[80] By the end of the embryonic period, the pattern of the mandible is complex and the mental foramen is present.

By the end of the embryonic period, the condensation for the entire mandible is clearly defined and ossification is spreading posteriorly. Early in the fetal period, there may be changes in Meckel's cartilage at a point between lateral incisor and canine tooth germs, suggesting incipient ossification. The expression of osteopontin is closely associated with calcifying foci in the extracellular matrix. Osteopontin might be expressed sequentially by chondrocytes and by cells that are differentiating further and exhibit an osteocytic phenotype.[81] Periosteal bone then forms on the upper and lateral aspects of the cartilage, following which, there is vascular penetration and endochondral ossification.

Intramembranous ossification extends rapidly into condylar and coronoid processes. Condylar growth involves a sequence of transitory stages uniquely defined by molecules that are intrinsically synthesized by cells in the condyles. During condylar growth, SOX 9 is expressed both by the proliferative cell layer and by chondrocytes. Vascular endothelial growth factor, VEGF, is expressed by hypertrophic chondrocytes, and its maximum level of expression precedes the maximum level of bone formation.[30] By about 11 weeks, a vesicular type of cartilage has formed on the articular surface of the condyle. Fibrocellular tissue separates the articular surface from the temporal bone. Between 11 and 12 weeks, cavitation begins in the temporomandibular joint and an intra-articular disc is indicated. Soon after, the secondary cartilage calcifies and is then vascularized. Endochondral ossification begins later and is well under way by 18–19 weeks. Vascular channels may pass from the fibrocellular layer through the cartilage into the marrow space. The entire process is quite similar to that which takes place at the ends of the clavicle. Secondary cartilage also appears along the anterior border of the coronoid process at 13–14 weeks. At term, the cartilage is reduced to a narrow zone.[82–85]

The part of Meckel's cartilage from the central incisor or the canine tooth germ becomes surrounded by a periosteal bone by the 14th week. The rest of the cartilage usually disappears. Small cartilaginous masses may form secondarily along both alveolar margins in the regions of the incisor teeth and along the lower border of the jaw in front, but these usually disappear completely by term.[83]

Each half of the mandible develops as a single skeletal element. Meckel's cartilage contributes little to the formation of the mandible. Condylar cartilages form independent of Meckel's cartilage. Throughout the prenatal period, the mandible is depressed and not opposed to the maxilla. The trabecular pattern is complex. An increase in width is the result of bone formation on lateral surfaces. Forward growth results from the deposition on the anterior surface of the symphysis, and backward growth results from accretion on the posterior border from the condyle to the angle. The main site of increase in height appears to be the alveolar border. The condyle grows up and back, and the remodeling mechanisms around the neck are similar to those in the long bones. Accretion occurs at the tip of the coronoid process. Posterior deposition and anterior absorption in tooth sockets allow the teeth to move forward and create space for molars.

Parietal and Frontal Bones

The vertebrate skull develops within a layer of mesenchyme between the embryonic brain and surface ectoderm. The outer skeletogenic membrane forms the skull bones and the inner meningeal layer forms the dura, arachnoid, and pia mater layers. The sutures between the membrane bones of the skull vault are growth centers in which proliferating osteogenic stem cells provide the source material for incremental growth of the calvarial bones. Osteogenic stem cells are located at the periphery of each membrane bone and in smaller numbers on the outer and inner surfaces. Continued growth of the skull vault depends on the maintenance of a balance between proliferation of the osteogenic stem cells and their differentiation. The coronal suture is a bilateral, vertical suture framed between the frontal and parietal bones. The suture is responsible for most of the growth of the skull in the front of the occipital plane.

The tissue distribution of *Fgfr* genes and *Twist* in the developing coronal suture indicates specific concentration, temporal, and spatial localization of the receptors. The parietal and frontal areas of *Fgfr* expression in the coronal suture area are separated by an area of midsutural mesenchyme in which no *Fgfr* gene expression can be detected. This separation of gene expression is maintained throughout the prenatal period during which the edges of the two bones show an increasing degree of overlap (Table 1.4). *Twist* expression appears essential for maintenance of osteogenic cell proliferation. *Twist* gene expression is down-regulated during differentiation of cultured mouse calvarial osteoprogenitor cells.

Proliferating osteoprogenitor cells are present at the suture edges of the bones and on the outer and inner surfaces — Fgf2 is expressed in these cells. The midsutural mesenchyme, in which only Twist expression is expressed, shows neither differentiation-related gene expression nor cell proliferation markers. The area appears to be a buffer zone between the frontal and parietal proliferating populations, in continuity at the periphery with cells co-expressing *Twist* and *Fgfr2*. Differential

Table 1.4
***Fgfr* and *Twist* Gene Expression in Mouse Embryo (E16)**

Receptor gene expression	Distribution
Fgfr1	Cells proximal to osteoid
Fgfr2	Frontal and parietal sutural cell populations, scattered localization in outer and inner surfaces of the osteoid plates
Fgfr3	Comparatively low expression, interspersed between *Fgfr1,2*; in a thin cartilaginous plate underlying the lower part of the coronal suture
Fgfr4	Muscle
Twist	Midsutural mesenchyme, overlapping with the frontal and parietal areas of *Fgfr2* expression

levels of FGF from high in the differentiated region to low in the suture ensure that sutural stem cell populations are maintained at the periphery of the growing bones. Relatively low levels of FGF are associated with osteogenic stem cell proliferation at the margins of the membrane bones that form the coronal suture and the inner and outer surfaces of these bones. Higher levels of FGF are associated with osteogenic differentiation. The mitogenic signal is mediated by FGFR2 and FGFR3. Higher levels of FGF2-stimulated FGFR2 signaling lead to down-regulation of *Fgfr2* and *Fgfr3* gene expression and up-regulation of *Fgfr1*. Signaling through FGFR1 does not have a mitogenic outcome, but leads to the expression of osteogenic differentiation genes. *Fgfr1* gene expression is down-regulated when differentiation is well established.

Differentiation is a multistage process involving the sequential expression of a number of different genes. Comparison of indicators of bone differentiation indicates that *Osteonectin* gene expression and alkaline phosphatase activity are present in cells close to the region of proliferation, in preosteoblast and osteoblasts at an early stage of differentiation. Osteopontin is expressed in more mature osteoblasts in contact with osteoid. Signaling through FGFR1 is associated with the onset of differentiation. FGFR1's role appears to be complete before osteoblasts are fully differentiated.

Each parietal bone develops from a pair of centers or occasionally from one appearing early in the fetal period. Ossification is initiated in the morphological center of the bone. The two centers fuse, and the trabeculae radiate from the growing central plate. The parietal bone grows rapidly, in constant increments in relation to sitting height, until about the 20th week, at which time it has reached its definitive form and area. It grows slowly until birth at about the same rate as the craniofacial complex. About week 20, the peripheral edges of the bone fuse to form a bony border.[86,87] The parietal bones are mesodermal derived, and the frontal bones are neural crest-derived with a layer of neural crest between the two parietal bones.

The two frontal bones that develop prenatally usually fuse after birth. Each bone is first indicated as a membranous condensation early in the embryonic period. Ossification begins in the late embryonic period or early fetal period. Bone first forms in the region of each supercilliary arch. Each such center simultaneously gives rise to a portion of both the pars frontalis where heavy islands of bone form. The islands fuse and form primary radiating trabeculae, following which secondary trabeculae are formed. Ossification in the pars orbitalis is less rapid, the radiating trabeculae are smaller, and an expanding network is maintained. The frontal bones grow rapidly, in constant increments, until about the 20th week, when the rate slows to keep pace with the rest of the skull. The initial rapid growth represents the expansion of bones until they cover the definitive area of the vault. The post frontal element is not a separate center of ossification. It is a region of slow growth in the posterior inferior angle of the frontal bone and becomes apparent only after the fifth fetal month.[86,88,89]

The zygomatic process and the orbital margin begin to thicken. At 12 weeks, the orbital rim is defined. Most of the orbital plate is completed by 13–14 weeks and has begun to fuse with the medial angular process. Reconstruction occurs from 20 weeks to term.[90]

The frontal sinuses form postnatally. A concurrent absorption of bone and ingrowth of lining tissue are evident, although the first indication of a sinus can be detected prenatally. By the end of the third fetal month, part of the middle nasal meatus starts to extend in a ventrocephalic direction. This is the beginning of the recessus frontalis, representing the first step in frontal sinus formation. Late in fetal life, this recess is complicated by the formation of frontal furrows or pits. The various rudiments of the sinus are well advanced at term. Occasionally the sinus itself may be present, although the final form may extend up to 12 months after birth.[91,92]

The onset and sequence of ossification have always been of practical importance, especially for postnatal periods, and many studies and tables have been published. There appears to be a definite sequence in the appearance of ossification centers.

Bone Remodeling

Remodeling entails the addition of new bone on the surface of preexisting bone. Concurrently, there is a simultaneous destruction of parts of the preexisting skeleton. Resorption of preexisting parts of the skeleton is followed by osteoblasts covering the inner aspects to the preosseous layer where delicate radiating striations are canaliculi. The cementing line joining the preosseous layer and the layers deposited previously is also called a reversal line. It marks the end of the destruction, which has hollowed out a cavity and the beginning of deposition. The new osteon's calcium content will increase at a variable pace and coincides with the local change in organic matrix. A three-stage process is proposed: deposition until the central canal in the osteon is narrowed, enlarging of the narrow central canal by osteoclasts and mineral deposition of a reversal line, and refilling of the enlarged canal by new bone deposition. Resting lines appear to be present due to a previous surface where osteoblastic activity had ceased and the superficial layer had become hypercalcified. Then, after a quiescent period, bone deposition resumed without an intervening phase of resorption.

In the growing skeleton, remodeling observed in the endochondral bone tissue of the metaphysic and the epiphysis is linked with morphogenesis. The intensity of renewal of compact bone is higher during the skeletal growth period than in the adult skeleton.

Remodeling appears symmetrical in similar bones, but remarkable differences in the process exist regarding temporal and spatial activity. The two main functions of bone remodeling enable the bone to be responsive to mechanical and metabolic demands placed upon the body. The two processes are integrated. A change in the microscopic structure brought about by some subtle mechanism triggered by a continuous adaptation to the mechanical function is also under metabolic control since it stores and releases calcium. Bone remodeling is a response to microdamage that is a normal part of daily living. Additionally and more subtly, genetic factors operate during development, defining the shape.

References

1. Lee, Y. S. and Chuong, C. M. (1992). Adhesion molecules in skeletogenesis: I. Transient expression of neural cell adhesion molecules (NCAM) in osteoblasts during endochondral and intramembranous ossification. *J. Bone Miner. Res.* 7, 1435–1446.
2. Eames, B. F., de la Fuente, L. and Helms, J. A. (2003). Molecular ontogeny of the skeleton. *Birth Defects Res. Part C Embryo Today* 69, 93–101.
3. Ferretti, M., Palumbo, C., Contri, M. and Marotti, G. (2002). Static and dynamic osteogenesis: two different types of bone formation. *Anat. Embryol. (Berl.)* 206, 21–29.
4. Caplan, A. I. (1988). Bone development. *Ciba. Found. Symp.* 136, 3–21.
5. Cobb, J. D. (1953). Relation of glycogen, phosphorylase, and ground substance to calcification of bone. *AMA Arch. Pathol.* 55, 496–502.
6. Scott, C. K. and Hightower, J. A. (1991). The matrix of endochondral bone differs from the matrix of intramembranous bone. *Calcif. Tissue Int.* 49, 349–354.
7. Aaron, J. E. and Skerry, T. M. (1994). Intramembranous trabecular generation in normal bone. *Bone Miner.* 25, 211–230.
8. Ashton, B. A., Allen, T. D., Howlett, C. R., Eaglesom, C. C., Hattori, A. and Owen, M. (1980). Formation of bone and cartilage by marrow stromal cells in diffusion chambers *in vivo*. *Clin. Orthop.* Issue 151, 294–307.
9. Scott, C. K., Bain, S. D. and Hightower, J. A. (1994). Intramembranous bone matrix is osteoinductive. *Anat. Rec.* 238, 23–30.
10. Lu, M. and Rabie, A. B. (2002). The effect of demineralized intramembranous bone matrix and basic fibroblast growth factor on the healing of allogeneic intramembranous bone grafts in the rabbit. *Arch. Oral Biol.* 47, 831–841.
11. Urabe, K., Jingushi, S., Ikenoue, T., Okazaki, K., Sakai, H., Li, C. and Iwamoto, Y. (2001). Immature osteoblastic cells express the pro-alpha2(XI) collagen gene during bone formation *in vitro* and *in vivo*. *J. Orthop. Res.* 19, 1013–1020.
12. Grave, B. (2000). Localization of TGF-Bs and perlecan in mouse skull development. *Ann. R. Australas Coll. Dent. Surg.* 15, 352–356.
13. Merida-Velasco, J. A., Sanchez-Montesinos, I., Espin-Ferra, J., Garcia-Garcia, J. D. and Roldan-Schilling, V. (1993). Developmental differences in the ossification process of the human corpus and ramus mandibulae. *Anat. Rec.* 235, 319–324.
14. Karp, N. S., McCarthy, J. G., Schreiber, J. S., Sissons, H. A. and Thorne, C. H. (1992). Membranous bone lengthening: a serial histological study. *Ann. Plast. Surg.* 29, 2–7.
15. Kagayama, M., Sasano, Y. and Akita, H. (1993). Lectin binding in bone matrix of adult rats with special reference to cement lines. *Tohoku J. Exp. Med.* 170, 81–91.

16. Tabata, Y., Hong, L., Miyamoto, S., Miyao, M., Hashimoto, N. and Ikada, Y. (2000). Bone formation at a rabbit skull defect by autologous bone marrow cells combined with gelatin microspheres containing TGF-beta1. *J. Biomater. Sci. Polym. Ed.* 11, 891–901.
17. Mathy, J. A., Lenton, K., Nacamuli, R. P., Fong, K. D., Song, H. M., Fang, T. D., Yang, G. P. and Longaker, M. T. (2003). FGF-2 stimulation affects calvarial osteoblast biology: quantitative analysis of nine genes important for cranial suture biology by real-time reverse transcription polymerase chain reaction. *Plast. Reconstr. Surg.* 112, 528–539.
18. Hajihosseini, M. K. and Heath, J. K. (2002). Expression patterns of fibroblast growth factors-18 and- 20 in mouse embryos is suggestive of novel roles in calvarial and limb development. *Mech. Dev.* 113, 79–83.
19. Francis-West, P. H., Robson, L. and Evans, D. J. (2003). Craniofacial development: the tissue and molecular interactions that control development of the head. *Adv. Anat. Embryol. Cell Biol.* 169, III–VI, 1–138.
20. Kiesler, J. and Ricer, R. (2003). The abnormal fontanel. *Am. Fam. Physician* 67, 2547–2552.
21. Hall, B. K. (1986). The role of movement and tissue interactions in the development and growth of bone and secondary cartilage in the clavicle of the embryonic chick. *J. Embryol. Exp. Morphol.* 93, 133–152.
22. Blair, H. C., Zaidi, M. and Schlesinger, P. H. (2002). Mechanisms balancing skeletal matrix synthesis and degradation. *Biochem. J.* 364, 329–341.
23. Liu, S. K. (2002). Metabolic disease in animals. *Semin. Musculoskelet. Radiol.* 6, 341–346.
24. Cohen, M. M., Jr. (2000). Merging the old skeletal biology with the new. I. Intramembranous ossification, endochondral ossification, ectopic bone, secondary cartilage, and pathologic considerations. *J. Craniofac. Genet. Dev. Biol.* 20, 84–93.
25. Bronckers, A. L., Sasaguri, K. and Engelse, M. A. (2003). Transcription and immunolocalization of Runx2/Cbfa1/Pebp2alphaA in developing rodent and human craniofacial tissues: further evidence suggesting osteoclasts phagocytose osteocytes. *Microsc. Res. Tech.* 61, 540–548.
26. Choi, K. Y., Lee, S. W., Park, M. H., Bae, Y. C., Shin, H. I., Nam, S., Kim, Y. J., Kim, H. J. and Ryoo, H. M. (2002). Spatio-temporal expression patterns of Runx2 isoforms in early skeletogenesis. *Exp. Mol. Med.* 34, 426–433.
27. Kundu, M., Javed, A., Jeon, J. P., Horner, A., Shum, L., Eckhaus, M., Muenke, M., Lian, J. B., Yang, Y., Nuckolls, G. H., Stein, G. S. and Liu, P. P. (2002). Cbfbeta interacts with Runx2 and has a critical role in bone development. *Nat. Genet.* 32, 639–644.
28. Yoshida, C. A., Furuichi, T., Fujita, T., Fukuyama, R., Kanatani, N., Kobayashi, S., Satake, M., Takada, K. and Komori, T. (2002). Core-binding factor beta interacts with Runx2 and is required for skeletal development. *Nat. Genet.* 32, 633–638.
29. Akiyama, H., Chaboissier, M. C., Martin, J. F., Schedl, A. and de Crombrugghe, B. (2002). The transcription factor Sox9 has essential roles in successive steps of the chondrocyte differentiation pathway and is required for expression of Sox5 and Sox6. *Genes Dev.* 16, 2813–2828.
30. Rabie, A. B. and Hagg, U. (2002). Factors regulating mandibular condylar growth. *Am. J. Orthod. Dentofacial Orthop.* 122, 401–409.
31. Nakamata, T., Aoyama, T., Okamoto, T., Hosaka, T., Nishijo, K., Nakayama, T., Nakamura, T. and Toguchida, J. (2003). *In vitro* demonstration of cell-to-cell interaction in growth plate cartilage using chondrocytes established from p53-/- mice. *J. Bone Miner. Res.* 18, 97–107.

32. MacLean, H. E., Kim, J. I., Glimcher, M. J., Wang, J., Kronenberg, H. M. and Glimcher, L. H. (2003). Absence of transcription factor c-maf causes abnormal terminal differentiation of hypertrophic chondrocytes during endochondral bone development. *Dev. Biol.* 262, 51–63.
33. Okazaki, K., Jingushi, S., Ikenoue, T., Urabe, K., Sakai, H. and Iwamoto, Y. (2003). Expression of parathyroid hormone-related peptide and insulin-like growth factor I during rat fracture healing. *J. Orthop. Res.* 21, 511–520.
34. Rabie, A. B., Tang, G. H., Xiong, H. and Hagg, U. (2003). PTHrP regulates chondrocyte maturation in condylar cartilage. *J. Dent. Res.* 82, 627–631.
35. Webster, S. V., Farquharson, C., Jefferies, D. and Kwan, A. P. (2003). Expression of type X collagen, Indian hedgehog and parathyroid hormone-related protein in normal and tibial dyschondroplastic chick growth plates. *Avian. Pathol.* 32, 69–80.
36. Iwamoto, M., Enomoto-Iwamoto, M. and Kurisu, K. (1999). Actions of hedgehog proteins on skeletal cells. *Crit. Rev. Oral Biol. Med.* 10, 477–486.
37. Reinhold, M. I., McEwen, D. G. and Naski, M. C. (2004). Fibroblast growth factor receptor 3 gene: regulation by serum response factor. *Mol. Endocrinol.* 18, 241–251.
38. Dailey, L., Laplantine, E., Priore, R. and Basilico, C. (2003). A network of transcriptional and signaling events is activated by FGF to induce chondrocyte growth arrest and differentiation. *J. Cell Biol.* 161, 1053–1066.
39. Makihira, S., Yan, W., Murakami, H., Furukawa, M., Kawai, T., Nikawa, H., Yoshida, E., Hamada, T., Okada, Y. and Kato, Y. (2003). Thyroid hormone enhances aggrecanase-2/ADAM-TS5 expression and proteoglycan degradation in growth plate cartilage. *Endocrinology* 144, 2480–2488.
40. Colnot, C., Thompson, Z., Miclau, T., Werb, Z. and Helms, J. A. (2003). Altered fracture repair in the absence of MMP9. *Development* 130, 4123–4133.
41. Drake, C. J. (2003). Embryonic and adult vasculogenesis. *Birth Defects Res. Part C Embryo Today* 69, 73–82.
42. Ortega, N., Behonick, D., Stickens, D. and Werb, Z. (2003). How proteases regulate bone morphogenesis. *Ann. N Y Acad. Sci.* 995, 109–116.
43. Nagel, A. (1993). The clinical significance of the nutrient artery. *Orthop. Rev.* 22, 557–561.
44. Kubota, A. (1966). [Studies on the nutrient arteries of the temporomandibular joint of the human fetus. 1. Arterial supply of the temporomandibular joint]. *Shikwa Gakuho* 66, 765–822.
45. Hallock, G. G., Anous, M. M. and Sheridan, B. C. (1993). The surgical anatomy of the principal nutrient vessel of the tibia. *Plast. Reconstr. Surg.* 92, 49–54.
46. Gavaghan, M. (1998). Vascular hemodynamics. *Aorn J.* 68, 212–26; quiz 227–228, 230, 233 passim.
47. Kronenberg, H. M. (2003). Developmental regulation of the growth plate. *Nature* 423, 332–336.
48. Nakajima, R., Inada, H., Koike, T. and Yamano, T. (2003). Effects of leptin to cultured growth plate chondrocytes. *Horm. Res.* 60, 91–98.
49. Nilsson, O., Chrysis, D., Pajulo, O., Boman, A., Holst, M., Rubinstein, J., Martin Ritzen, E. and Savendahl, L. (2003). Localization of estrogen receptors-alpha and-beta and androgen receptor in the human growth plate at different pubertal stages. *J. Endocrinol.* 177, 319–326.
50. Claassen, H., Kampen, W. U. and Kirsch, T. (1996). Localization of collagens and alkaline phosphatase activity during mineralization and ossification of human first rib cartilage. *Histochem. Cell Biol.* 105, 213–219.

51. Vaananen, K., Morris, D. C., Munoz, P. A. and Parvinen, E. K. (1987). Immunohistochemical study of alkaline phosphatase in growth plate cartilage, bone, and fetal calf isolated chondrocytes using monoclonal antibodies. *Acta Histochem.* 82, 211–217.
52. Chang, W. C., Christensen, T. M., Pinilla, T. P. and Keaveny, T. M. (1999). Uniaxial yield strains for bovine trabecular bone are isotropic and asymmetric. *J. Orthop. Res.* 17, 582–585.
53. Ashman, R. B., Rosinia, G., Cowin, S. C., Fontenot, M. G. and Rice, J. C. (1985). The bone tissue of the canine mandible is elastically isotropic. *J. Biomech.* 18, 717–721.
54. Ito, M., Nishida, A., Koga, A., Ikeda, S., Shiraishi, A., Uetani, M., Hayashi, K. and Nakamura, T. (2002). Contribution of trabecular and cortical components to the mechanical properties of bone and their regulating parameters. *Bone* 31, 351–358.
55. Rivas, R. and Shapiro, F. (2002). Structural stages in the development of the long bones and epiphyses: a study in the New Zealand white rabbit. *J. Bone Joint Surg. Am.* 84-A, 85–100.
56. Wrathall, A. E., Bailey, J. and Hebert, C. N. (1974). A radiographic study of development of the appendicular skeleton in the fetal pig. *Res. Vet. Sci.* 17, 154–168.
57. Kerr, G. R., Wallace, J. H., Chesney, C. F. and Waisman, H. A. (1972). Growth and development of the fetal rhesus monkey. 3. Maturation and linear growth of the skull and appendicular skeleton. *Growth* 36, 59–76.
58. Panattoni, G. L., D'Amelio, P., Di Stefano, M. and Isaia, G. C. (2000). Ossification centers of human femur. *Calcif. Tissue Int.* 66, 255–258.
59. Bradbeer, J. N., Lindsay, P. C. and Reeve, J. (1994). Fluctuation of mineral apposition rate at individual bone-remodeling sites in human iliac cancellous bone: independent correlations with osteoid width and osteoblastic alkaline phosphatase activity. *J. Bone Miner. Res.* 9, 1679–1686.
60. Nazario, A. C., Tanaka, C. I., Santana, R. M. and Juliano, Y. (1992). Sex variation at the time of sonographic appearance of the proximal tibial epiphyseal ossification center. *Rev. Paul Med.* 110, 26–28.
61. Holtrop, M. E. (1967). The potencies of the epiphyseal cartilage in endochondral ossification. *Proc. K Ned. Akad. Wet. C* 70, 21–28.
62. Newman, B., Gigout, L. I., Sudre, L., Grant, M. E. and Wallis, G. A. (2001). Coordinated expression of matrix Gla protein is required during endochondral ossification for chondrocyte survival. *J. Cell Biol.* 154, 659–666.
63. Yagami, K., Suh, J. Y., Enomoto-Iwamoto, M., Koyama, E., Abrams, W. R., Shapiro, I. M., Pacifici, M. and Iwamoto, M. (1999). Matrix GLA protein is a developmental regulator of chondrocyte mineralization and, when constitutively expressed, blocks endochondral and intramembranous ossification in the limb. *J. Cell Biol.* 147, 1097–1108.
64. Rubin, C. T., Gross, T. S., McLeod, K. J. and Bain, S. D. (1995). Morphologic stages in lamellar bone formation stimulated by a potent mechanical stimulus. *J. Bone Miner. Res.* 10, 488–495.
65. Jaworski, Z. F. and Wieczorek, E. (1985). Constants in lamellar bone formation determined by osteoblast kinetics. *Bone* 6, 361–363.
66. Brown, R. A., Blunn, G. W., Salisbury, J. R. and Byers, P. D. (1993). Two patterns of calcification in primary (physeal) and secondary (epiphyseal) growth cartilage. *Clin. Orthop.* Issue 294, 318–324.
67. Syftestad, G. T., Weitzhandler, M. and Caplan, A. I. (1985). Isolation and characterization of osteogenic cells derived from first bone of the embryonic tibia. *Dev. Biol.* 110, 275–283.

68. Navagiri, S. S. and Dubey, P. N. (1976). The ossification centers in the lower end of tibiotarsus in domestic fowl. *Z. Mikrosk. Anat. Forsch.* 90, 360–367.
69. Morrison, E. H., Ferguson, M. W., Bayliss, M. T. and Archer, C. W. (1996). The development of articular cartilage: I. The spatial and temporal patterns of collagen types. *J. Anat.* 189 (Part 1), 9–22.
70. Schmeling, A., Schulz, R., Reisinger, W., Muhler, M., Wernecke, K. D. and Geserick, G. (2004). Studies on the time frame for ossification of the medial clavicular epiphyseal cartilage in conventional radiography. *Int. J. Legal Med.* 118, 5–8.
71. Iseki, S., Wilkie, A. O. and Morriss-Kay, G. M. (1999). Fgfr1 and Fgfr2 have distinct differentiation-and proliferation-related roles in the developing mouse skull vault. *Development* 126, 5611–5620.
72. Naski, M. C. and Ornitz, D. M. (1998). FGF signaling in skeletal development. *Front. Biosci.* 3, D781–794.
73. Atchley, W. R. and Hall, B. K. (1991). A model for development and evolution of complex morphological structures. *Biol. Rev. Cambridge Philos. Soc.* 66, 101–157.
74. Li, Z. L., Chisaka, O., Koseki, H., Akasaka, T., Ishibashi, M. and Shiota, K. (1997). Heat shock-induced homeotic transformations of the axial skeleton and associated shifts of Hox gene expression domains in mouse embryos. *Reprod. Toxicol.* 11, 761–770.
75. Rabie, A. B., She, T. T. and Harley, V. R. (2003). Forward mandibular positioning upregulates SOX9 and type II collagen expression in the glenoid fossa. *J. Dent. Res.* 82, 725–730.
76. Zhao, Q., Brauer, P. R., Xiao, L., McGuire, M. H. and Yee, J. A. (2002). Expression of parathyroid hormone-related peptide (PthrP) and its receptor (PTH1R) during the histogenesis of cartilage and bone in the chicken mandibular process. *J. Anat.* 201, 137–151.
77. Bontemps, C., Cannistra, C., Hannecke, V., Michel, P., Fonzi, L. and Barbet, J. P. (2001). [The first appearance of Meckel's cartilage in the fetus]. *Bull. Group Int. Rech. Sci. Stomatol. Odontol.* 43, 94–99.
78. Orliaguet, T., Darcha, C., Dechelotte, P. and Vanneuville, G. (1994). Meckel's cartilage in the human embryo and fetus. *Anat. Rec.* 238, 491–497.
79. Orliaguet, T., Dechelotte, P., Scheye, T. and Vanneuville, G. (1993). Relations between Meckel's cartilage and the morphogenesis of the mandible in the human embryo. *Surg. Radiol. Anat.* 15, 41–46.
80. Morimoto, K., Hashimoto, N. and Suetsugu, T. (1987). Prenatal developmental process of human temporomandibular joint. *J. Prosthet. Dent.* 57, 723–730.
81. Ishizeki, K., Nomura, S., Takigawa, M., Shioji, H. and Nawa, T. (1998). Expression of osteopontin in Meckel's cartilage cells during phenotypic transdifferentiation *in vitro*, as detected by *in situ* hybridization and immunocytochemical analysis. *Histochem. Cell Biol.* 110, 457–466.
82. Petersen, W., Tsokos, M. and Pufe, T. (2002). Expression of VEGF121 and VEGF165 in hypertrophic chondrocytes of the human growth plate and epiphyseal cartilage. *J. Anat.* 201, 153–157.
83. Thompson, Z., Miclau, T., Hu, D. and Helms, J. A. (2002). A model for intramembranous ossification during fracture healing. *J. Orthop. Res.* 20, 1091–1098.
84. Zimmermann, B. (1992). Degeneration of osteoblasts involved in intramembranous ossification of fetal rat calvaria. *Cell Tissue Res.* 267, 75–84.
85. Katchburian, E. (1978). Early mineral deposition in intramembranous ossification [proceedings]. *J. Anat.* 126, 426–428.
86. Jiang, X., Iseki, S., Maxson, R. E., Sucov, H. M. and Morriss-Kay, G. M. (2002). Tissue origins and interactions in the mammalian skull vault. *Dev. Biol.* 241, 106–116.

87. Cohen, M. M., Jr. and Kreiborg, S. (1996). Suture formation, premature sutural fusion, and suture default zones in Apert syndrome. *Am. J. Med. Genet.* 62, 339–344.
88. Bradley, J. P., Levine, J. P., Roth, D. A., McCarthy, J. G. and Longaker, M. T. (1996). Studies in cranial suture biology: IV. Temporal sequence of posterior frontal cranial suture fusion in the mouse. *Plast. Reconstr. Surg.* 98, 1039–1045.
89. Roth, D. A., Bradley, J. P., Levine, J. P., McMullen, H. F., McCarthy, J. G. and Longaker, M. T. (1996). Studies in cranial suture biology: part II. Role of the dura in cranial suture fusion. *Plast. Reconstr. Surg.* 97, 693–699.
90. Plavcan, J. M. and German, R. Z. (1995). Quantitative evaluation of craniofacial growth in the third trimester human. *Cleft Palate Craniofac. J.* 32, 394–404.
91. Shah, R. K., Dhingra, J. K., Carter, B. L. and Rebeiz, E. E. (2003). Paranasal sinus development: a radiographic study. *Laryngoscope* 113, 205–209.
92. Honig, J. F., Merten, H. A., Schutte, R., Grohmann, U. A. and Cassisis, A. (2002). Experimental study of the frontal sinus development on Goettingen miniature pigs. *J. Craniofac. Surg.* 13, 418–426.
93. Hulth, A., Johnell, O., Lindberg, L. and Heinegard, D. (1993). Sequential appearance of macromolecules in bone induction in the rat. *J. Orthop. Res.* 11, 367–378.
94. Poole, A. R. and Rosenberg, L. C. (1986). Chondrocalcin and the calcification of cartilage. A review. *Clin. Orthop.* 114–118.
95. Poole, A. R., Pidoux, I., Reiner, A., Choi, H. and Rosenberg, L. C. (1984). Association of an extracellular protein (chondrocalcin) with the calcification of cartilage in endochondral bone formation. *J. Cell Biol.* 98, 54–65.
96. Roy, M. E. and Nishimoto, S. K. (2002). Matrix Gla protein binding to hydroxyapatite is dependent on the ionic environment: calcium enhances binding affinity but phosphate and magnesium decrease affinity. *Bone* 31, 296–302.
97. Zebboudj, A. F., Imura, M. and Bostrom, K. (2002). Matrix GLA protein, a regulatory protein for bone morphogenetic protein-2. *J. Biol. Chem.* 277, 4388–4394.
98. Bostrom, K., Tsao, D., Shen, S., Wang, Y. and Demer, L. L. (2001). Matrix GLA protein modulates differentiation induced by bone morphogenetic protein-2 in C3H10T1/2 cells. *J. Biol. Chem.* 276, 14044–14052.
99. Fusetti, F., Pijning, T., Kalk, K. H., Bos, E. and Dijkstra, B. W. (2003). Crystal structure and carbohydrate-binding properties of the human cartilage glycoprotein-39. *J. Biol. Chem.* 278, 37753–37760.
100. Steenbakkers, P. G., Baeten, D., Rovers, E., Veys, E. M., Rijnders, A. W., Meijerink, J., De Keyser, F. and Boots, A. M. (2003). Localization of MHC class II/human cartilage glycoprotein-39 complexes in synovia of rheumatoid arthritis patients using complex-specific monoclonal antibodies. *J. Immunol.* 170, 5719–5727.
101. Burton-Wurster, N., Liu, W., Matthews, G. L., Lust, G., Roughley, P. J., Glant, T. T. and Cs-Szabo, G. (2003). TGF beta 1 and biglycan, decorin, and fibromodulin metabolism in canine cartilage. *Osteoarthr. Cartilage* 11, 167–176.
102. Theocharis, A. D., Karamanos, N. K., Papageorgakopoulou, N., Tsiganos, C. P. and Theocharis, D. A. (2002). Isolation and characterization of matrix proteoglycans from human nasal cartilage. Compositional and structural comparison between normal and scoliotic tissues. *Biochim. Biophys. Acta* 1569, 117–126.
103. Connor, J. R., Dodds, R. A., Emery, J. G., Kirkpatrick, R. B., Rosenberg, M. and Gowen, M. (2000). Human cartilage glycoprotein 39 (HC gp-39) mRNA expression in adult and fetal chondrocytes, osteoblasts and osteocytes by *in-situ* hybridization. *Osteoarthr. Cartilage* 8, 87–95.

104. Demoor-Fossard, M., Redini, F., Boittin, M. and Pujol, J. P. (1998). Expression of decorin and biglycan by rabbit articular chondrocytes. Effects of cytokines and phenotypic modulation. *Biochim. Biophys. Acta* 1398, 179–191.
105. Giannoni, P., Siegrist, M., Hunziker, E. B. and Wong, M. (2003). The mechanosensitivity of cartilage oligomeric matrix protein (COMP). *Biorheology* 40, 101–109.
106. Ramstad, V. E., Franzen, A., Heinegard, D., Wendel, M. and Reinholt, F. P. (2003). Ultrastructural distribution of osteoadherin in rat bone shows a pattern similar to that of bone sialoprotein. *Calcif. Tissue Int.* 72, 57–64.
107. Di Cesare, P. E., Chen, F. S., Moergelin, M., Carlson, C. S., Leslie, M. P., Perris, R. and Fang, C. (2002). Matrix–matrix interaction of cartilage oligomeric matrix protein and fibronectin. *Matrix Biol* 21, 461–470.
108. Shibata, S., Fukada, K., Suzuki, S., Ogawa, T. and Yamashita, Y. (2002). *In situ* hybridization and immunohistochemistry of bone sialoprotein and secreted phosphoprotein 1 (osteopontin) in the developing mouse mandibular condylar cartilage compared with limb bud cartilage. *J. Anat.* 200, 309–320.
109. Qabar, A., Derick, L., Lawler, J. and Dixit, V. (1995). Thrombospondin 3 is a pentameric molecule held together by interchain disulfide linkage involving two cysteine residues. *J. Biol. Chem.* 270, 12725–12729.
110. Tucker, R. P., Adams, J. C. and Lawler, J. (1995). Thrombospondin-4 is expressed by early osteogenic tissues in the chick embryo. *Dev. Dyn.* 203, 477–490.
111. Chen, J., Shapiro, H. S. and Sodek, J. (1992). Development expression of bone sialoprotein mRNA in rat mineralized connective tissues. *J. Bone Miner. Res.* 7, 987–997.
112. Miller, R. R. and McDevitt, C. A. (1988). Thrombospondin is present in articular cartilage and is synthesized by articular chondrocytes. *Biochem. Biophys. Res. Commun.* 153, 708–714.
113. Karaplis, A. C., Luz, A., Glowacki, J., Bronson, R. T., Tybulewicz, V. L., Kronenberg, H. M. and Mulligan, R. C. (1994). Lethal skeletal dysplasia from targeted disruption of the parathyroid hormone-related peptide gene. *Genes Dev.* 8, 277–289.
114. Lanske, B., Karaplis, A. C., Lee, K., Luz, A., Vortkamp, A., Pirro, A., Karperien, M., Defize, L. H., Ho, C., Mulligan, R. C., Abou-Samra, A. B., Juppner, H., Segre, G. V. and Kronenberg, H. M. (1996). PTH/PTHrP receptor in early development and Indian hedgehog-regulated bone growth. *Science* 273, 663–666.
115. Weir, E. C., Philbrick, W. M., Amling, M., Neff, L. A., Baron, R. and Broadus, A. E. (1996). Targeted overexpression of parathyroid hormone-related peptide in chondrocytes causes chondrodysplasia and delayed endochondral bone formation. *Proc. Natl. Acad. Sci. USA* 93, 10240–10245.
116. Schipani, E., Langman, C. B., Parfitt, A. M., Jensen, G. S., Kikuchi, S., Kooh, S. W., Cole, W. G. and Juppner, H. (1996). Constitutively activated receptors for parathyroid hormone and parathyroid hormone-related peptide in Jansen's metaphyseal chondrodysplasia. *N. Engl. J. Med.* 335, 708–714.
117. Vortkamp, A., Lee, K., Lanske, B., Segre, G. V., Kronenberg, H. M. and Tabin, C. J. (1996). Regulation of rate of cartilage differentiation by Indian hedgehog and PTH-related protein. *Science* 273, 613–622.

2 Bone Physiology Dynamics

Bruce Doll and Hannjörg Koch

Contents

Normal bone formation is a prolonged process that is carefully regulated and involves sequential expression of growth-regulatory factors. The physiology of bone involves a complex interrelationship between the cellular, molecular, and systemic components. The skeletal system is continuously remodeled — a balanced response of mineral resorption and deposition consistent with mechanical and molecular influences.

BONE REMODELING

The idea that bone remodeling is controlled by mechanical as well as metabolic factors, popularized by Wolff a century ago, has been increasingly intensely studied in recent decades. Current thinking about Wolff's law revolves around several key concepts and observations.[1, 2] First, it is postulated that bone contains sensor cells that monitor mechanical strain (or another load-related variable), compare it to a physiologically desirable range of values, and activate corrective biological processes when the sensed variable falls outside this range. Several investigators have developed computational simulations of how the bone adapts to mechanical loading based on this general scheme. Such models routinely assume that, when the mechanical stimulus is too low, remodeling removes bone and, when it is too high, remodeling adds bone.

Second, many investigators have suggested that osteocytes, distributed throughout the bone matrix, are the bone's mechanosensing cell. There has been considerable speculation that osteocytes produce a signal proportional to mechanical loading

0-8493-1621-9/05/$0.00+$1.50
© 2005 by CRC Press LLC

by sensing strain on bone surfaces through stretch-activated ion channels, flow of interstitial fluid, electrical potentials, or some other phenomenon. This idea arises from the observation that these cells are ideally located for this function, and apparently communicate with one another, as well as with osteoblasts and bone lining cells, through dendritic processes and gap junctions, forming a functional syncytium. Cell-to-cell communication of electrical signals and small molecules through gap junctions has been demonstrated in osteoblasts. There is evidence for similar gap junctions in osteocytes, and it seems likely that they participate in such communication with osteoblasts, and bone-lining cells as well.[3]

Third, it has been suggested that osteocytes also sense fatigue damage and transmit signals that activate remodeling to remove the damage.[4] Microdamage in cortical bone is associated with an increased activation of remodeling, and it is generally assumed that the same is true in cancellous bone.[5] This remodeling is postulated to remove damaged bone and prevent the occurrence of fatigue fracture.[6,7]

A fourth key concept suggests that cells of the osteoblast lineage control the initiation of remodeling.[8] Subsequently, many investigators have adopted the modified hypothesis that these "retired" osteoblasts, known as bone lining cells, are responsible for activating basic multicellular units to remodel bone in response to signals from osteocytes or hormones. Taken together, these four concepts form an attractive model in which osteocytes sense mechanical changes and initiate remodeling to modify bone structure accordingly. Many investigators subscribe to this general model, even though they may have differing views regarding its details.[9]

A theory has been developed to resolve several inconsistencies between current concepts and observations about bone remodeling. For example, the observation that remodeling increases both when mechanical loading is excessively low in a disuse state, and when it is excessively high, producing substantial fatigue damage, is contrary to the widely held assumption that a signal generated by osteocytes in proportion to mechanical loading stimulates bone lining cells to activate remodeling. The new theory resolves this disparity by assuming that lining cells are inclined to activate remodeling unless restrained by an inhibitory signal, and that the mechanically provoked osteocytic signal serves this inhibitory function. Consequently, remodeling is elevated when signal generation declines due to reduced loading, or when signal generation or transmission is interrupted by damage due to excessive loading. Otherwise, remodeling is maintained at a relatively low level by inhibitory signals produced through physiologic loading. Furthermore, the inhibitory signal is postulated to be identical to that proposed by Marotti as the mechanism for conversion of osteoblasts to osteocytes and responsible for the diminishment of apposition rate during refilling of osteonal basic multicellular units. This theory postulates that, when a previously formed osteocyte is sufficiently covered by new bone and osteoid, it sends an inhibitory signal through its dendritic processes to the neighboring osteoblasts which reduces their individual apposition rates. The osteoblast most affected by this inhibition becomes buried by its neighbors, and becomes one of the next layer of osteocytes. Consequently, a single, mechanically derived signal produced in the osteocytic syncytium may control osteoblast and bone-lining cell functions, and thereby a variety of important phenomena in bone biology.[2,6]

Remodeling could be activated by mechanical damage in a variety of ways. Several different kinds of damage have been identified in histological sections from cyclically loaded bone specimens.[10] These include microcracks and more diffuse disturbances of the calcified matrix structure. It seems clear from observing such damage that it could easily interfere with both the normal flow of fluids over osteocyte processes within canaliculae (signal generation) and with the passage of signals from one cell to another through these processes (signal transmission). It is also possible that diffuse damage to the calcified matrix could release Ca ions or cytokines that could act to reduce the inhibitory signals produced by the osteocytes, or their transmission from cell-to-cell. Furthermore, it has been hypothesized that microdamage results in osteocyte apoptosis, again interrupting both signal generation and transmission. By one or more of these mechanisms, microdamage would be expected to reduce the osteocytic signals to the bone-lining cells, allowing activation of increased remodeling.[10, 11]

Disuse and damage produced by overload are the two situations in which substantial departures from the normal range of mechanical loading—low or high, respectively—are known to activate remodeling. Another instance in which remodeling is known to be activated is local bone matrix necrosis—that is, when osteocytes die. This may occur due to microdamage or when the haversian blood supply is interrupted following fracture. Osteocyte death would necessarily reduce the generation and transmission of inhibitory signals to nearby bone lining cells and activate increased remodeling; the desirable result of this would, of course, be replacement of the dead bone with new bone structural units containing viable osteocytes. More gradual and diffuse osteocyte death has been associated with aging, which, along with the resulting diminishment of the inhibitory signal, would lead to increased remodeling and bone loss. But at this point, the pace of change would be subtle, in keeping with senile osteoporosis.

Unification of these concepts is addressed by Martin's proposal that a single kind of signal transmitted through the bone's cellular syncytium of osteocytes, osteoblasts, and lining cells controls remodeling in response to extremes of mechanical loading and such exigencies as osteocyte death and hormonal fluctuations. When BMU refilling is complete, and the remaining osteoblasts become bone-lining cells, they continue to receive signals from the osteocytes in the completed bone structural unit. Now, however, instead of regulating osteocyte formation, the signals are hypothesized to inhibit bone-lining cells from activating a new remodeling cycle. This extension of Marotti's hypothesis, while contrary to the current paradigm of osteocytes as activators of remodeling, can eliminate inconsistency in current concepts and serve as the basis for a unifying theory of bone remodeling.

The theory recognized that these three kinds of bone cells (osteoblasts, osteoclasts, and osteocytes) are inexorably linked not only by gap junctions, but through a common differentiation pathway as well. Osteoblasts change into osteocytes and bone-lining cells seamlessly, and none of the three can realize their full function in remodeling without the other two. Such communality, consistent with the existence of a common signal that osteocytes use to guide osteoblasts into the bone matrix, serves to gauge its burden of mechanical stress and damage, and signals bone-lining cells when it is time to remove and replace the tissue.

In such a system, the activation of remodeling would be increased when either the generation or the transmission of the inhibiting signals diminishes. This could occur in several ways, consistent with both the general model and the mechanostat theory. First, remodeling would increase in a disuse state because strain-generated inhibitory signals would be diminished. Furthermore, this response would occur not only in a global disuse situation but also in cases of local disuse. Thus, for example, severing of a trabecular strut would unload it and lead to activation of remodeling by those lining cells on its remaining surfaces because they would no longer receive inhibiting signals from their connecting osteocytes.

Other factors clearly play roles in modulating the activation of remodeling.[12] For example, daily injections of biosynthetic parathyroid hormone activate remodeling in cortical bone, and intact parathyroid and thyroid glands seem to be necessary for the activation of cortical remodeling in a disuse state.[13] Marrow cellularity is another element that may affect the activation of remodeling, with more hematopoietic marrow being associated with higher turnover rates. If the concepts introduced here are valid, there should be pathways by which such factors affect the generation, transmission, or interpretation of the inhibitory signal in the osteocyte-lining cell syncytium.

The response of the osteocytes facilitates bone growth and adaptation to the body's mechanical needs for strength with lightness. Osteocytes utilize some molecular signaling pathways such as the generation of nitric oxide and prostaglandins as well as directing cell–cell communication via gap junctions.[14–17] They may also direct the removal of damaged or redundant bone through mechanisms linked to their own apoptosis or via the secretion of specialized cellular attachment proteins such as osteopontin. Osteocytes possess receptors for parathyroid hormone/parathyroid hormone related peptide and both estrogen receptors alpha and beta.[18] They also express glutamate/aspartate transporter, GLAST-1, which are involved with glutamate neurotransmission in nerve cells. Extracellular binding of glutamate to bone cells causes increase in osteoblast number. The regulation of the expression and distribution of GLAST-1 occurs by mechanical stimuli. Extracellular glutamate, a substrate of GLAST-1, in osteocytes may regulate glutamate signaling in bone, consistent with its operation in the central nervous system.[19] At least some of these receptors and their ligands may regulate osteocyte apoptosis and modulate osteocyte signaling.[20] In conclusion, osteocytes modulate signals arising from mechanical loading, subsequently contributing to the appearance and disappearance of mineralized tissue.

AGING AND BONE INTEGRITY

Osteocyte density can be expressed as the mean value for cells measured from different areas. However, the integrity of the osteocyte network depends not only on mean changes but also on the pattern of cell distribution. If osteocytes are distributed unevenly in the bone, the morphology of the cell network could be different from one area to another. A poor cellular network would be expected where there are fewer osteocytes, as has been demonstrated by confocal microscopy. There may be a critical value for osteocyte density, below which network signal transmission is impaired, but in cancellous bone there are minimal data on osteocyte or lacunar distribution.

Osteocyte density in iliac cancellous bone (cell number per unit section area of bone) declines with age in bone >45 μm from the surface, most of which is interstitial bone, but does not decline with age in bone <25 μm from the surface. Deep bone is less frequently remodeled and so is older than superficial bone; this suggests that one function of cancellous bone remodeling is to maintain osteocyte density in superficial bone. If bone remodeling is driven solely by hormonal and mechanical factors, and is not influenced by osteocytes, there should be a positive correlation between the bone formation rate and osteocyte density. This would reflect more frequent replacement of bone, in which some osteocytes may have died, by new bone, and would correspond to the higher values for both bone turnover and osteocyte density in superficial than in deep bone. Alternatively, if osteocytes are able to inhibit bone remodeling, there should be either absent or negative correlation between the rate of bone formation and osteocyte density.

There is a significant decline in osteocyte density with age in the deep regions of normal human iliac cancellous bone, beginning at or before skeletal maturity and without postmenopausal acceleration. There was no change with age in variability between regions, but the decline did not occur in the surface bone, presumably because of more frequent bone remodeling. The spontaneous death of osteocytes appears to be a normal phenomenon, which may contribute to fatigue microdamage accumulation with age.[6, 20–23]

Osteocytes exert a restraining effect on bone remodeling, presumably by release of an inhibitory signal molecule. The lack of effect of osteocyte density in deep bone suggests that the concentration of this putative molecule at the bone surface declines with increasing distance of its cell of origin from the surface, an effect augmented by the paucity of canaliculi that cross a cement line. The additional positive effect of empty lacunae, which are sites of previous osteocyte death, suggests that the loss of inhibitory effect depends not only on the absolute number but also on the relative decline in osteocyte density. The hormonal, mechanical, and damage repair effects on bone remodeling remain dominant, but can be mediated in part by changing signal release by osteocytes without a change in their number.

A decline in osteocyte viability with age has been observed in cortical bone at various sites. The proportion of rib bone lacunae containing osteocytes declines from 95% at adolescence to about 70% by 35 years of age with minimal further reduction.[24] A decline in stainable osteocytes was demonstrated both absolutely and as a proportion of total lacunae with increasing age. Beginning before menopause, healthy women demonstrated about 0.5% of the value at age 20 per year in central cancellous bone.[25] However, the data suggest an exponential approach to an asymptotic value of total osteocytes rather than a linear decline. A similar decline was found in the ileum in a recent autopsy study. The sample was small and predominantly male. Their linear regression slope for osteocytes disappearance was –1.04/year, compared with –1.31/year in a separate study; thus, the decline with age may be faster in women than in men. The assessment of total osteocytes decline using lacuna with vs. without osteocytes is confounded by the probability of not detecting lacunae obliterated by micropetrosis. An empty lacuna is strong evidence of cell death, but the decline in total lacunae suggests that some become filled with mineralized debris after osteocyte death in a process defined as micropetrosis, and are therefore no longer detectable.[25]

The internal structure of compact bone changes with age and mirrors its functional state. Mandibular bone was evaluated at different ages in rats. The basal bone was compared with the alveolar compact bone in the rat mandible. Large irregular central vascular canals and lacunae were more concentric in the basal than the alveolar bone for young rats. Canaliculi within the mandibular compact bone thinned with extensive branching and then decreased in size and number with advanced age. The lacunae appeared to proceed from the large circular structures of youth to the flat forms of the aged.[26] In rat long bones, the bone volume usually decreases with progression of time after a growth period. Alveolar bone turnover in rats after undergoing ovarectomy declined one year after ovary removal. The bone volume decreased, and trabecular changes were brought about by the fragmentation of each trabecula.[27]

Iliac cancellous bone has a mean volume-based turnover bone-forming rate/bone volume (BFR/BV) of 15–20%/year, much higher than in the peripheral cortical bone where it may be no more than 2–4%/year; however, these calculations do not take into account the distance from the surface, where remodeling begins. The probability that a particular moiety of bone will be remodeled decreases, and its estimated age increases with increasing distance from the surface. At normal values for surface-based bone formation rate, the mean age would be <2 years in superficial bone and >10 years in deep bone. Observations that osteocyte density declines with age only in deep older bone and not in younger superficial bone imply that the age of the bone and not that of the subject is important to the status of the osteocytes. One function of bone remodeling in iliac cancellous bone may be to prevent spontaneous osteocyte death, which can be accomplished in two different ways. With advancing age of the bone, a possibility is that osteocyte death leads to the targeted replacement of a specific region of bone, similar to the postulated role of osteocyte death for the targeted replacement of fatigue-damaged cortical bone. Significant osteocyte death may also be minimized by nontargeted bone remodeling, thereby maintaining a younger population of osteocytes.

Marotti et al. indirectly estimated the number of processes that come into contact with osteoblasts, by counting the number of canalicular openings present on the osteogenic surfaces viewed under the scanning electron microscope. They observed a uniform distribution of canalicular openings ranging between 9.4 and 20.9 μm beneath one rat osteoblast. In another study, the mean of the number of processes that came into contact with one osteoblast was 4.8 μm. The distribution of osteocyte processes to osteoblasts was not uniform. Osteoblasts directly beneath an osteoid–osteocyte contained numerous processes, but osteoblasts somewhat further away constituted only very few processes. These differences from Marotti's study suggest that not all canaliculi are filled with osteocyte processes, but this seems unlikely, as the distribution pattern of canaliculi generally coincides with that of the osteocyte processes. Therefore, species, age, and/or anatomical differences between the two studies likely account for the different observations. Interestingly, it was rare to find an osteoblast that had connections with several osteocytes at one time, that is, usually one osteoblast was connected to only one osteoid–osteocyte. This suggests that an osteocyte might have its territory of osteoblasts contacted by osteocyte processes. Furthermore, it is reported that only a limited number of osteoblasts can become osteocytes, and that this unique selection is done by committed osteocytes. However, the precise mechanism of the differentiation from osteoblast to osteocyte is not known. The heterogeneous distribution of osteocyte processes to osteoblasts might be a key to understanding why only a certain number of osteoblasts become osteocytes. In addition, such selection might be done in the territory of an osteoid–osteocyte.

Contribution of Cellular Components and Alterations in Gene Expression

Decreased bone formation during bone remodeling and fracture healing in the elderly patient can be caused by either a reduced number of recruited osteoblast precursors, a decline in proliferative activity of osteogenic precursor cells, a reduced maturation of osteoblast precursors, and/or a reduced functional activity of mature osteoblast during bone formation.[28] Advancing age is associated with changes in the bone mineral, bone matrix, and osteogenic cell population.[29, 30] Additionally, it is a frequent clinical experience that fractures heal faster in children than in adults.[31] Mechanistic explanations for the differences in bone-healing response reflect a general functional decline in the homeostatic mechanisms during aging and senescence.[32] A comprehensive understanding of bone-healing mechanisms in the elderly patient will enable therapies to resolve diminished healing capability.[33–35] For example, differences in fracture healing in the elderly population may be attributable to local or systemic changes in hormonal and growth factor secretion, altered receptor levels, changes in the extracellular matrix composition, or an uncoupling of osteoblastic and osteoclastic activity.

Predominant research in the aged skeleton has focused on clinical studies involving osteoporotic and osteopenic patients. Osteoporosis is an expanding area of pathology due to the increasing lifespan of our population. Osteoporotic weakness in the skeleton associated with diminished mineralized content is a consequence of presently theorized imbalances in the normally coupled activity of osteoblasts and osteoclasts. However, there is no unanimity in the literature about the definition of the terms osteoporosis and osteopenia. In 1994, the World Health Organization Osteoporosis Study Group established diagnostic categories based on bone mineral density using young adult women as the reference group.

The standard deviation from the young–adult mean, "T-score," defines the diagnostic categories: Normal: T-score ±1 SD, osteopenia: T-score –1 to –2.5 SD, osteoporosis: T-score –2.5 or less, and Severe osteoporosis: T-score –2.5 or less and fragility fracture. Cellular processes *in vitro* and *in vivo* dealing with cellular aging of mature osteoblasts, committed osteoprogenitor cells, and uncommitted mesenchymal stem cells, proliferative and differentiation activity characterized by the expression of osteoblast specific genes, matrix synthesis, and mineralization were investigated in these two pathological models. General cellular and biochemical processes of fracture repair in the elderly, healthy patient are less available, perhaps because of the rarity of an aged, healthy skeletal system or because we need a refined definition for an aged, healthy skeletal system.[36, 37]

Significant research has investigated the coupling between bone resorption and bone formation and its imbalances, leading to osteopenia and osteoporosis. Emphasis on pathological conditions compromising bone healing has overshadowed research focused on defining normal age-related effects on bone healing. There is a need for further investigations on processes affecting fracture healing under nonpathologic conditions in the elderly, healthy patient. At the moment, less information is available regarding the similarities and changes in the process of fracture healing in the elderly patient. Furthermore, definition of skeletal age groups including molecular and cellular parameters are required. A standardized fracture repair model for comparisons of components involved in the fracture-healing process in elderly individuals is also required. Presently, it is difficult to make general conclusions on fracture healing in the elderly, healthy patient due to the different animal models and fracture techniques used. The data obtained in animal fracture-healing models (rat, rabbit) must be carefully applied to the human physiologic fracture repair process in the elderly patient, because the relative ages of the animals may be quite different compared with the human subjects. Relevant studies include a molecular analysis of age-related differences in healing.

Molecular and Cellular Activity in Bone During Aging

The functional activity of mature osteoblasts *in vitro* has been examined. Serially passaged cultures of human trabecular osteoblasts exhibited limited proliferative activity and underwent cellular aging *in vitro*.[38, 39] Several changes were evident during serial passaging of human trabecular osteoblasts. Alterations in morphology and cytoskeleton organization, an increase in cell size and higher levels of senescence associated β-galactosidase activity, a reduction in macromolecular synthesis, and reduced mRNA levels of alkaline phosphatase (AP), osteocalcin (OC) and collagen type I (Col I) were noted.[33, 39] The gene products are important for bone matrix formation, matrix maturation, and mineralization (Table 2.1).

Genes important to the osteoblast phenotype are under transcriptional control. Specific alterations in transcriptional control may lead to an altered osteoblast phenotype. Runx2, also known as Cbfa1 and topoisomerase I, are involved in transcriptional control and maintenance of chromosomal integrity. Changes of Cbfa1 and topoisomerase I levels during cellular aging of human trabecular osteoblasts have been detected.[40] Ducy and co-workers reported an age-related progressive and significant decline in steady-state mRNA levels of these genes in human bone cells undergoing cellular aging *in vitro*. Reduced osteoblast function during cellular aging was evident. Runx2 is a transcription factor known to bind to osteoblast-specific *cis*-acting elements (OSE2) in genes expressed during osteoblast differentiation (alkaline phosphatase, OC, collagen type I). A progressive age-related decrease of Runx2 expression could modify the expression of these genes. A decrease in osteoblast functions, such as a decreased formation of extracellular matrix and reduced mineralization are possible outcomes.[41,42]

Similar molecular events are apparent during fracture healing. The capacity to facilitate bone healing in an elderly patient may be influenced by altered cellular transcriptional capability. A reduced proliferative capacity and function of osteoblasts could contribute to the development of delayed fracture healing in the elderly healthy patient. A correlation between donor age and the impairment of osteoblastic functions (Col I, OC, and extracellular matrix synthesis) as well as a reduced AP activity during *in vitro* culture of human mature osteoblasts could contribute to problematic clinical challenges involving bone healing in the elderly.[42–45]

Martinez and co-workers examined the cell proliferation rate and the secretion of C-terminal type I procollagen (PICP), AP, and OC in primary cultures of osteoblastic cells from human trabecular bone in relation to skeletal site and donor age. They noticed a lower proliferation rate and OC secretion in osteoblastic cells from the older donors compared with younger subjects.[46] AP secretion was higher in the older subjects, whereas PICP secretion was unchanged. Significant differences

Table 2.1
Function of Molecular Proteins in the Osteocyte

Molecular marker	Cell: osteocyte/bone-lining cell
Osteocalcin	Secreted by mature osteocytes, incorporated into lacunar matrix[67, 68]
Osteopontin	Secreted by mature osteocytes, incorporated into lacunar matrix[67]
Collagen I	Secreted by mature osteocytes, incorporated into lacunar matrix[67]
Osteoblast/osteocyte factor 45 (OF45)	OF45 mRNA is transiently expressed by mature osteoblasts and subsequently expressed by osteocytes throughout ossification in the skeleton and this protein represents an important marker of the osteocyte phenotype and most likely participates in regulating osteocyte function[69]
Connexin	Terminally differentiated osteocyte-like MLO-Y4 cells respond to PEMF with changes in local factor production and reduced Cx43, suggesting decreased gap junctional signaling[14]
Alkaline phosphatase	Express lower levels of alkaline phosphatase (vs osteoblast) when cultured with (1-34) PTH and stretched[70]
Osteoprotegerin (OPG)	RANKL and OPG may play a role in osteocyte signaling, OPG and M-CSF as soluble factors and RANKL as a surface molecule that is functional in osteocytes or along their exposed dendritic processes[71]
Macrophage colony stimulating factor (M-CSF)	MLO-Y4 cells express and secrete large amounts of M-CSF[71]
CD44	Express CD44, adhesion molecule, in accordance with their morphological changes from osteoblasts into osteocytes[72]
Connective tissue growth factor (CTGF)	Regulation of osteocyte function during the mechanical stimulation of bone[73]
Transforming growth factor beta (TGF-β)	Secrete significant amounts of TGF-beta, which inhibits bone resorption and is modulated by estrogen[71]

in these expression patterns existed in relation to the skeletal site of origin.[47] Precursor callus cells (CFC) at the fracture site that differentiate into chondrocytes or directly into osteoblasts by intramembranous ossification were shown to proliferate at a greater rate in the 6-week rats than in 6-month-old rats using the measurement of total DNA. The theoretical basis of these studies and their importance in the understanding of the process of bone healing in the elderly patient are useful for evaluating osteoblastic alterations associated with aging.

Additional evidence demonstrated that human bone-derived cells undergo a decrease in their proliferative capacity with donor age. These cells also have a reduced proliferative rate after stimulation with various growth factors. The gender and age-related changes in iliac crest cortical bone and serum OC in humans have been studied. A significant age-related decline of bone and serum OC content with increasing patient age was apparent. Furthermore, a parallel decrease in OC was noted in age-matched groups. However, generally higher concentrations of bone and serum OC were found in men than in women.[48]

OSTEOGENIC ACTIVITY IN AGED ANIMALS

Animal models have indicated differences in inductive capacity relative to the age of the source. Bone from old rats was found to be less osteogenic than younger ratsamples.[49,50] Demineralized bone matrix (DBM) prepared from younger donors was more osteoinductive than preparations from older animals.[50,51] The osteogenic response was earlier and more complete in 6-month-old rats with a 4 mm rat calvarial defect compared to 24-month-old rats. Osteoblasts and woven bone appeared at 7 days in the younger animals, but were not detected in the older animals 21 days after creation of the calvarial defect. Differentiation of osteoblastic cells in older animals decreased.

AGE-RELATED BONE HEALING

Age-related bone healing may be a consequence of a lack of osteoblastic formation and/or function. Working with serially passaged adult rat bone cells, Williams and co-workers investigated the potential for mineralization and response to hormones. They observed a diminution of the response to PTH and PGE_2 with serial passage. No significant decrease in the potential to mineralize *in vitro* after extended serial passages was detectable.[52] Martinez demonstrated that both age and site of osteoblast cell harvest were determinants for the osteoblast response to 1,23 (OH) vitamin D_3. Using age-matched controls, OC secretion was lower in hip osteoblast cells compared with those from the knee in the older group. OC secretion was similar in cell cultures from both skeletal sites in the younger group. 1,25(OH) 2 vitamin D_3 stimulation of both cells from either age group was not different when knee osteoblast cells were tested. Cells from the hip in the older group were less responsive.[53] In an ultrastructural analysis of trabecular bone and bone marrow stromal cells, Roholl and co-workers were able to examine distinct changes in the number of osteoblasts per unit bone surface in aging rats. The study determined whether the spontaneous trabecular bone loss with aging is due to the loss of functional osteoblasts caused by a deficient maturation of pre-osteoblasts into osteoblasts or due to an earlier stage defect leading to a diminished number of pre-osteoblasts. The proliferation potential of bone marrow mesenchymal stromal cells, considered a source of osteoprogenitor cells, was examined. They found a significant reduction by more than 10-fold in the number of osteoblasts. The population of pre-osteoblasts, inactive bone-lining cells of the trabecular bone, pre-osteoclasts, and osteoclasts were considered age independent. Important for the interpretation of these data is the number of cells related to units of bone surface. There was a significant decrease in total bone volume and bone surface of more than twofold found in the process of aging. The absolute number of osteoblasts was decreased twofold. A significant age-related decline of the number of stromal cells in the bone marrow was reported. In conclusion the results suggest a diminished maturation of pre-osteoblasts into osteoblasts during aging in rats, based on a more than 10-fold increase of the ratio of pre-osteoblasts to osteoblasts with age, a constant number of pre-osteoblasts and a lowered number of osteoblasts per unit bone. The data indicated an age-independent stimulation of osteoblasts in analyzing the appearance of osteoblasts in the process of bone repair of femoral fractures 3 and 16 days after fracturing. Apparently, the pre-osteoblasts in aged rats underwent osteoblast differentiation under the extreme stimulatory conditions of fracture healing. Under normal aging conditions, there may be a lack of appropriate stimuli for this maturation to take place. The number of colony-forming cells were significantly lower in aged compared to adult and young rats. Diminished cell population was accompanied by an altered differentiation of pre-osteoblast to osteoblast. The number of osteoblasts was low in the face of a constant number of pre-osteoblasts per unit of bone. Ultrastructural appearance of the osteoblasts suggested less functional activity. The diminished maturation of pre-osteoblasts to osteoblasts could be a consequence of deficient maturation stimuli in the microenvironment and may not be intrinsic to aged animal pre-osteoblasts.[54] Sekine and co-workers evaluated the effect of aging on the recovery of cellular proliferative ability in the experimentally fractured condylar model. Bromodeoxyruridine (BrdU) immunohistochemistry was employed to determine cell proliferative activity in the intermediate cell layer of the condyle. 3-6-and 36-month-aged Sprague–Dawley rats were used to study this effect. A significant age-related decline in the recovery of cell proliferation following fracture was determined, indicating that the stimulus of bone fracture could not promote osteoprogenitor cell recruitment and proliferation to levels comparable with levels reached in a fracture model in young rats.[28] These results seem to contradict former results of Roholl but differences in origin of the cell pool, the staging of the cells in the osteoblastic lineage hierarchy, and differences in the age range of animals may explain these inconsistencies. The findings underscore a need for uniform standardized fracture models to study the effects of increasing age on the mechanisms of fracture repair in an elderly, healthy population.

In an experimental implantation model, demineralized bone matrix (DBM), containing bone morphogenetic proteins (BMP), and bone marrow was transplanted to an ectopic intramuscular site directly or after enclosure in filter chambers. Bone marrow produced more bone in young than in old animals following implantation directly into muscle or after enclosure in filter chambers. Similar results were found with the implantation of DBM directly to muscle. No bone formation was stimulated by implantation of DBM in filter chambers, suggesting that direct cell contact or interaction with the DBM may be required for the release of the BMP, present in the matrix. Furthermore, a significant decrease in the total calcium content and a reduced AP activity per cell in implants of DBM was detected in the old animals in comparison to implants in the young animals. These results are in agreement with those of Reddi and Nishimoto, who found an age-dependent decline in bone formation capacity by subcutaneous implantation of DBM in rats. This was assessed histologically and by AP activity, ^{45}Ca uptake, and absolute calcium- and vitamin K-dependent protein (osteocalcin, serum level) content.[49,55] Using a similar model, Irving and co-workers have shown that 6-week-old rats form bone in 14 days while 2-year old rats form bone in 23 days.[56] Syftestad and Urist demonstrated that older rat bone was less effective in generating an osteogenic response.[57] A decrease in available BMP may occur while aging.[57,58] However, responsiveness by osteogenic cells to BMP is retained in senescent rats. In another rat study recombinant BMP-2 combined with fibrous collagen matrix has the ability to induce new bone formation continuous with original bone. Responsiveness to BMP was independent of age when generating new bone in the palate; however, less bone was formed in the older rat model.[59] The number of bone marrow stromal cells undergoing differentiation and/or the capacity of bone marrow stromal cells to form new bone in an appropriate environment appears to decrease with aging as indicated by animal studies.

In contradiction to the former hypothesis supporting a decrease in the osteoinductive potential of DBM with increasing donor age,[60] it was found that there was a significant increase in the potential of DBM to induce bone formation at the recipient site with increasing donor age. These results were inconsistent with studies reporting a decrease in BMP content in the bone matrix with aging.[61] Contradictory results may reflect differences in the age range of study animals or in techniques of DBM preparation, implantation, and evaluation methods.

Basil and co-workers tested the osteogenic response of 1- and 16-month-old rats to intramuscularly implanted DBM. They observed that the implants in the older animals had less calcium and alkaline phosphatase activity compared to younger animals. This may be due to failure of differentiation of stem cells into osteoblastic lineage, failure of the osteoblasts to calcify a matrix, and the reduced population of stem cells in an older animal. DBM prepared from old animals, was found to produce less bone than that from young animals when implanted into ectopic sites of the allogenic host. DBM from 4-month-old rats were placed at subcutaneous pockets of 1-, 3-, 10-, and 16-month-old rats and bone-forming capacity was observed. The metabolism of the bone cells decreased with age with a decline in the total calcium deposition and alkaline phosphatase activity. Interestingly, the decline was significant between 1- and 3-month-old animals. Similar to other studies, less amounts of inductive factors in "old" bone may account for the diminished response of the elderly animals to bone trauma.[62] Additional animal studies demonstrated that the AP activity in various bone cells from aged animals is reduced.[63]

BLOOD FLOW TO THE BONE DECREASES WITH AGE

Along with factors related to the bone cell number and metabolism, there are other factors that may affect the rate of bone healing in the aged, the most important being blood supply to the bone. A decrease in blood flow has been hypothesized to be one of the reasons for decline in bone mass and increased risk of fracture. This has been shown in elderly women with lower extremity arterial disease.[64,65] In an animal model the blood flow to the femur and hind limb was lower in 24-month-old rats than in 2- and 6-month-old rats. However, this decline was not observed in other skeletal sites in the rat. The decline in blood flow may be related to increased vascular resistance and fatty replacement of the marrow, especially in the shafts of long bones. Fatty marrow has approximately 70% less blood flow than the hematopoietic marrow.[66] Age-related decreases in blood flow to the bone might indirectly diminish bone-healing capacity. The relationship between cellular, molecular events and angiogenesis to bone healing is a crucial supporting activity. The relationship exemplifies a systemic interdependence culminating in a functional, healed skeletal defect.

In summary, bone physiology dynamics involve intra- and intercellular molecular activity. Age, environment, and systemic health status further complicate the reactive nature of the bone to accommodate change via modifications in mechanosensation and molecular signaling. Using gene array technology, additional efforts to understand the molecular cascade culminating in bone formation will identify gene expression patterns consistent with health bone metabolism.

Bibliography

1. Lee, T. C. and Taylor, D. (1999). Bone remodelling: should we cry Wolff? *Ir. J. Med. Sci.* 168, 102–105.
2. Burr, D. B. and Martin, R. B. (1989). Errors in bone remodeling: toward a unified theory of metabolic bone disease. *Am. J. Anat.* 186, 186–216.
3. Tami, A. E., Nasser, P., Verborgt, O., Schaffler, M. B. and Knothe Tate, M. L. (2002). The role of interstitial fluid flow in the remodeling response to fatigue loading. *J. Bone Miner. Res.* 17, 2030–2037.
4. Noble, B. S., Peet, N., Stevens, H. Y., Brabbs, A., Mosley, J. R., Reilly, G. C., Reeve, J., Skerry, T. M. and Lanyon, L. E. (2003). Mechanical loading: biphasic osteocyte survival and targeting of osteoclasts for bone destruction in rat cortical bone. *Am. J. Physiol. Cell Physiol.* 284, C934–C943.
5. Frank, J. D., Ryan, M., Kalscheur, V. L., Ruaux-Mason, C. P., Hozak, R. R. and Muir, P. (2002). Aging and accumulation of microdamage in canine bone. *Bone* 30, 201–206.
6. Burr, D. B. (2002). Targeted and nontargeted remodeling. *Bone* 30, 2–4.
7. Burr, D. B. (1992). Estimated intracortical bone turnover in the femur of growing macaques: implications for their use as models in skeletal pathology. *Anat. Rec.* 232, 180–189.
8. Rodan, G. A. and Martin, T. J. (2000). Therapeutic approaches to bone diseases. *Science* 289, 1508–1514.
9. Hernandez, C. J., Hazelwood, S. J. and Martin, R. B. (1999). The relationship between basic multicellular unit activation and origination in cancellous bone. *Bone* 25, 585–587.
10. Lee, T. C., Staines, A. and Taylor, D. (2002). Bone adaptation to load: microdamage as a stimulus for bone remodelling. *J. Anat.* 201, 437–446.
11. O'Brien, F. J., Taylor, D. and Lee, T. C. (2003). Microcrack accumulation at different intervals during fatigue testing of compact bone. *J. Biomech.* 36, 973–980.
12. Stanislaus, D., Yang, X., Liang, J. D., Wolfe, J., Cain, R. L., Onyia, J. E., Falla, N., Marder, P., Bidwell, J. P., Queener, S. W. and Hock, J. M. (2000). *In vivo* regulation of apoptosis in metaphyseal trabecular bone of young rats by synthetic human parathyroid hormone (1–34) fragment. *Bone* 27, 209–218.
13. Rubin, M. R. and Bilezikian, J. P. (2003). The anabolic effects of parathyroid hormone therapy. *Clin. Geriatr. Med.* 19, 415–432.
14. Lohmann, C. H., Schwartz, Z., Liu, Y., Li, Z., Simon, B. J., Sylvia, V. L., Dean, D. D., Bonewald, L. F., Donahue, H. J. and Boyan, B. D. (2003). Pulsed electromagnetic fields affect phenotype and connexin 43 protein expression in MLO–Y4 osteocyte-like cells and ROS 17/2.8 osteoblast-like cells. *J. Orthop. Res.* 21, 326–334.
15. Loveridge, N., Fletcher, S., Power, J., Caballero-Alias, A. M., Das-Gupta, V., Rushton, N., Parker, M., Reeve, J. and Pitsillides, A. A. (2002). Patterns of osteocytic endothelial nitric oxide synthase expression in the femoral neck cortex: differences between cases of intracapsular hip fracture and controls. *Bone* 30, 866–871.
16. Nomura, S. and Takano-Yamamoto, T. (2000). Molecular events caused by mechanical stress in bone. *Matrix Biol.* 19, 91–96.
17. Kato, Y., Windle, J. J., Koop, B. A., Mundy, G. R. and Bonewald, L. F. (1997). Establishment of an osteocyte-like cell line, MLO-Y4. *J. Bone Miner. Res.* 12, 2014–2023.
18. Vidal, O., Kindblom, L. G. and Ohlsson, C. (1999). Expression and localization of estrogen receptor-beta in murine and human bone. *J. Bone Miner. Res.* 14, 923–929.

19. Huggett, J. F., Mustafa, A., O'Neal, L. and Mason, D. J. (2002). The glutamate transporter GLAST-1 (EAAT-1) is expressed in the plasma membrane of osteocytes and is responsive to extracellular glutamate concentration. *Biochem. Soc. Trans.* 30, 890–893.
20. Noble, B. S. and Reeve, J. (2000). Osteocyte function, osteocyte death and bone fracture resistance. *Mol. Cell. Endocrinol.* 159, 7–13.
21. Weinstein, R. S., Nicholas, R. W. and Manolagas, S. C. (2000). Apoptosis of osteocytes in glucocorticoid-induced osteonecrosis of the hip. *J. Clin. Endocrinol. Metab.* 85, 2907–2912.
22. Vashishth, D., Verborgt, O., Divine, G., Schaffler, M. B. and Fyhrie, D. P. (2000). Decline in osteocyte lacunar density in human cortical bone is associated with accumulation of microcracks with age. *Bone* 26, 375–380.
23. Noble, B. S., Stevens, H., Loveridge, N. and Reeve, J. (1997). Identification of apoptotic changes in osteocytes in normal and pathological human bone. *Bone* 20, 273–282.
24. Seeman, E. (2002). Pathogenesis of bone fragility in women and men. *Lancet* 359, 1841–1850.
25. Qiu, S., Rao, D. S., Palnitkar, S. and Parfitt, A. M. (2002). Age and distance from the surface but not menopause reduce osteocyte density in human cancellous bone. *Bone* 31, 313–318.
26. Okada, S., Yoshida, S., Ashrafi, S. H. and Schraufnagel, D. E. (2002). The canalicular structure of compact bone in the rat at different ages. *Microsc. Microanal.* 8, 104–115.
27. Tanaka, M., Toyooka, E., Kohno, S., Ozawa, H. and Ejiri, S. (2003). Long-term changes in trabecular structure of aged rat alveolar bone after ovariectomy. *Oral Surg. Oral Med. Oral Pathol. Oral Radiol. Endod.* 95, 495–502.
28. Sekine, J., Sano, K. and Inokuchi, T. (1995). Effect of aging on the rat condylar fracture model evaluated by bromodeoxyuridine immunohistochemistry. *J. Oral Maxillofac. Surg.* 53, 1317–1321; discussion 1322–1323.
29. O'Driscoll, S. W., Saris, D. B., Ito, Y. and Fitzimmons, J. S. (2001). The chondrogenic potential of periosteum decreases with age. *J. Orthop. Res.* 19, 95–103.
30. Tonna, E. A. (1978). Electron microscopic study of bone surface changes during aging. The loss of cellular control and biofeedback. *J. Gerontol.* 33, 163–177.
31. Mears, D. C. (1999). Surgical treatment of acetabular fractures in elderly patients with osteoporotic bone. *J. Am. Acad. Orthop. Surg.* 7, 128–141.
32. Horan, M. A. and Clague, J. E. (1999). Injury in the aging: recovery and rehabilitation. *Br. Med. Bull.* 55, 895–909.
33. Kveiborg, M., Rattan, S. I., Clark, B. F., Eriksen, E. F. and Kassem, M. (2001). Treatment with 1,25-dihydroxyvitamin D3 reduces impairment of human osteoblast functions during cellular aging in culture. *J. Cell. Physiol.* 186, 298–306.
34. Barton, R. N., Horan, M. A., Rose, J. G. and Clague, J. E. (2001). Cortisol kinetics in elderly women with persistently high cortisol concentrations after proximal femur fracture. *J. Orthop. Trauma* 15, 321–325.
35. Baltzer, A. W., Lattermann, C., Whalen, J. D., Wooley, P., Weiss, K., Grimm, M., Ghivizzani, S. C., Robbins, P. D. and Evans, C. H. (2000). Genetic enhancement of fracture repair: healing of an experimental segmental defect by adenoviral transfer of the BMP-2 gene. *Gene Ther.* 7, 734–739.
36. Heaney, R. P. (1998). Bone mass, bone fragility, and the decision to treat. *JAMA* 280, 2119–2120.
37. World Health Organization. (1991). Osteoporosis Consensus Development Conference. Group, W. H. O. O. S.

38. Ankersen, L. (1994). Aging of human trabecular osteoblasts in culture. *Arch. Gerontol. Geriatr.* 3, 5–12.
39. Kassem, M., Ankersen, L., Eriksen, E. F., Clark, B. F. and Rattan, S. I. (1997). Demonstration of cellular aging and senescence in serially passaged long-term cultures of human trabecular osteoblasts. *Osteoporos. Int.* 7, 514–524.
40. Christiansen, M., Kveiborg, M., Kassem, M., Clark, B. F. and Rattan, S. I. (2000). CBFA1 and topoisomerase I mRNA levels decline during cellular aging of human trabecular osteoblasts. *J. Gerontol. A Biol. Sci. Med. Sci.* 55, B194–B200.
41. Ducy, P., Zhang, R., Geoffroy, V., Ridall, A. L. and Karsenty, G. (1997). Osf2/Cbfa1: a transcriptional activator of osteoblast differentiation. *Cell* 89, 747–754.
42. Chavassieux, P. M., Chenu, C., Valentin-Opran, A., Merle, B., Delmas, P. D., Hartmann, D. J., Saez, S. and Meunier, P. J. (1990). Influence of experimental conditions on osteoblast activity in human primary bone cell cultures. *J. Bone Miner. Res.* 5, 337–343.
43. Fedarko, N. S., Vetter, U. K., Weinstein, S. and Robey, P. G. (1992). Age-related changes in hyaluronan, proteoglycan, collagen, and osteonectin synthesis by human bone cells. *J. Cell. Physiol.* 151, 215–227.
44. Sutherland, M. S., Rao, L. G., Muzaffar, S. A., Wylie, J. N., Wong, M. M., McBroom, R. J. and Murray, T. M. (1995). Age-dependent expression of osteoblastic phenotypic markers in normal human osteoblasts cultured long-term in the presence of dexamethasone. *Osteoporos. Int.* 5, 335–343.
45. Pfeilschifter, J., Diel, I., Pilz, U., Brunotte, K., Naumann, A. and Ziegler, R. (1993). Mitogenic responsiveness of human bone cells *in vitro* to hormones and growth factors decreases with age. *J. Bone Miner. Res.* 8, 707–717.
46. Martinez, M. E., del Campo, M. T., Medina, S., Sanchez, M., Sanchez-Cabezudo, M. J., Esbrit, P., Martinez, P., Moreno, I., Rodrigo, A., Garces, M. V. and Munuera, L. (1999). Influence of skeletal site of origin and donor age on osteoblastic cell growth and differentiation. *Calcif. Tissue Int.* 64, 280–286.
47. Marie, P. J. (1995). Human endosteal osteoblastic cells: relationship with bone formation. *Calcif. Tissue Int.* 56, S13–S16.
48. Vanderschueren, D., Gevers, G., Raymaekers, G., Devos, P. and Dequeker, J. (1990). Sex- and age-related changes in bone and serum osteocalcin. *Calcif. Tissue Int.* 46, 179–182.
49. Nishimoto, S. K., Chang, C. H., Gendler, E., Stryker, W. F. and Nimni, M. E. (1985). The effect of aging on bone formation in rats: biochemical and histological evidence for decreased bone formation capacity. *Calcif. Tissue Int.* 37, 617–624.
50. Urist, M. R., Hudak, R. T., Huo, Y. K. and Rasmussen, J. K. (1985). Osteoporosis: a bone morphogenetic protein auto-immune disorder. *Prog. Clin. Biol. Res.* 187, 77–96.
51. Hollinger, J. (1993). Factors for osseous repair and delivery. Part I. *J. Craniofac. Surg.* 4, 102–108.
52. Williams, D. C., Boder, G. B., Toomey, R. E., Paul, D. C., Hillman Jr. C. C., King, K. L., Van Frank, R. M. and Johnston Jr. C. C. (1980). Mineralization and metabolic response in serially passaged adult rat bone cells. *Calcif. Tissue Int.* 30, 233–246.
53. Martinez, M. E., Medina, S., Sanchez, M., Del Campo, M. T., Esbrit, P., Rodrigo, A., Martinez, P., Sanchez-Cabezudo, M. J., Moreno, I., Garces, M. V. and Munuera, L. (1999). Influence of skeletal site of origin and donor age on 1,25(OH)2D3-induced response of various osteoblastic markers in human osteoblastic cells. *Bone* 24, 203–209.
54. Roholl, P. J., Blauw, E., Zurcher, C., Dormans, J. A. and Theuns, H. M. (1994). Evidence for a diminished maturation of preosteoblasts into osteoblasts during aging in rats: an ultrastructural analysis. *J. Bone Miner. Res.* 9, 355–366.

55. Reddi, A. H. (1985). Age-dependent decline in extracellular matrix-induced local bone differentiation. *Isr. J. Med. Sci.* 21, 312–313.
56. Irving, J. T., LeBolt, S. A. and Schneider, E. L. (1981). Ectopic bone formation and aging. *Clin. Orthop*. Issue 154, 249–253.
57. Syftestad, G. T. and Urist, M. R. (1982). Bone aging. *Clin Orthop*. 288–297.
58. Kakiuchi, M., Hosoya, T., Takaoka, K., Amitani, K. and Ono, K. (1985). Human bone matrix gelatin as a clinical alloimplant. A retrospective review of 160 cases. *Int. Orthop.* 9, 181–188.
59. Matsumoto, A., Yamaji, K., Kawanami, M. and Kato, H. (2001). Effect of aging on bone formation induced by recombinant human bone morphogenetic protein-2 combined with fibrous collagen membranes at subperiosteal sites. *J. Periodontal Res.* 36, 175–182.
60. Urist, M. R., Nillsson, O. S., Hudak, R., Huo, Y. K., Rasmussen, J., Hirota, W. and Lietze, A. (1985). Immunologic evidence of a bone morphogenetic protein in the milieu interieur. *Ann. Biol. Clin.* 43, 755–766.
61. Fleet, J. C., Cashman, K., Cox, K. and Rosen, V. (1996). The effects of aging on the bone inductive activity of recombinant human bone morphogenetic protein-2. *Endocrinology* 137, 4605–4610.
62. Nishida, S. (1992). Properties of osteoprogenitor cells in bone marrow of aged and ovarectomized rats. *J. Bone Miner. Res.* 7, 113.
63. Quarto, R., Thomas, D. and Liang, C. T. (1995). Bone progenitor cell deficits and the age-associated decline in bone repair capacity. *Calcif. Tissue Int.* 56, 123–129.
64. Dolan, N. C., Liu, K., Criqui, M. H., Greenland, P., Guralnik, J. M., Chan, C., Schneider, J. R., Mandapat, A. L., Martin, G. and McDermott, M. M. (2002). Peripheral artery disease, diabetes, and reduced lower extremity functioning. *Diabetes Care* 25, 113–120.
65. Smith, J. J., Toogood, G. J. and Galland, R. B. (1999). Reconstruction for lower limb occlusive disease in the elderly. *Cardiovasc. Surg.* 7, 58–61.
66. Bloomfield, S. A., Hogan, H. A. and Delp, M. D. (2002). Decreases in bone blood flow and bone material properties in aging Fischer-344 rats. *Clin. Orthop*. Issue 396, 248–257.
67. Aarden, E. M., Wassenaar, A. M., Alblas, M. J. and Nijweide, P. J. (1996). Immunocytochemical demonstration of extracellular matrix proteins in isolated osteocytes. *Histochem. Cell Biol.* 106, 495–501.
68. van der Plas, A., Aarden, E. M., Feijen, J. H., de Boer, A. H., Wiltink, A., Alblas, M. J., de Leij, L. and Nijweide, P. J. (1994). Characteristics and properties of osteocytes in culture. *J. Bone Miner. Res.* 9, 1697–1704.
69. Igarashi, M., Kamiya, N., Ito, K. and Takagi, M. (2002). *In situ* localization and *in vitro* expression of osteoblast/osteocyte factor 45 mRNA during bone cell differentiation. *Histochem. J.* 34, 255–263.
70. Sekiya, H., Mikuni-Takagaki, Y., Kondoh, T. and Seto, K. (1999). Synergistic effect of PTH on the mechanical responses of human alveolar osteocytes. *Biochem. Biophys. Res. Commun.* 264, 719–723.
71. Zhao, S., Zhang, Y. K., Harris, S., Ahuja, S. S. and Bonewald, L. F. (2002). MLO-Y4 osteocyte-like cells support osteoclast formation and activation. *J. Bone Miner. Res.* 17, 2068–2079.
72. Nakamura, H., Kenmotsu, S., Sakai, H. and Ozawa, H. (1995). Localization of CD44, the hyaluronate receptor, on the plasma membrane of osteocytes and osteoclasts in rat tibiae. *Cell Tissue Res.* 280, 225–233.
73. Yamashiro, T., Fukunaga, T., Kobashi, N., Kamioka, H., Nakanishi, T., Takigawa, M. and Takano-Yamamoto, T. (2001). Mechanical stimulation induces CTGF expression in rat osteocytes. *J. Dent. Res.* 80, 461–465.

3 Cell Biology of the Skeletal System

Contents

0-8493-1621-9/05/$0.00+$1.50
© 2005 by CRC Press LLC

Osteoblastic and Osteocytic Biology and Bone Tissue Engineering

Henry J. Donahue, Christopher A. Siedlecki and Erwin Vogler

The successful engineering of bone tissue requires (1) a mechanical environment that stimulates bone cell differentiation, (2) a scaffold made from a biomaterial that supports bone cell proliferation and differentiation, (3) appropriate cells, including pluripotent stem cells as well as supporting cells, and (4) appropriate osteogenic molecules, including extracellular matrix proteins, incorporated into the scaffold biomaterial in a distribution pattern that optimizes bone cell proliferation and differentiation. In this chapter, we will discuss the osteoblastic and osteocytic cell biology relevant to bone tissue engineering. This will include a review of the regulation of osteoblastic differentiation with special emphasis on mechanical regulation of osteoblastic differentiation. We will also discuss how the biomaterials used in tissue engineering can also function to regulate osteoblastic behavior.

OSTEOBLAST PROLIFERATION, DIFFERENTIATION, AND FUNCTION

Osteoblasts are the bone cells that form bone. Mature osteoblastic cells are highly polarized with a prominent Golgi apparatus typical of highly secretory cells.[1] The main secretory product of osteoblasts is type 1 collagen, but osteoblasts secrete other noncollagenous proteins including osteopontin, osteocalcin, and bone sialoprotein. Osteoblasts form bone by facilitating mineralization but the mechanism by which this occurs is not well understood. One possibility is that lipid matrix vesicles that bud off of bone cells create a microenvironment where calcium and phosphate are concentrated in a ratio allowing for optimized crystallization. These crystals then align on secreted collagen and form a nucleation site that facilitates subsequent mineralization and hydroxyapatite formation. Another possibility is that mineralization is initiated by components of the collagen molecule[1] in a manner not requiring matrix vesicles. It is also possible, indeed likely, that matrix vesicle-dependent and-independent mineralization occur concurrently.[2,3] In any case, osteoblasts are critically important in mineralization because they both secrete collagen and produce matrix vesicles.

Osteoblastic cells arise from pluripotent mesenchymal progenitor cells that can also develop into adipocytes, myocytes, and chondrocytes.[4] The differentiation pathway taken by the progenitor cell is regulated by tissue-specific transcription factors.[5] Runx2 (Cbfa1) enables differentiation of the progenitor cell into osteoblastic cells, the MyoD family of transcription factors enable differentiation to myocytes, Sox5/6/9 enables chondrocytic cell development, and PPARγ enables differentiation into adipocytes. Once the pluripotent progenitor cells have committed to the osteoblastic lineage, they progress through three developmental stages of differentiation: proliferation, matrix maturation, and mineralization.[6] While these stages and the genes expressed during these stages have been identified *in vitro,* they are believed to reflect *in vivo* maturation of osteoblasts (Figure 3.1).

In general, type 1 collagen and histone H4 peak during the proliferation phase and decline thereafter, alkaline phosphatase peaks during the matrix formation phase and declines thereafter, while osteopontin and osteocalcin peak during the mineralization phase. It is important to note that this differentiation sequence may actually be more complicated. Aubin et al.[4] have identified at least seven transitional stages rather than the three just described. Whether there are three, seven, or more transitional stages in osteoblastic differentiation, the fact that osteoblastic cells express phenotypically characteristic genes as they differentiate *in vitro* has important

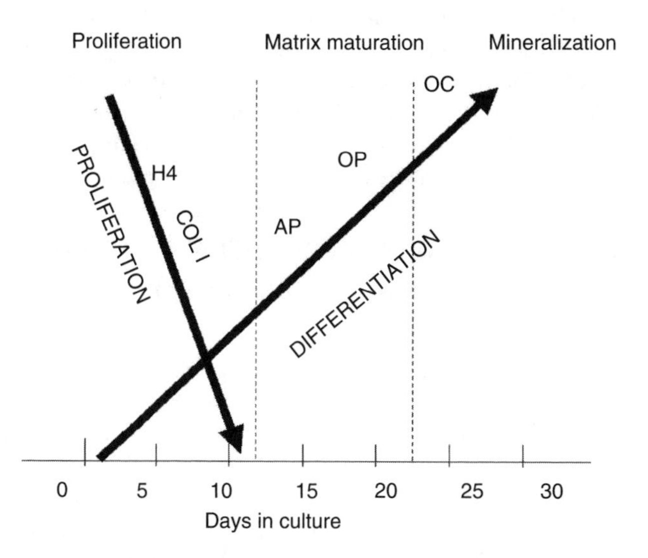

Figure 3.1 Temporal expression of genes characteristic of osteoblastic differentiation. Osteoblastic cells in culture go through stages of proliferation, maturation, and mineralization. Type 1 collagen (COL I) and histone H4 peak during the proliferative phase and decline thereafter. Alkaline phosphatase (AP) peaks during the matrix maturation phase, while osteopontin (OP) and osteocalcin (OC) peak in the late maturation or early mineralization phases and decline thereafter. Adapted from Lian *et al.*[6]

implications for bone tissue engineering. For instance, one can examine the effect of various biomaterials on osteoblastic differentiation and identify materials that support osteoblastic differentiation. Similarly, one can identify bimolecular and biophysical signals that stimulate bone cell differentiation. These biomaterials and signals could then be exploited in bone tissue engineering.

While osteoblastic cells are perhaps the most widely studied bone cells, by far the most abundant bone cell is the osteocyte. At least 90% of all cells in bone are osteocytes.[7] However, because they are relatively difficult to isolate and culture *in vitro*, osteocytes are the least studied of the bone cells. Osteocytes arise from osteoblasts that have become entrapped in the mineralized matrix they have formed. As osteocytes transition from osteoblasts, they develop long dendritic-like processes that exist in small channels within the bone known as canuliculi. The processes of individual osteocytes connect with the processes of other osteocytes, as well as with osteoblasts, and communicate via gap junctions, creating a highly communicative network of cells within the bone matrix ideally situated to integrate and amplify extracellular signals.

While the regulation of osteoblastic proliferation and differentiation has been widely studied, very little is known regarding the regulation of the transition from osteoblast to osteocyte, the terminal step in osteoblastic differentiation. However, recent studies have shown that osteoblastic MC3T3-E1 cells will differentiate into a cell with an osteocytic morphology, that is, stellate with interconnected processes, when cultured in a three-dimensional-collagen gel. This does not occur when the cells are cultured on tissue culture polystyrene.[8] Transition from osteoblastic morphology to osteocytic morphology, in three-dimensional gels, was blocked by the matrix metalloproteinase (MMP) inhibitor GM6001, suggesting the involvement of MMPs in the transition of osteoblasts to osteocytes. Furthermore, exposure to active transforming growth factor beta (TGF-β) restored transition from osteoblasts to osteocytes in the presence of the MMP inhibitor. Thus, MMP-activated TGF-β, as well as a three-dimensional environment, contributes to osteoblastic transition to osteocytes.

REGULATION OF OSTEOBLAST PROLIFERATION AND DIFFERENTIATION BY BIOLOGICAL FACTORS

An understanding of which growth and transcription factors regulate osteoblastic cell proliferation and differentiation is critical for optimized bone tissue engineering. For instance, incorporation of appropriate growth factors into biomaterials, creating biomimetic materials, could greatly enhance osteoblastic proliferation and differentiation. Perhaps the most widely studied growth factors in this context are the bone morphogenic proteins (BMPs). Over 15 BMPs have been identified in vertebrates and all, except BMP-1, are members of the TGF-β superfamily. BMPs are secreted by osteoprogenitor cells and mature osteoblasts and are potent activators of bone cell differentiation *in vitro*, especially BMP-2, 4, and 7,[5,9] and ectopic bone formation *in vivo*.[10] However, BMP gene ablation studies have not provided definitive evidence for a role of BMPs in osteoblastic differentiation. This may be due to functional redundancy within the BMP family. BMP-2 has been demonstrated to be effective in healing critical-sized defects in several animal species.[9] Interestingly, relatively high doses of BMPs are required for efficacy, raising issues regarding cost and safety.[9,11]

This, together with the observation that in order to be effective *in vivo*, BMPs must be delivered via carriers including collagen matrix, demineralized bone matrix, and synthetic polymers,[9] emphasizes the need to optimize molecular incorporation of BMPs into biomaterials for use in bone tissue engineering.

TGF-β itself, which is secreted largely by platelets and interacts with specific receptors on osteoblastic cells, has also been demonstrated, under certain circumstances, to stimulate osteoblastic differentiation *in vitro*,[12,13] suggesting it may be applicable to bone tissue engineering. However, the situation is complicated by the finding that osteocalcin promoter-driven expression of TGF-β in mice results in the development of a complex low bone mass phenotype with an overall increase in bone resorption.[14] Furthermore, when examined within the context of bone regeneration, TGF-β does not produce bone when applied ectopically[12] and when applied to healing fractures, only induces bone regeneration at very high concentrations given frequently.[15,16] Thus, at this point, TGF-β has limited potential for use in biomimetic biomaterials designed for bone tissue engineering.

Other growth factors that may regulate osteoblastic differentiation include fibroblast growth factor (FGF), insulin-like growth factor-1 (IGF), and platelet-derived growth factor (PDGF). FGFs, especially FGF-2 and 18, stimulate osteoblastic cell proliferation and inhibit osteoblastic cell differentiation,[17] as does PDGF.[18] IGF-1, on the other hand, stimulates both osteoblastic differentiation and proliferation.[19] FGF, IGF-1,[9] and PDGF[20] have all shown the potential to regenerate bone *in vivo* utilizing various delivery mechanisms, and therefore show promise in bone tissue engineering.

REGULATION OF OSTEOBLAST PROLIFERATION AND DIFFERENTIATION BY BIOPHYSICAL FACTORS

Bone is able to sense and respond to mechanical signals. However, the mechanisms by which this occurs are only partly understood. A better understanding of how mechanical signals affect bone cell metabolism would provide clues as to how such signals could be exploited in bone tissue engineering. For instance, knowing which, if any, mechanical signals stimulate bone cell proliferation and differentiation would suggest which mechanical signals should be incorporated into bioreactors for bone tissue engineering.

Matrix deformation and interstitial fluid flow are the two most likely mechanisms by which mechanical loads are transduced to bone cells *in vivo*. Matrix deformation can potentially affect bone cell activity via cellular deformation or piezoelectric effects. There is considerable *in vitro* evidence that substrate stretch within the physiological range (0.34–1.3%) affects bone cell activity.[21] Indeed, at least two putative stretch-activated channels have been identified in bone cells.[22,23] Additionally, *in vitro* studies suggest that substrate deformation activates these channels. Thus, substrate deformation may be a candidate for a mechanical signal that enhances bone cell differentiation and proliferation, and thus would be useful in bone tissue engineering.

Matrix deformation also induces endogenous electric fields. When bone is functionally loaded, electric currents are generated as a result of the piezoelectric properties of collagen.[24] Furthermore, relatively large electrokinetic currents are produced by the boundary interaction effects of charged constituents of fluids passing the mineral phase of the extracellular matrix in response to strain in the bone tissue (the so-called streaming potentials).[25] Both *in vivo* and *in vitro* studies suggest that electric fields can increase bone cell activity, and thus could be exploited in bone tissue engineering.[26–30]

Interstitial fluid flow, a direct effect of loading bone, is another mechanism by which mechanical signals could be transduced to bone cells. The responsiveness of bone cells to fluid flow has been well established. Osteoblastic cells exposed to fluid flow display increased PGE_2 production,[31] release of nitric oxide,[32] cAMP production,[33] osteopontin mRNA,[34,35] TGF-β1 mRNA[36] and Erg1 mRNA.[37] Fluid flow also increases osteocytic prostaglandins[38] and nitric oxide.[39]

The majority of studies published to date on the effects of fluid flow on bone cells have applied either steady flow or pulsatile flow (dynamic flow that never reverses). This type of flow would occur as a result of arterial pressure. In contrast, physiologic fluid flow resulting from the repetitive loading of daily activities is oscillatory in nature (reversing flow that occurs in alternating directions). This distinction is important when considering bone tissue engineering protocols in that the two signals may have different effects on bone cells. For instance, we have demonstrated that steady flow, while a less physiological signal, is a more potent stimulator of cytosolic Ca^{2+} mobilization in osteoblastic cells than oscillating flow.[40] However, it remains to be determined whether this means that steady flow would be more stimulatory of osteoblastic proliferation and differentiation than oscillating fluid flow. Interestingly, results from our laboratory demonstrate that physiological levels of oscillating fluid flow induced cytosolic Ca^{2+} oscillations in a greater percentage of cells than did physiological levels of substrate stretch and increased steady-state levels of osteopontin mRNA, indicative of increased osteoblastic differentiation, while substrate deformation did not.[41] This would suggest that if substrate stretch were used in a bone tissue engineering protocol, it would need to be at levels that do not occur under normal physiological conditions.

The Role of Gap Junctional Intercellular Communication in Bone Cell Differentiation

Gap junctional intercellular communication (GJIC) contributes to bone cell responsiveness to diverse extracellular signals, including hormonal,[42] mechanical,[43] and electrical.[44] Therefore, it is important to briefly consider gap junction physiology in the context of bone tissue engineering. Gap junctions are membrane-spanning channels that allow passage of small molecules (<1kDa) such as calcium ions (Ca^{2+}), inositol phosphates, and cyclic nucleotides, from one cell to another. Gap junctions directly connect (exist between) adjoining osteoblasts and osteocyte,[45,46] and are the key couplers in the osteocyte syncytium.[47,48] Additionally, bone cells are functionally coupled both *in vivo*[49] and *in vitro*[50–54] where osteoblasts are coupled with osteoblastic as well as osteocytic cells[55] (Figure 3.2).

Accumulating data suggest that gap junctions contribute to bone cell differentiation. *In vivo* studies suggest that gap junctions may be involved in cell signaling processes important to limb bud differentiation and skeletogenesis in embryonic mice[56] and contribute to cellular differentiation and intramembranous bone forma-

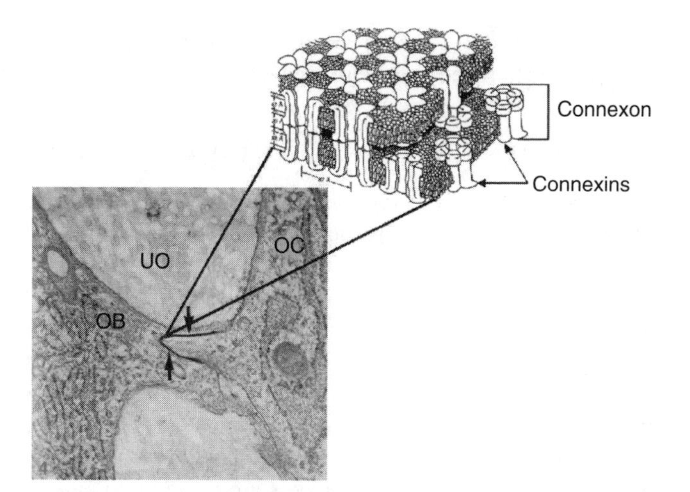

Figure 3.2 Gap junctions in bone. Electron microscopy was first used to identify gap junctions between bone cells. Gap junctions can be seen as electron-dense regions between osteoblasts (OB) and osteocytes (OC). Unmineralized osteoid (UO) surrounds the processes. The insert depicts a schematic model of gap junctional channels. Each channel is formed from two hexameric hemichannels (connexons) contributed by each of two adjacent cells. Each connexon is in turn composed of six protein subunits called connexins. Intercellular signaling molecules less than 1 kDa in size can pass from cell to cell via these aqueous pores. Additionally, these channels can gate (open and close) in response to various cellular regulators. (Micrograph from Doty, S.B. (1981) morphological evidence of gap junctions between bone cells. *Calcif Tissue Int.,* 33, 509–512, © 1981 Springer-Verlag. Insert reproduced from *The Journal of Cell Biology,* 1977, vol. 74, by copyright permission of The Rockefeller University Press.)

tion in developing chick mandible.[57,58] Additionally, rat tendon cells vary in their expression of connexins according to their state of differentiation, suggesting a role in differentiation.[59] Finally, Lecanda et al. reported impaired intramembranous bone formation in Cx43 null mice, suggesting a defect in osteoblastic maturation.[60]

In vitro studies also suggest a role for gap junctions in bone cell differentiation. Chiba et al.[61] have demonstrated that Cx43 expression parallels alkaline phosphatase activity and osteocalcin secretion in differentiating human osteoblastic cells. Schiller et al.[62] showed that Cx43 mRNA expression increased in osteoblastic MC3T3-E1 cells as they differentiated in culture and blocking GJIC forces osteoblastic cells toward an adipocytic phenotype.[63] PTH responsiveness, a characteristic of mature osteoblasts, is also decreased in gap junction-deficient osteoblastic cells.[64] Additionally, osteoblastic ROS 17/2.8 cells expressing Cx43 antisense cDNA sequences or Cx45 are less well coupled than normal ROS 17/2.8 cells, and they display a decreased expression of markers of fully differentiated osteoblasts.[65,66] Furthermore, calvarial osteoblasts from Cx43 null mice display decreased GJIC and alkaline phosphatase activity and an inability to form mineralized nodules *in vitro*.[67] Cx43 expression and function parallels differentiation of the human fetal osteoblastic cell line hFOB 1.19, and blocking GJIC decreases alkaline phosphatase activity in these cells.[21] Taken together, both *in vivo* and *in vitro* data strongly suggest that GJIC is critical for optimal osteoblastic differentiation. In the context of bone tissue engineering, this suggests that factors that increase GJIC could be exploited in developing bone tissue engineering protocols.

Effect of Biomaterials on Bone Cell Differentiation

Whether engineering skeletal constructs *in vitro*, delivering osteogenic factors *in vivo*, or implanting artificial bone constructs, the surface characteristics of the substratum on which bone cells are growing are extremely important. It has long been observed that the substratum is not a passive anchoring surface but rather participates in a complex relationship with cells (and the proteinaceous fluid phase) in such a way that certain materials stimulate adhesion and proliferation, whereas others are inhibitory.[68] Adherent cells typically condition the substratum over exposure time by the extrusion of the extracellular matrix. In the particular case of osteoblasts, the substratum can become mineralized through enzyme and cell-mediated processes.[69] These complex cell–protein–surface interactions are essential to understand for the purpose of prospective design of biomaterials used in bone tissue engineering, and much remains to be learnt in this pursuit.

Surface properties are of paramount importance in controlling cytocompatibility because experiments show that not only do different materials induce different cell-adhesion outcomes but also the same material with different surface treatments can profoundly affect cell behavior.[70,71] Surface properties may be roughly divided into chemical and morphological categories, although it is not always easy to clearly differentiate one property from another. This is because topological features, natural or manmade, can have distinct chemistry (such as at an edge), even if composed of an apparently homogenous material. Surface chemistry gives rise to surface energy that, for the purpose of biomaterial applications in which water is the biological solvent phase,[72] can be understood most easily in terms of water wetting and measured with water contact angles.[70,73] Surface morphology (topography), chemistry, and energy effects on bone cell behavior are briefly reviewed below, but the reader should bear in mind that these are not always clearly differentiated in the literature.

Surface roughness is a topological feature that can affect bone cell behavior. Many studies have focused on the influence of titanium surface roughness on bone cell proliferation and differentiation. In summary, these studies demonstrate that increased surface roughness increases osteoblast attachment, decreases proliferation, and increases alkaline phosphatase, osteocalcin, TGF-β and prostaglandin production.[29–77] This would suggest that increased surface roughness favors osteoblastic differentiation. Furthermore, osteoblastic cells cultured on rougher titanium surfaces display increased responsiveness to $1{,}25(OH)_2$ vitamin D[78] and estradiol.[29] However, most of these studies were carried out with transformed cell lines. There have been relatively few studies examining the effect of surface roughness of biomaterials, other than titanium, on primary bone-cell proliferation and differentiation, and these results have not been generally consistent with the studies using titanium as the substratum. For example, Deligianni et al.[79] found that the roughness of hydroxyapatite did not affect alkaline phosphatase activity of human bone marrow cells, but the proliferation of these cells was increased on the rougher surfaces. Montanaro et al.[80] showed that alkaline phosphatase activity and osteocalcin production in MG63 cells were not affected by surface roughness of fluorohydroxyapatite-coated titanium alloys, and Laczka-Osyczka[81] found no significant differences in alkaline phosphatase, type 1 collagen, osteopontin, or bone sialoprotein production by bone marrow cells cultured on glass-ceramics with variable roughness. Thus, while the surface roughness of titanium clearly affects osteoblastic cell differentiation, this has not been demonstrated for other biomaterials that many be used in bone tissue engineering *in vivo* or *in vitro*. The lack of consensus in the literature is perhaps related to the difficulty in clearly separating surface chemistry and roughness effects. In any event, further research on the effects of substratum topography is clearly indicated.

In addition to topography at the microscale, some recent literature points to the importance of topography at the nanometer scale. For example, Dalby et al.[82] examined fibroblasts cultured on nanometer high islands manufactured by polymer demixing of polystyrene (PS) and poly(4-bromostyrene) (PBrS) creating a PSP/BrS blend. Fibroblasts cultured on this material with 13 nm high islands spread out more, displayed a better-developed microtubule system, and more focal adhesions than cells cultured on polysyrene with no islands. Additionally, micro array analysis revealed that 68 of 584 genes examined displayed an average 36% upregulation in cells cultured on 13 nm high islands relative to those cultures on control biomaterial. While these changes are rather modest and not confirmed by other mRNA detecting techniques, these studies do suggest that fibroblast responds to nanoscale topography. Dalby et al. also demonstrated that endothelial cell morphology was different on a surface with 13 nm high islands compared to 35 nm or 95 nm high islands.[83] We are aware of only one study that examined the effect of nanotopography on osteoblastic-cell behavior. Riehle et al.[84] demonstrated that, relative to PS/PBrS with 13 nm high islands or flat PS, osteoblastic cells cultured on PS/PBrS with 35 and 95 nm high islands adhered and spread out more. However, subsequent filipodia extension, spreading, and proliferation were similar on all the materials. It is clearly premature to draw conclusions from these early studies but, taken at face value, the results suggest that osteoblastic cells do not respond to nanotopography to the same degree as endothelial cells do. Additional studies on the effect of nanotopography on osteoblastic cell proliferation and differentiation are necessary to determine whether biomaterial nanotopography can be exploited in bone tissue engineering.

In addition to topography, surface chemistry and energy have clearly observable effects on cell adhesion and proliferation. A good example of how surface chemistry can affect osteoblast behavior comes from studies of bioactive glasses. One widely studied bioactive glass composition, known as Bioglass 45S5 (U.S. Biomaterials Corporation, Alachua, FL, USA), containing 45% SiO_2, 24.5% NaO_2, 24.5% CaO, and 6% P_2O_5, undergoes a series of surface reactions when placed in appropriate aqueous buffer solutions. This results in the formation of a layer of carbonated hydroxyapatite on the surface. By contrast, a bioinert composition designated 60S, containing 60.1% SiO_2, 17.7% NaO_2, 19.6% CaO, and 2.6% P_2O_5 does not form hydroxyapatite on the surface. Furthermore, it has been shown that osteoblastic cells mineralize disks fabricated from 45S5[85] but not on the 60S surface.[86] It has also been demonstrated that ionic products released as a result of the dissolution of Bioglass, especially silicon, stimulate the proliferation of human osteoblastic cells.[87] Thus, the surface chemistry of Bioglass clearly affects osteoblastic behavior.

Surface chemistry and energy can also affect cell behavior through effects on protein adsorption and activity, which in turn profoundly affects cell adhesion and proliferation. An example relevant to bone-analog materials is that incorporation of fluorapatite into hydroxyapatite surfaces changes the orientation of calcium-binding proteins and thus cellular attachment.[88] Another example is illustrated by the binding of $\alpha_5\beta_1$ integrin molecules on osteoblastic MC3T3-1 cells to adsorbed fibronectin, resulting in differences in osteoblast differentiation measured by alkaline phosphatase activity, matrix mineralization, and osteoblastic-specific gene expression.[89] It appears that this is a result of substrate-dependent differences in the conformation of fibronectin as measured by antibody binding.[90] Furthermore, the density of ligands for attachment seems to play an important role. Rezania and Healy demonstrated that when the density of a bone sialoprotein peptide fraction containing the integrin binding amino acid sequence arginine–glycine–aspartate (RGD) was ≥ 0.62 pmol/cm^2, mineralization of the extracellular matrix after culture with rat calvaria osteoblast-like cells was increased at time points greater than 2 weeks compared to lower densities of the same peptide.[91] The distribution of surface chemistry and subsequent protein function also appears to play a role in culture of human bone-derived cells. In this case, spatial confinement of cells, by micropatterning changes in material surface chemistry, affected focal adhesion formation and cytoskeletal organization.[92] Taken together, these observations suggest that materials used in bone tissue engineering must present the appropriate signals at the appropriate length scale, the density must be sufficient to stimulate focal adhesion formation and cytoskeletal organization, and yet the distribution of surface chemistry on the underlying materials must be appropriately homogenous to allow bone cells to spread so that they may undergo proper morphogenesis, differentiation growth, and function.

Even preparation and sterilization techniques commonly used for orthopedic implant materials can alter surface chemistry and affect cell–material interactions. For instance, plasma glow discharge of titanium/aluminum/vanadium alloys (Ti6Al4V) can result in increased concentration of aluminum oxides that can be toxic to cells. The surface chemistry of minerals (such as hydroxyapatite) and mineral/polymer composites is particularly complex due to solution-phase dissolution/precipitation reactions that occur at the hydrated interface, as already mentioned in the context of Bioglass. In pure water or buffer, these reactions are dictated by solubility constants,[93–97] but can be mediated by enzymes and/or cells *in vitro* and *in vivo*.[98]

Relating cell adhesion to surface chemistry, as resolved by high-vacuum surface spectroscopies, for example, may be difficult or impossible due to hydration reactions and protein adsorption events that effectively obscure the original surface to contacting cells.[70,99] However, some progress has been made in relating cell adhesion and proliferation to surface energetics as measured by water contact angles.[68,70,100–103] Contact angles measure the extent of physiochemical interactions of a fluid (e.g., water) with a material surface, which is termed "wetting." Low (acute) angles, as measured through a drop resting on a surface (Figure 3.3),

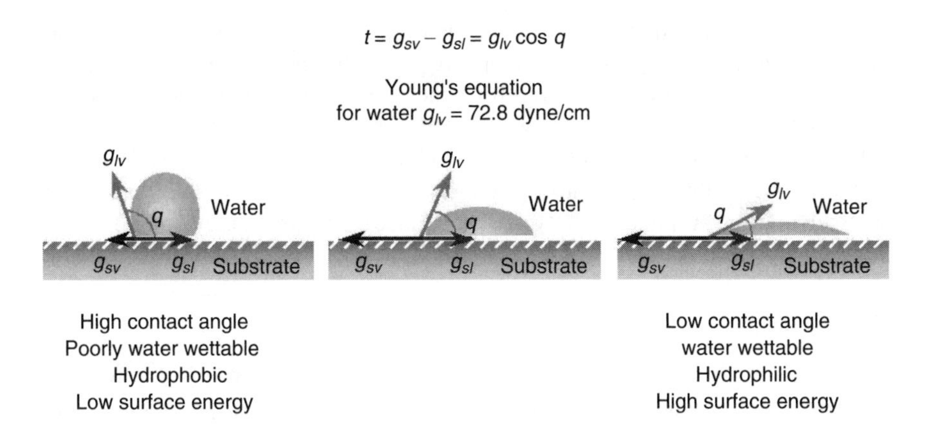

Figure 3.3 Characterization of substrate surface energy. The surface energy of a substrate can be determined from contact angles of water droplets placed on the substrate.

signal relatively strong fluid–surface interactions whereas high (obtuse) angles correspond to relatively weaker interactions.[73] Material surfaces exhibiting low water-contact angles are frequently termed "hydrophilic" or "high surface energy" and those exhibiting high water-contact angles are termed "hydrophobic" or "low surface energy," although these relative and somewhat arbitrary designations have introduced considerable confusion in the biomaterials literature.[104] Relationships among surface chemistry, surface energy (wettability), adhesion, and (protein) adsorption are quite complex and topics of current research in biomaterials surface science are beyond the scope of this review. However, it is very clear from decades of research into cell adhesion that surface chemistry/energy profoundly affects cell attachment and proliferation,[71,100,101,105] only a few of which are mentioned below.

Studies with human fibroblasts and endothelial cells demonstrated that cells adhere to a greater degree on substrata with a higher surface energy.[102,106,107] Studies with osteoblastic cells have been inconsistent with respect to the importance of surface energy effects on cell behavior, possibly due to the complexities in characterizing the surface energy of mineral-based materials or biodegradable polymers. Daw et al.[108] reported that surface energy did not affect osteoblastic ROS 17/2.8 cell attachment. Redey et al.[109] found that human primary culture osteoblastic cells displayed greater attachment and collagen production when cultured on stoichiometric hydoxyapatite relative to cells cultured on type A carbonate appetite. These results were attributed to differences in surface energy of the two materials. Cai et al.[110] examined the behavior of rat calvarial osteoblasts cultured on poly (D,L,-lactic acid) (PDLLA) or chitosan-modified PDLLA. Cell attachment, proliferation, and alkaline phosphatase activity were higher on the chitosan-modified PDLLA, which also had a higher surface energy. More recently, we demonstrated that attachment and proliferation of osteoblastic hFOB1.19 cells were highly correlated with surface energy, with cell compatibility being the highest for hydrophilic materials including monolithic quartz or glass slides and decreasing smoothly with surface energy to poorly cytocompatible with hydrophobized counterparts.[111] Correlations of this kind strongly suggest that surface energy/water wettability is an important consideration in identifying biomaterials optimized for bone tissue engineering, but much more work is required to demonstrate causal (rather than casual) relationships. Toward this goal, it will be necessary to separate the complex cell–material interactions into component pieces that help identify secure cause-and-effect relationships because it is unlikely that any a single property such as surface chemistry/energy dominates the entire process culminating in cytocompatibility.

SUMMARY

Recent advances in our understanding of bone cell and molecular biology have provided important insights into which molecules regulate bone cell proliferation and differentiation and the mechanisms by which this occurs. This information can now be exploited in the rationale design of bone tissue engineering protocols. Furthermore, accumulating evidence suggests that certain biophysical factors, for example, fluid flow, are important regulators of bone cell proliferation and differentiation, which is perhaps not surprising since bone cells exist *in vivo* in a very mechanically active environment. It is logical, then, to consider the mechanical environment when designing bone tissue engineering protocols. The substrate with which bone cells interact is also of paramount importance in any application of bone tissue engineering. Indeed, how cells interact with different biomaterials can affect how biophysical or biomolecular signals regulate cell activity. Therefore, future development of bone tissue engineering protocols should emphasize biomolecular, biophysical, and biomaterial signaling characteristics combined optimally for stimulation of bone cell proliferation and differentiation.

ABBREVIATIONS:

TRAP, tartrate-resistant acid phosphatase
PTH, parathyroid hormone
PTHrP, parathyroid hormone related peptide
PGE_2, prostaglandin E_2
IL-1/IL-4/IL-6/IL-8/IL-11, Interleukin-1, etc.
M-CSF, macrophage colony-stimulating factor
CSF-1, synonymous with M-CSF
RANK, receptor-activated nuclear-factor kappa beta (on osteoclasts)
RANKL, RANK Ligand (on osteoblasts)
OPG, osteoprotegerin (a soluble form of RANK)
TNFα, tumor necrosis factor α
LIF, leukemia inhibitory factor
PKA/PKC, protein kinase A or C
RGD, arginine–Glycine–Aspartate sequence
PI3K, phosphatidylinositol 3-kinase
p-NPPase, para-nitrophenyl phosphatase
MMP-9, matrix metaloproteinase-9
MIP-1α, macrophage inflammatory protein-1α
Pyk2, a focal adhesion kinase

The Osteoclast

Carol V. Gay

INTRODUCTION

Bone resorption is balanced by bone formation in normal adult bone, so that bone mass is maintained. Normally, bone-resorbing osteoclasts are present in appropriate numbers and, when mature, they cycle between active and inactive states in a tightly regulated manner. When in balance, the remodeling process involves equal and linked participation of both osteoclasts and bone-forming osteoblasts. Excessive bone resorption is a component of osteopenic conditions including osteoporosis, osteoarthritis, rheumatoid arthritis, orthopedic implant-induced osteolysis, peridontitis, and metastatic bone disease. Mature osteoclasts are believed to exist for ~2 weeks; however, a nearly inexhaustible source of osteoclast precursors is present in the bone marrow.

The life span of the osteoclast can be divided into distinct stages: (1) homing of precursor cells of hemopoietic origin to bone surfaces; (2) development of preosteoclasts into mononuclear osteoclasts; (3) fusion of mononuclear cells into multinucleate osteoclasts; (4) development of osteoclast polarity through attraction and adherence to bone surfaces; (5) assembly of an acidifying resorptive apparatus; (6) cycling through activation and inactivation phases, and (7) detachment and final inactivation that ends in apoptosis. Regulation occurs at all stages of the osteoclast life span.

Osteoclasts were identified as bone-resorbing cells by microscopy many years ago by Robin[113] and Kölliker,[114] but it was not until the 1960s that detailed insights into osteoclast function became possible. Research in the 1960s focused on identifying bone dissolving acids secreted by osteoclasts. Using ^{3}H-acetazolamide-specific inhibitor autoradiography, Gay and Mueller[115] showed conclusively that carbonic anhydrase is abundantly present in osteoclasts. Congenital absence of carbonic anhydrase isoenzyme II results in osteopetrosis, a disorder characterized by defective osteoclast activity.[116,117] The discovery of a central role for carbonic anhydrase in osteoclasts led to studies that defined the mechanism of acid secretion and to the discovery that carbonic anhydrase supplies H^+ to the proton-ATPase. This included showing that there is a proton-pumping vacuolar ATPase at the ruffled border[118–120], that inhibition of carbonic anhydrase markedly slows bone resorption in an organ culture model,[121] and that acid production by osteoclasts depends on carbonic anhydrase activity.[122,123] Investigations into the origin of osteoclasts paralleled studies of function. This was followed by considerable strides being made in elucidating mechanisms of the regulation of osteoclast development and activity, as methods to manipulate genes have evolved.

MORPHOLOGICAL CHARACTERISTICS

The mature osteoclast is multinucleate, containing 6–8 nuclei typically, but it can have many more. The cross-sectional diameter varies from around 20 to several hundred microns. When actively resorbing bone, an osteoclast may be flattened against the bone surface or it may be highly contorted as it seals itself over the end of a spicule. The ruffled border is the most obvious organelle and it has received considerable attention. The ruffled border, so termed by Scott and Pease[124] in the first electron microscopic study of osteoclasts, is completely surrounded by a clear zone, the structure by which the osteoclast adheres to the bone surface. The clear zone is so named because of its appearance in light microscope preparations.[125] This region is packed with actin filaments.[126] The prevalence of actin in the clear zone accounts for the absence of organelles in this region. The adherence of osteoclasts to bone surfaces is now known to involve the interaction of membrane-associated $\alpha_v\beta_3$ integrin in clear zone membrane with osteopontin in exposed bone matrix surfaces.[127]

The ruffled border consists of cytoplasmic extensions that are finger-like or sheet-like protrusions. The protrusions may be several microns long, are widely variable in thickness and length, and usually longer than they are wide. The ruffled border provides an extensive surface area of specialized plasma membrane through which acid is secreted and hydrolytic enzymes are released. The size of the ruffled border varies according to cell activity;[128] inactive osteoclasts have little or no ruffled border.[129,130] Inactive, detached osteoclasts migrate between the perivascular space and bone surface at a rate of 30–250 μm/hr.[131] Large osteoclasts may have more than one ruffled border region, each of which can resorb at different rates. Osteoclasts are phagocytic, and bits of resorbed bone appear to be internalized.

There appear to be two reservoirs of ruffled border membrane. In some instances, extensive interdigitations are seen along the lateral surfaces of inactive osteoclasts.[132] Even more pronounced is the occurrence of numerous cytoplasmic vesicles that give the cells a foamy appearance in histological sections. It is clear that these vesicles are a reservoir for ruffled border formation. As shown by Fukushima et al.,[130] a unique enzyme, alkaline *p*-nitrophenyl phosphatase, is present in both the vesicles and ruffled border membrane, but not in the nearby plasma membrane. The amount of vesicles is inversely proportional to the extent of ruffled border present. A comparison of the distribution of enzymes in osteoclasts implicates unique functions of organelles in the osteoclast, particularly with respect to distinct regions of the plasma membrane.[133]

Many of the vacuoles and vesicles present are involved in either endocytosis or exocytosis.[129,134] A hallmark of osteoclasts, tartrate-resistant acid phosphatase (TRAP), is found within some vesicles, lysosomes, Golgi, and matrix.[130] Polarized secretion of TRAP as well as proteolytic enzymes has been reported.[135]

Ribosomes are present individually and in small clusters, suggesting some capacity for renewal of intracellular proteins.[125] Rough endoplasmic reticulum and Golgi are scant, relative to the entire cytoplasmic volume, indicating that synthesis of protein for export, that is, proteolytic enzymes, is conservative.[125]

Mitochondria are notably abundant, indicating that osteoclasts have a high level of aerobic metabolism. Mitochondria in osteoclasts contain numerous granules rich in calcium phosphate. Such granules are considered to serve as a buffer for surges in intracellular Ca^{2+}.[136] Enhanced uptake and release of Ca^{2+} by mitochondria occur in tissues that handle large amounts of Ca^{2+}.[137] It is estimated that osteoclasts contain more mitochondria than any other cell in the body.[125]

OSTEOCLAST GENERATION

ROLE OF OSTEOBLASTS AND STROMAL CELLS IN OSTEOCLAST DIFFERENTIATION

Multinucleate osteoclasts are derived from fusion of hemopoietic cells of the monocyte–macrophage lineage. The process requires contact with osteoblasts or stromal cells in the marrow (pre-osteoblasts). Osteotropic factors, including PTH, PGE_2, $1,25(OH)_2$ vitamin D_3, and IL-11, orchestrate osteoclast formation through specific receptors on osteoblasts and stromal cells. This phenomenon was discovered by Suda and colleagues, who developed a co-culture system of mouse calvarial osteoblasts and spleen-derived hemopoietic cells as a source of pre-osteoclasts.[138–140] The multinucleate cells formed in the co-culture system exhibited major characteristics of osteoclasts, that is, tartrate-resistant acid phosphatase (TRAP) activity, expression of calcitonin receptors, p60c-Src, $\alpha_v\beta_3$ integrin, and the ability to form resorption pits on bone or dentine slices.

Knockout and transgenic technology has played a decisive role in elucidating the mechanisms involved. Osteoblasts isolated from PTH receptor knockout mice (PTHR1-/-) co-cultured with wild-type hemopoietic cells did not support osteoclast formation; the converse experiment supported osteoclast formation.[141] Similarly, when the receptor for PGE_2 (EP4 subtype) was absent in osteoblasts, osteoclastogenesis did not occur under the influence of PGE_2.[142] The same result was found using the vitamin D receptor knockout mouse.[143] These studies prove that functional osteoblasts are required for initiating osteoclastogenesis.

Other osteotrophic factors are also important. Using the *op/op* mouse model, in which there is an extra base (thymidine) in the M-CSF gene, Yoshida and colleagues[144] established that M-CSF is necessary for osteoclast formation. In addition, Niida et al.[145] report that human recombinant vascular endothelial growth factor (VEGF) will induce osteoclast formation in the *op/op* mouse, apparently by substituting for M-CSF.

In summary, osteoclastogenesis requires cell–cell contact and signaling by osteotrophic factors through specific receptors on osteoblasts and/or stromal cells. This discovery was key for elucidating the next phase of investigations into the mechanism of osteoclast formation, that is, the RANK–RANKL interaction.

THE RANK–RANKL INTERACTION IN OSTEOCLAST FORMATION

Discovery of RANK, RANKL, and OPG

RANK (receptor-activated nuclear-factor kappa β) is expressed on both osteoclast progenitors and mature osteoclasts. Its receptor RANK Ligand (RANKL) is expressed by osteoblasts and stromal cells in response to osteotrophic factors such as $1,25(OH_2)D_3$, PTH, PGE_2, and IL-11 (Figure 3.4). This discovery resulted from investigations by several laboratories in a number of fields, including skeletal metabolism and immunology. An historical account has been presented by Takahashi and colleagues.[146] As the story developed, several regulatory proteins were identified and named. These were as follows: osteoclast differentiation factor (ODF), osteoprotegerin ligand (OPGL), tumor

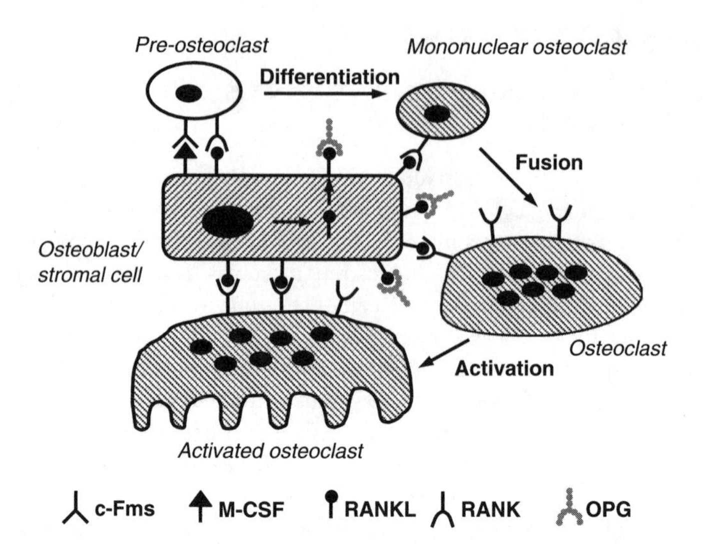

Figure 3.4 Diagram of the generation of osteoclasts from pre-osteoclasts as directed by M-CSF and RANKL on osteoblasts or marrow stromal cells. Both osteoclast progenitors and mature osteoclasts express surface bound RANK, the receptor for RANKL. RANKL expression by osteoblasts is stimulated by the bone-regulatory factors, including PTH, IL-11, 1,25 vitamin D_3, and PGE_2. These factors stimulate differentiation, multinucleation, fusion, and activation of osteoclasts. OPG, a soluble decoy receptor of RANKL, is produced by osteoblasts and stromal cells and blocks the RANK–RANKL induction of osteoclastogenesis.

necrosis factor-related activation-induced cytokine (TRANCE), and RANKL. Through cloning techniques, these four proteins were found to be identical. In addition, a soluble form of RANK, called osteoprotegerin (OPG), is produced by osteoblasts. OPG is a decoy receptor for RANKL and is a powerful inhibitor of osteoclast differentiation.[147] The terms 'RANKL,' 'RANK,' and 'OPG' are recommended by the American Soceity for Bone and Mineral Research Committee on Nomenclature.[148]

Current evidence indicates that RANK (on pre-osteoclasts) acts as the sole signaling receptor for RANKL (on osteoblasts) for inducing differentiation of pre-osteoclasts to fuse and develop into osteoclasts. The binding of cross-linking antibodies to the extracellular domain of RANK caused cell surface clustering of RANK, by-passing the need to interact with RANKL; a spleen culture model osteoclast formation was stimulated.[149,150] On the other hand, when Fab fragments of anti-RANK antibody were introduced, clustering was prevented, RANK–RANKL interaction was blocked, and osteoclasts did not develop.[150] Subsequently, it was found that soluble RANK, the extracellular domain of RANK, now called OPG, also blocked osteoclast formation.[147] Additionally, transgenic mice overexpressing OPG developed osteopetrosis, a symptom of absent or nonfunctioning osteoclasts.[149]

As mentioned, both RANKL and M-CSF have been established as essential for osteoclast formation. It is interesting that both RANKL and M-CSF are rather widely distributed in nature.[151–153] This raises the question of why osteoclasts are confined to bone. Through a series of co-culture experiments with hemopoietic cells on a lawn of an osteoblast cell line, SaOS-4/3, both M-CSF and RANKL were found to be deployed on osteoblast surfaces,[154] as illustrated in Figure 3.4. The simultaneous presence of both proteins on osteoblast surfaces may explain why osteoclast formation occurs only in the bone environment.

Regulation of RANKL and OPG expression in osteoblasts and stromal cells

RANKL expression in osteoblasts is inducible through a variety of signal pathways. These include the following: PTH/PTHrP, TNFα, IL-1, and PGE_2 signaling through the cyclic AMP pathway;[142,155,156] LIF, oncostatin M, IL-6 + sIL-6R, and IL-11 signaling through the gp130/STAT130 pathway;[157] vitamin D receptor (VDR) occupancy by 1,25 (OH_2) vitamin D_3 signaling VDR binding domains found in the 5′-flanking promoter region of the RANKL gene;[158] and compounds such as ionomycin and phorbolmyristatic acid (PMA) stimulating the Ca^{2+}/PKC pathway.[159,160] OPG expression is often suppressed by the factors that stimulate RANKL expression. For example, forskolin, an activator of cAMP/PKA signaling, suppressed OPG mRNA expression while enhancing RANKL mRNA expression.[160] Cross-talk between pathways that lead to coordinated up- and downregulation of RANKL, RANK, and OPG are currently under investigation.

Signaling through RANK in osteoclasts

When RANKL of osteoblasts engages RANK on pre-osteoclasts or mature osteoclasts, signal cascades are initiated, which lead to differentiation, survival, or activation of the osteoclast. In a study of RANK–/– mice, which were treated with powerful osteotropic factors (1,25$(OH_2)D_3$, PTHrP, and IL-11) neither TRAP-positive cell formation nor hypercalcemia was induced.[161] This study indicates that osteoclasts cannot form and mature in the absence of RANK. A number of RANK signaling pathways have been identified. The cytoplasmic tail of RANK interacts with TNF receptor-associated factor 6 (TRAF6). TRAF6 interacts with TRAF1, 2, 3, and/or 5 to activate NFκB, JNK, p38, or ERK pathways.[162–166] TRAF6 appears to be the crucial link between RANK and the other TRAFs since TRAF6 –/– mice exhibited severe osteopetrosis.[167]

Further supporting the central importance of TRAF6 is the report that lymphocyte T-cell production of interferon-γ (INFγ) blocks osteoclastogenesis.[168] In this study, it was shown that RANK–RANKL signaling was blocked due to INFγ-induced degradation of TRAF6.

NKκB activation leads to increased resorption as detected by the pit assay.[169] JNK signaling through AP-1, the ubiquitous activator protein (comprised of fos and jun transcription factors), supports osteoclast differentiation[170] as does the p38 MAP kinase pathway.[149,171] ERK signaling via MEK prolonged survival, but did not affect cell activity.[169] Greater details on these pathways can be found in Takahasi et al.[146]

RANKL/RANK/TRAF6 also signals through c-Src.[166] Soriano et al.[172] were the first to report that targeted disruption of the c-Src gene causes osteopetrosis. Osteoclasts were shown to be present in c-Src –/– mice, but ruffled borders were absent.[173] Spleen cells from c-Src-deficient mice could differentiate into TRAP-positive multinucleated cells, but they did not form resorption pits.[174]

To date, many signaling pathways through RANK in osteoclast have been established. Substantial evidence supports the conclusion that RANK–RANKL interaction is required for osteoclasts to form and differentiate.

Regulation of RANKL expression in lymphocytes

Lymphocyte expression of RANKL is an important consideration because bone resorption can increase at sites of inflammation. Activated T-cells, in which RANKL expression is upregulated, have been shown to stimulate osteoclast formation and activation.[175] The signaling pathways involved in inducing RANKL expression are PKC, phosphoinositol-3 kinase, and calcineurin. In a T-cell-dependent model of adjuvant arthritis in rats, cartilage destruction was prevalent, but was prevented by OPG treatment. $Cd44^{(+)}$ T-cells, stimulated by microbial challenge and introduced into a mouse model, caused periodontitis and alveolar bone loss; treatment with OPG blocked RANKL signaling and reduced the loss of bone.[176]

OSTEOCLAST ACTIVITY

Adherence and Development of Polarity

Once osteoclasts have been induced to form by fusion of pre-osteoclasts present among hemopoietic cells, as orchestrated by interaction with osteoblasts, a number of morphologic and functional changes occur. First, polarity develops in such a manner that recognition and binding of the osteoclast to bone surfaces occur. The ruffled border that develops becomes completely surrounded by the clear zone, whose membrane is enriched in RGD-peptide-recognizing proteins, especially $\alpha_v\beta_3$ integrin.[177] This particular integrin recognizes matrix peptides and has a high affinity for osteopontin, but it also recognizes other peptides including bone sialoprotein and fibronectin. Osteoclasts bind to matrix proteins when bone mineral crystals are present; osteoclasts neither adhere to nor resorb a nonmineralized matrix.[178] Osteopontin and other RGD-peptides are believed to bind to crystal surfaces in a manner that allows presentation of the RGD sequence to osteoclasts. Osteopontin has a particularly high density of negative charges due to phosphorylation; this fosters binding of calcium phosphate and calcium carbonate crystals.[179]

The first studies indicating that osteoclast activation is adherence-dependent and that binding leads to intracellular signaling and induction of bone resorption came from Horton and colleagues in 1985.[180,181] Echistatin blocks binding of $\alpha_v\beta_3$ integrin to its ligand, causing cell inactivation and detachment, as detected by morphological changes and the pit assay. Detachment is accompanied by redistribution of the $\alpha_v\beta_3$ integrin. It appears that $\alpha_v\beta_3$ integrins contribute to osteoclast function in at least two ways: support of cell migration and maintenance of the sealing zone during resorption.[182] Less is known of the roles of other integrins in osteoclasts.

The ability of the osteoclast to detach from bone surfaces and move to other locales is an important aspect of osteoclast function. Migration is under the control of *c-fms*, which encodes the CSF-1 receptor; CSF-1R is expressed in abundance in mature osteoclasts.[183] Osteoclasts migrate along CSF-1 gradients with cytoskeletal rearrangements involving phosphatidylinositol 3-kinase (PI3K) and Src.[184,185] The interaction of CSF-1 with its receptor leads to the redistribution of PI3K to the cell periphery and disassembly of the ruffled border. When migration ceases, ruffled borders reassemble. PI3K is a pivotal enzyme in the process; its inhibition results in actin ring disruption,[186] impaired resorption,[186,187] and impaired migration.[184] Osteoclast binding to osteopontin on the bone surface induces actin polymerization and resorption.[188] Migration and polarity development in osteoclasts are under tightly regulated signaling pathways. Signaling is mediated through interaction of c-Src and other effectors, namely PI3K, Pyk2 or Cb1.[189]

Assembly of the Ruffled Border and other Components of the Mature Osteoclast

To become functional, osteoclasts must form a ruffled border, wherein resides a unique assembly of molecules for secreting an acidified solution enriched in proteolytic enzymes. Carbonic anhydrase and vacuolar-type H^+-ATPase are the key players in acidification. The tight seal surrounding the ruffled border creates a space so that the resorption fluid is directed onto the bone surface in a precise and regulatable manner. The ruffled border develops from the fusion of intracellular vesicles.[118,130,190] Some of the vesicles are derived from Golgi and containproteolytic enzymes. Constituents of the ruffled border appear to be expressed during the development of osteoclasts.[191–193] Figure 3.5 outlines the major mechanisms in osteoclasts that support bone degradation.

Carbonic anhydrase

Microscopic studies established the presence of carbonic anhydrase in osteoclasts. [194] Autoradiographic localization revealed that the enzyme was present in as great abundance as in other tissues where the enzyme plays a key role (e.g., pancreas and kidney).[115] Immunocytochemical studies showed the major isoform present to be carbonic anhydrase II (CAII).[115,195] Electron microscopic immunocytochemistry revealed carbonic anhydrase in the cytosol, in vesicles, and along the membrane of active ruffled borders; calcitonin treatment caused a shift of carbonic anhydrase off the ruffled border in both *in vivo* and *in vitro* systems.[196,197]

Carbonic anhydrase is well established as being necessary for bone resorption. *In vivo* inhibitor studies by Kenny and colleagues[198–200] revealed the physiologic importance of carbonic anhydrase in bone resorption. When procedures for isolating osteoclasts were developed, it became possible to design mechanistic studies to show that osteoclast acidification is dependent on carbonic anhydrase activity.[123,201–205] Carbonic anhydrase deficiency due to a congenital condition results in osteopetrosis.[116,117]

Both H^+ and HCO_3^- are products of carbonic anhydrase activity, that is, CO_2 hydration (Figure 3.5). The extrusion of H^+ through vacuolar H^+-ATPase action is balanced by Cl^-/HCO_3^- exchange.[206] This prevents internal alkalinization due to loss of H^+ extrusion. Intracellular Ca^{2+} levels are also linked to carbonic anhydrase activity since blocking the enzyme causes increased influx of Ca^{2+}.[207]

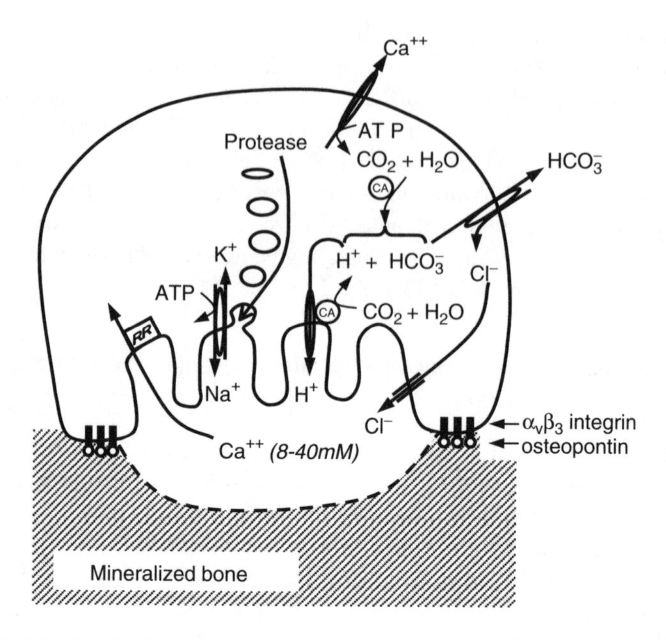

Figure 3.5 Matrix acidification and proteolysis by osteoclasts. Binding of the osteoclast to mineralized matrix through the osteopontin-$\alpha_v\beta_3$ integrin initiates ruffled border formation and positioning of ion-translocating proteins. Carbonic anhydrase (CA) in cytosol and at the ruffled border supplies H^+ to a vaculolar-type H^+-ATPase. The resulting ionic imbalance is corrected by HCO_3^- efflux through a Cl^-/HCO_3^- exchanger. The excess intracellular chloride enters the resorption pit through a chloride channel in the ruffled border. Ryanodine receptors (RR), purported to be in the ruffled border serve as a negative feedback Ca^{2+} sensor. An active Na^+, K^+-ATPase is present in the ruffled border. Ca^{2+}-ATPase is responsible for calcium efflux on the nonresorptive side of the cell; its role is likely to maintain low cytosolic Ca^{2+} concentrations. Matrix metaloproteinases, formed by the Golgi, are secreted into the resorption pit.

Carbonic anhydrase is required for osteoclast differentiation. The formation of TRAP-positive multinuclear cells was decreased while the number of TRAP-positive mononuclear cells increased, indicating that interference with CAII expression interferes with precursor fusion.[207,208]

Hormones involved in the regulation of bone growth affect CAII expression. The CAII promoter in chicken osteoclasts contains a vitamin D response element, and the addition of vitamin D to cultured osteoclasts increases the level of CAII mRNA expression.[209,210] Further, vitamin D-induced stimulation of osteoclast differentiation requires CAII due, in part, to the requirement for regulation of intracellular pH.[207] This suggests that CAII is expressed in osteoclast precursors. PTH can also stimulate CAII mRNA expression.[210] Retinoic acid may have an antagonistic role to that of vitamin D during osteoclast differentiation, and this affects CAII expression.[211] Retinoic acid stimulates the proliferation of pre-osteoclasts and numbers of multinucleated cells in osteoclast cultures but has no effect on CAII expression, whereas vitamin D added to these same cultures along with retinoic acid stimulates proliferation and increases CAII expression. A retinoic acid response element has recently been identified in the CAII promoter.[212]

Vacuolar ATPase

Because carbonic anhydrase facilitates H^+ and HCO_3^- production but cannot act as a pump, efforts were made to identify a proton-translocating system in osteoclasts. Using antibodies to gastric H^+, K^+-ATPase, immunocytochemical cross-reactivity was detected along the ruffled border;[213] however the role of the H^+, K^+-ATPase is still not clear.[214] A proton-ATPase was also detected in ruffled border membranes by an enzyme histochemical staining procedure activity.[215] Subsequent to these initial investigations, three laboratories independently demonstrated the H^+-translocating function of this enzyme in osteoclasts and identified it as a vacuolar-type ATPase (V-ATPase).[118–120] Plating isolated osteoclasts onto a resorbable surface, such as slices of dentine, causes cell activation and release of H^+ into resorption pits.[216]

When osteoclasts are cultured on RGD-containing peptide (vitronectin, fibronectin, collagen I)-coated plastic, the V-ATPase is distributed to the ruffled membranes and is activated, as shown by the capacity of the cells to secrete acid. V-ATPase in activated osteoclasts associates with actin filaments, a process believed to move the enzyme from diffuse cytosolic distribution to focal localization in the ruffled border.[216] Myosin II, which interacts with actin filaments, is also involved, since microinjection of antibodies to myosin II diminishes bone resorption as assessed by the pit assay.[217] RNA transcripts of V-ATPase have also been shown to shift from diffuse distribution to the ruffled border area when osteoclasts become activated.[218] When fully formed osteoclasts become inactivated, the ruffled border vesiculates internally as shown using alkaline p-NPPase, a ruffled border marker enzyme.[130] V-ATPase is expressed in mononuclear osteoclasts along with other markers of mature osteoclasts, such as tartrate-resistant acid phosphatase and calcitonin receptors. The expression of these proteins occurs before the cells bind to the substrate.[192,193]

V-ATPase is a membrane-integrated protein. Its mRNA is localized in the region of cytoplasm between the nuclei and membrane where osteoclasts are attached to the bone.[218] Reagents that disrupt the cytoskeleton or inhibit V-ATPase activity also inhibit polarization of the mRNA.[218] Antisense nucleic acids to V-ATPase decrease levels of message and cause impaired acidification and bone resorption.[207,208] V-ATPase consists of several complexes: V_i located in the cytoplasm, a membrane-integrated subunit, and V_0 that contains the H^+ pump. Isoform variants of subunit 'a' in the V_0 complex have been identified; the expression of mRNA for each isoform is tissue-specific.[219] One of these isoforms, a3, appears to be osteoclast-specific and is induced during osteoclast differentiation.[220] Mutations in the a3 subunit of human vacuolar-type H^+-ATPase causes osteopetrosis.[221] In addition, mice carrying a disrupted form of the gene encoding this subunit[222] or a deletion of a portion of this gene[223] exhibit osteopetrosis due to loss of osteoclast function.

Other plasma membrane enzymes and channels that support osteoclast activity

Ouabain-inhibitable Na/K-ATPase activity has been found in the ruffled border membrane at greater densities than elsewhere in the plasma membrane,[224] and immunocytochemistry has revealed Na/K-ATPase in the entire plasma membrane.[225] The difference between the two observations likely reflects methodology; the former reflects enzyme activity and the latter reflects the position of protein subunits, whether active or not. The role of Na/K-ATPase in the ruffled border has not been defined, but it is likely involved with maintaining an electrolyte balance. Spectrin, known to be associated with sites of Na/K-ATPase activity, has been localized to the ruffled membrane.[226]

Phosphatidylinositol-3 kinase (PI3K) is involved in ruffled border formation in osteoclasts. This has been shown by specifically inhibiting PI3K and finding that vesicles derived from ruffled borders appeared in abundance in the cytosol; once the inhibitor was removed, the number of vesicles diminished as a result of fusion with the ruffled border.[227]

Plasma membrane Ca^{2+}-ATPase (PMCA) is abundant on the nonresorptive surface of active osteoclasts[228] and it pumps Ca^{2+} to the cell exterior.[229] RGD-containing peptides stimulate Ca^{2+} efflux in osteoclasts through the PMCA.[230] The specific role of PMCA has not been defined but is likely involved with restoring cytosolic Ca^{2+} levels following Ca^{2+} signaling events as it does in other cells. High extracellular Ca^{2+}, that is, 8–40 mM in resorption pits,[231] stimulates IL-6 production, which in turn inhibits Ca^{2+} sensing,[232] constituting a negative feedback mechanism. Ryanodine receptor type 2, postulated to be in the ruffled border membrane, appears to serve as both a Ca^{2+} sensor and an inwardly directed Ca^{2+} channel.[233]

The substantial secretion of protons into resorption lacunae is accompanied by Cl^- through an electrically coupled chloride channel.[234,235] Acid–base balance within the osteoclast is maintained by a Cl^-/HCO_3^- exchanger on the basolateral plasma membrane.[236]

The GTPases, Rac1 and Rac2, are intimately involved with the cytoskeleton and influence ruffled border development. Introduction of antibodies to these enzymes into osteoclasts causes disruption of the actin ring, osteoclast detachment, loss of ruffled border, and reduced resorption.[237]

Src is a plasma membrane-anchored tyrosine kinase that elicits responses through many pathways. Src has many substrates, including actin and integrin binding proteins (e.g., tensin, paxillin, vinculin, cortactin, talin).[238] While complete understanding of Src activity in osteoclasts awaits discovery, it is clear that activation of Src leads to alterations in the cytoskeleton in ways that lead to cell activation. Further discussion of Src is presented in the next section.

Proteolytic enzymes

Osteoclasts secrete tartrate-resistant acid phosphatase and thiol proteinases (e.g., acidic collagenase and cathepsin K) into resorption pits.[135,239] These serve to break down the matrix once bone mineral deposits have been dissolved. Of the several matrix metalloproteinases found in osteoclasts, MMP-9 is abundantly expressed; however, MMP-9 knockout mice experience only a transient disturbance in bone resorption.[240,241] Among the thiol proteases, cathepsin K is the most abundant. Knockout of cathepsin K retards matrix degradation and results in osteopetrosis, whereas overexpression causes the opposite problem.[242] Degraded collagen fragments are translocated through the osteoclast cytoplasm by vacuolar transcytosis.[243,244]

Signal Pathways in Mature Osteoclasts

It is well established that osteoclasts are activated indirectly by osteotropic factors that upregulate RANKL expression in osteoblasts and then stimulate osteoclasts through RANK. There are some exceptions to this rule, however.

Direct activators of osteoclast activity

IL-1 directly stimulates osteoclasts. Cultures of mature osteoclasts express IL-1 receptors and form pits in response to IL-1;[245] this response was inhibited by IL-1 receptor antagonist (IL-1ra) but not by blocking the RANK–RANKL interaction with OPG.[147]

TNFα can stimulate osteoclast differentiation and pit-forming activity in the absence of RANK–RANKL interaction. This was shown in bone marrow cultures from RANK –/– mice treated with TNFα.[246,247] When antibody to TNFα was present, the response was blocked; OPG had no effect. In another study using RANK –/– mice, TNFα infusion led to the formation of TRAP-positive osteoclasts.[161]

PTH/PTHrP receptors are present in mature osteoclasts. Using *in situ* hybridization, mRNA for PTHR-1 has been detected in osteoclasts in sections of iliac crest biopsies. In samples from normal individuals, 65.7% of osteoclasts contained PTHR-1 mRNA and in patients with secondary hyperparathyroidism, 98% of osteoclasts contained message.[248] Immunostaining has revealed PTH/PTHrP receptor in osteoclasts from rat bone,[249–251] deer antler,[252] human bone,[248] and the RAW 264.7 osteoclast cell line.[251] Rapid responses of osteoclasts to PTH or PTHrP have been reported.[253] The studies showing rapid responses utilized full-length PTH (1–84) and so it is not clear whether the N terminus or the C terminus was responsible for the effect. Most studies indicate that levels of PTH/PTHrP receptors in osteoclasts are low and occur intermittently. It is interesting that PTHrP is secreted by both osteoblasts[254] and osteoclasts.[255] Consequently, PTHrP may act as a paracrine or autocrine factor to influence mature osteoclast function. PTH has both anabolic and catabolic effects on the bone; low and intermittent doses elicit bone apposition, while high and continuous administration stimulates resorption.[256] One way in which these dichotomous effects could occur might be through receptors on both osteoblasts and osteoclasts. For example, the osteoblasts expressing PTH receptors constitutively and in abundance could react well to intermittent stimulation. The pauses in stimulation would provide time for intracellular signals (e.g., Ca^{2+}, cAMP) to return to baseline. On the other hand, the osteoclast, which rapidly clears its surface of occupied PTH receptors,[257] would be able to respond to continuous doses of PTH by recycling unoccupied receptors, to the cell surface. Two PTH-responsive cells have been shown to rapidly clear occupied PTH receptors by endocytosis, namely, cultured kidney cells[258,259] and osteoclasts.[257]

Estradiol influences osteoclast formation, differentiation, and activity indirectly through the regulation of cytokine synthesis by osteoblasts. Estrogen also acts directly on osteoclasts at both genomic and nongenomic levels.[253,260] Estradiol reduces mRNA expression of the type I IL-1 receptor in osteoclasts while increasing expression of decoy-type II IL-1 receptor.[261] Several studies have shown that estradiol reduces resorption activity, through an estrogen response element, in human and avian osteoclast-like cells,[262–264] rabbit osteoclasts,[265] and mouse osteoclast-like cells.[266] Other studies report the lack of an estrogen effect on osteoclast activity.[190,267–269] The difference in these reports may reflect the stage osteoclasts are in when isolated. For example, Sugiyama et al.[270] found that estrogen did not influence adherence of osteoclasts isolated from avian medullary bone in the resorptive phase, but caused a reduction in adherence of osteoclasts during bone formation. Estrogen reduces mRNA and protein synthesis of several lysosomal enzymes in osteoclasts, including cathepsin L, β-glucuronidase, lysozyme, and a cathepsin unique to osteoclasts, cathepsin K.[263–265,271]

The main means by which nongenomic effects are distinguished from genomic effects is through the detection of rapid signaling events with a membrane-impermeable form of estrogen. Cultures of mature osteoclasts bind and respond to 17β-estradiol. A fluorescent estradiol–BSA complex bound to osteoclasts and was displaceable by estradiol monomer.[272] Estradiol depolarizes osteoclast membrane potential, and stimulates Src activity and Src-dependent actin ring formation.[272,273] The effects occur within 60 sec and so are believed to represent nongenomic regulation at the plasma membrane. Membrane-impermeable estrogen also has rapid effects on intracellular pH, cyclic AMP, cyclic GMP, and intracellular Ca^{2+} in human pre-osteoclastic cells.[274] PTH-stimulated acid production by isolated osteoclasts is blocked by 17β-estradiol.[122] Membrane depolarization occurs following estradiol treatment as a consequence of the opening of inwardly directed K^+ channels.[275] Signaling events following occupancy of plasma membrane estrogen receptors involves a rapid (within 2 min) phosphorylation of tyrosines at the plasma membrane. Src is activated and translocated to the plasma membrane; it is likely the main phosphorylating enzyme involved.[273] Superoxide anion generation in osteoclast is inhibited by 17β-estradiol within 5–45 sec.[276]

Inhibitors of osteoclast activity

Calcitonin rapidly blocks resorption through a now well-defined receptor present on osteoclasts. The physiological role of calcitonin is believed to be prevention of bone loss at times of stress on calcium reserves, for example, during pregnancy, lactation, and growth. Stress on calcium reserves may be why calcitonin receptors evolved to be under tight control and are rapidly endocytosed following hormone binding.[277] Osteoclasts respond rapidly to calcitonin, losing their attachment sites as well as their ruffled borders through vesiculation and cell rounding.[196] Calcitonin causes a shift in the cytoskeletal elements, vimentin, and tubulin.[278] Calcitonin also abolishes superoxide anion (O_2^-) production[279] and inhibits osteopontin expression in osteoclasts.[280] The calcitonin receptor exists in several isoforms and signals through several pathways, including cAMP, Ca^{2+}, and MAP1 kinase. The binding affinity and capacity of the calcitonin receptor is influenced by 17β-estradiol and progesterone in laying hens.[281] Continued exposure of osteoclasts to calcitonin results in the down-regulation of the calcitonin receptor. This effect occurs at the transcriptional level and renders osteoclasts insensitive to prolonged calcitonin exposure.[282,283]

Nitric oxide (NO) decreases osteoclast activity through a NO-dependent guanosyl cyclase and cGMP-dependent protein kinase.[284] Osteoblasts responding to stretch are the likely source of NO,[285] although osteoclasts can also produce nitric oxide.[286] At high concentrations, NO and Ca^{2+} can initiate apoptosis.[287,288]

Other inhibitors of osteoclast activity, under some conditions, include high phosphate and Ca^{2+} concentrations, glucocorticoids, prostaglandins, IL-4, IL-8, M-CSF, and MIP-1α.[289]

Other factors influencing osteoclasts

Fluid shear has a number of effects on osteoblasts, causing them to secrete IL-11, which in turn influences osteoclasts.[290] Cell–cell contact (osteoblast–osteoclast; stromal cell–osteoclast) is a likely means of direct control; osteoclasts have been shown to express connexin 43, elements of gap junctions.[291] Stem cell factor is expressed on osteoblast surfaces, but not secreted; therefore, its ability to activate existing osteoclasts as well as enhance osteoclast differentiation requires cell–cell contact.[292] As indicated earlier, control of osteoclast activity involves multiple mechanisms.

Involvement of the tyrosine kinase, Src, in osteoclast regulation

The protein pp60 derived from the oncogene *c-src* has tyrosine kinase activity. This enzyme is critical for osteoclast activity and it has many roles. Its importance in bone was first recognized when deleted by knockout technology.[172] Mutations in pp60 result in osteopetrosis.[173] Src protein associates with microtubules in osteoclast-like cells,[293] and it supports ruffled border formation.[173] Therefore, osteoclasts are prepared for activity, at least in part, by an Src-mediated process. Src-dependent phosphorylation of Pyk2, a kinase in the focal adhesion kinase complex, appears to foster adhesion-induced formation of the sealing zone[294] and gelsolin-associated PI3-kinase activity, which facilitates actin filament formation and osteoclast motility.[295] The latter activity is initiated through binding of osteopontin to the $\alpha_v\beta_3$ integrin. Since this effect was shown to occur through plasma membrane estrogen receptors, it is possible that it is mediated through altered membrane potential.[275] In addition, a downregulation of osteoclast activity also occurs through Src-mediated tyrosine kinase phosphorylation through 17β-estradiol receptor occupancy.[273] This may occur through a reduced interaction between actin and cortactin, another Src-mediated effect.[296] Yet another role for Src in osteoclast function is linked with intracellular pH. When intracellular pH becomes more acidic, tyrosine kinase pathways are activated, increasing Src activity.[297] Src appears to have a pivotal role in the regulation of osteoclast activity.

END OF LIFE: APOPTOSIS

Apoptosis is a form of cell death that provides a means of containment of intracellular lytic enzymes that, if released, would damage neighboring cells. Apoptosis was first characterized as a series of morphological changes.[298,299] Cells undergoing apoptosis become detached from adjacent cells, condense, and finally break into readily phagocytosed fragments.

Initially, apoptosis in osteoclasts was discovered when M-CSF was withdrawn from culture media.[300,301] A number of apoptotic stimuli are now known. Loss of adhesion by interference with integrin–matrix interaction causes apoptosis, as shown by introducing small interfering RGD-peptides[302,303] or antisense oligonucleotides to the α_v gene.[304] Vitamin K_2 stimulates apoptosis presumably because of free radical production.[305,306] Inhibiting vacuolar H^+-ATPase results in apoptotic cell death.[307] Estrogen, testosterone, glucocorticoids, and bisphosphonate treatment lead to osteoclast apoptosis.[308] High extracellular calcium is apoptotic.[288]

Antiapoptotic agents include factors that stimulate bone resorption, such as PTH and 1,25 (OH_2) vitamin D_3.[308] These signal through RANKL and so it is not surprising that these antiapoptotic effects are countered by NFκB inhibition[309] and OPG.[310] Inflammatory cytokines IL-1 and IL-6 are antiapoptotic.[308] Calcitonin promotes osteoclast survival and opposes nitric oxide-induced apoptosis.[287,311,312] It is interesting that calcitonin is antiapoptotic even though calcitonin causes cell detachment.[313]

Controlled apoptosis of osteoclasts at remodeling sites is considered to be an important feature of the advancing resorption front at sites of bone remodeling.[314]

Bibliography

1. Marks, S. C. and Odgren, P. R. (2002). Structure and development of the skeleton. in *Principles of Bone Biology,* Bilezikian, J. P., Raisz, L. G. and Rodan, G. A., Eds., Vol. 1,– 2 vols. Academic Press, New York, pp. 3–15.
2. Goldberg, M. and Boskey, A. L. (1996). Lipids and biomineralizations. *Prog. Histochem.and Cytochemistry* 31, 1–187.
3. Boskey, A. L. (1998). Biomineralization: conflicts, challenges, and opportunities. *J. Cell. Biochem. — Suppl.* 30–31, 83–91.
4. Aubin, J. E. (2001). Regulation of osteoblast formation and function. *Rev. Endocr. and Metab. Disorders* 2, 81–94.
5. Katagiri, T. and Takahashi, N. (2002). Regulatory mechanisms of osteoblast and osteoclast differentiation. *Oral Dis.* 8, 147–159.
6. Lian, J. B., Stein, G. S., Stein, J. L. and van Wijnen, A. J. (1998). Transcriptional control of osteoblast differentiation. *Biochem. Soc. Trans.* 26, 14–21.
7. Parfitt, A. M. (1977). The cellular basis of bone turnover and bone loss: a rebuttal of the osteocytic resorption—bone flow theory. *Clin. Orthop. and Relat. Res.*, 236–247.
8. Karsdal, M. A., Larsen, L., Engsig, M. T., Lou, H., Ferreras, M., Lochter, A., Delaisse, J. M. and Foged, N. T. (2002). Matrix metalloproteinase-dependent activation of latent transforming growth factor-beta controls the conversion of osteoblasts into osteocytes by blocking osteoblast apoptosis. *J. Biol. Chem.* 277, 44061–44067.
9. Lieberman, J. R., Daluiski, A. and Einhorn, T. A. (2002). The role of growth factors in the repair of bone. Biology and clinical applications. *J. Bone and Miner. Res.* 17, 977–978.
10. Wang, E. A., Rosen, V., D'Alessandro, J. S., Bauduy, M., Cordes, P., Harada, T., Israel, D. I., Hewick, R. M., Kerns, K. M., LaPan, P. and et al. (1990). Recombinant human bone morphogenetic protein induces bone formation. *Proc. Nat. Acad. Sci. USA* 87, 2220–2224.
11. Schmitt, J. M., Hwang, K., Winn, S. R. and Hollinger, J. O. (1999). Bone morphogenetic proteins: an update on basic biology and clinical relevance. *J. Orthop. Res.* 17, 269–278.

12. Bonewald, L. F. (2002). Transforming growth factor beta. In *Principles of Bone Biology* (Bilezikian, J. P., Raisz, L. G. and Rodan, G. A., eds.), Vol. 2, pp. 903–918. 2 vols. Academic Press, New York.
13. Karsenty, G. (1999). The genetic transformation of bone biology. *Genes and Dev.* 13, 3037–3051.
14. Erlebacher, A. and Derynck, R. (1996). Increased expression of TGF-Beta2 in osteoblasts results in an osteoporosis-like phenotype. *J. Cell Biol.* 132, 195–210.
15. Nielsen, H. M., Andreassen, T. T., Ledet, T. and Oxlund, H. (1994). Local injection of TGF-beta increases the strength of tibial fractures in the rat. *Acta Orthop. Scand.* 65, 37–41.
16. Critchlow, M. A., Bland, Y. S. and Ashhurst, D. E. (1995). The effect of exogenous transforming growth factor-beta 2 on healing fractures in the rabbit. *Bone* 16, 521–527.
17. Shimoaka, T., Ogasawara, T., Yonamine, A., Chikazu, D., Kawano, H., Nakamura, K., Itoh, N. and Kawaguchi, H. (2002). Regulation of osteoblast, chondrocyte, and osteoclast functions by fibroblast growth factor (FGF)-18 in comparison with FGF-2 and FGF-10. *J. Biol. Chem.* 277, 7493–7500.
18. Kubota, K., Sakikawa, C., Katsumata, M., Nakamura, T. and Wakabayashi, K. (2002). Platelet-derived growth factor BB secreted from osteoclasts acts as an osteoblastogenesis inhibitory factor. *J. Bone and Miner. Res.* 17, 257–265.
19. Zhang, J. T. and Nicholson, B. J. (1989). Sequence and tissue distribution of a second protein of hepatic gap junctions, Cx26, as deduced from its cDNA. *J. Cell Biol.* 109, 3391–3401.
20. Lee, J. Y., Nam, S. H., Im, S. Y., Park, Y. J., Lee, Y. M., Seol, Y. J., Chung, C. P. and Lee, S. J. (2002). Enhanced bone formation by controlled growth factor delivery from chitosan-based biomaterials. *J. Control. Release* 78, 187–197.
21. Donahue, H. J. (2000). Gap junctions and biophysical regulation of bone cell differentiation. *Bone* 26, 417–422.
22. Duncan, R. L., Kizer, N., Barry, E. L., Friedman, P. A. and Hruska, K. A. (1996). Antisense oligodeoxynucleotide inhibition of a swelling-activated cation channel in osteoblast-like osteosarcoma cells. *Proc. Nat. Acad. Sci. USA* 93, 1864–1869.
23. Kizer, N., Guo, X. L. and Hruska, K. (1997). Reconstitution of stretch-activated cation channels by expression of the alpha-subunit of the epithelial sodium channel cloned from osteoblasts. [erratum appears in Proc. Natl. Acad. Sci. USA 1997 Apr 15;94(8):4233]. *Proc. Nat. Acad.Sci. USA* 94, 1013–1018.
24. Marino, A. A., Becker, R. O. and Soderholm, S. C. (1971). Origin of the piezoelectric effect in bone. *Calcif. Tissue Res.* 8, 177–180.
25. Hastings, G. W. and Mahmud, F. A. (1988). Electrical effects in bone. *J. Biomed. Eng.* 10, 515–521.
26. Rubin, J., McLeod, K. J., Titus, L., Nanes, M. S., Catherwood, B. D. and Rubin, C. T. (1996). Formation of osteoclast-like cells is suppressed by low frequency, low intensity electric fields. *J. Orthop. Res.* 14, 7–15.
27. Luben, R. A., Cain, C. D., Chen, M. C., Rosen, D. M. and Adey, W. R. (1982). Effects of electromagnetic stimuli on bone and bone cells *in vitro*: inhibition of responses to parathyroid hormone by low-energy low-frequency fields. *Proc. Nat. Acad. Sci. USA* 79, 4180–4184.
28. McLeod, K. J., Donahue, H. J., Levin, P. E., Fontaine, M. A. and Rubin, C. T. (1993). Electric fields modulate bone cell function in a density-dependent manner. *J. Bone and Miner. Res.* 8, 977–984.
29. Lohmann, C. H., Tandy, E. M., Sylvia, V. L., Hell-Vocke, A. K., Cochran, D. L., Dean, D. D., Boyan, B. D. and Schwartz, Z. (2002). Response of normal female

human osteoblasts (NHOst) to 17beta-estradiol is modulated by implant surface morphology. *J. Biomed. Mater. Res.* 62, 204–213.
30. Rubin, C. T., Donahue, H. J., Rubin, J. E. and McLeod, K. J. (1993). Optimization of electric field parameters for the control of bone remodeling: exploitation of an indigenous mechanism for the prevention osteopenia. *J. Bone and Miner. Res.* 8, 573–581.
31. Reich, K. M., McAllister, T. N., Gudi, S. and Frangos, J. A. (1997). Activation of G proteins mediates flow-induced prostaglandin E2 production in osteoblasts. *Endocrinology* 138, 1014–1018.
32. Johnson, D. L., McAllister, T. N. and Frangos, J. A. (1996). Fluid flow stimulates rapid and continuous release of nitric oxide in osteoblasts. *Am. J. Physiol. Endocrinol. and Metab.* 34, E 205–E 208.
33. Reich, K. M., Gay, C. V. and Frangos, J. A. (1990). Fluid shear stress as a mediator of osteoblast cyclic adenosine monophosphate production. *J. Cell. Physiol.* 143, 100–104.
34. You, J., Reilly, G. C., Zhen, X., Yellowley, C. E., Chen, Q., Donahue, H. J. and Jacobs, C. R. (2001). Osteopontin gene regulation by oscillatory fluid flow via intracellular calcium mobilization and activation of mitogen-activated protein kinase in MC3T3-E1 osteoblasts. *J. Biol. Chem.* 276, 13365–13371.
35. Owan, I., Burr, D. B., Turner, C. H., Qiu, J., Tu, Y., Onyia, J. E. and Duncan, R. L. (1997). Mechanotransduction in bone: osteoblasts are more responsive to fluid forces than mechanical strain. *Am. J. Physiol. Cell Physiol.* 42, C810–C815.
36. Sakai, K., Mohtai, M. and Iwamoto, Y. (1998). Fluid shear stress increases transforming growth factor beta 1 expression in human osteoblast-like cells: modulation by cation channel blockades. *Calcif. Tissue Int.* 63, 515–520.
37. Ogata, T. (1997). Fluid flow induces enhancement of the Egr-1 mRNA level in osteoblast-like cells: involvement of tyrosine kinase and serum. *J. Cell. Physiol.* 170, 27–34.
38. Ajubi, N. E., Klein-Nulend, J., Alblas, M. J., Burger, E. H. and Nijweide, P. J. (1999). Signal transduction pathways involved in fluid flow-induced PGE2 production by cultured osteocytes. *Am. J. Physiol.* 276, E171–E178.
39. Klein-Nulend, J., Semeins, C. M., Ajubi, N. E., Nijweide, P. J. and Burger, E. H. (1995). Pulsating fluid flow increases nitric oxide (NO) synthesis by osteocytes but not periosteal fibroblasts—correlation with prostaglandin upregulation. *Biochem. and Biophys. Res. Commun.* 217, 640–648.
40. Jacobs, C. R., Yellowley, C. E., Davis, B. R., Zhou, Z. and Donahue, H. J. (1998). Differential effect of steady versus oscillating flow on bone cells. *J. Biomech.* 31, 969–976.
41. You, J., Yellowley, C. E., Donahue, H. J., Zhang, Y., Chen, Q. and Jacobs, C. R. (2000). Substrate deformation levels associated with routine physical activity are less stimulatory to bone cells relative to loading-induced oscillatory fluid flow. *J. Biomech. Eng.* 122, 387–393.
42. Vander Molen, M. A., Rubin, C. T., McLeod, K. J., McCauley, L. K. and Donahue, H. J. (1996). Gap junctional intercellular communication contributes to hormonal responsiveness in osteoblastic networks. *J. Biol. Chem.* 271, 12165–12171.
43. Saunders, M. M., You, J., Trosko, J., Tamasaki, H., Donahue, H. J. and Jacobs, C. R. (2001). The role of gap junctions and gap junctional intercellular communication in cell ensemble responsiveness to oscillatory fluid flow in osteoblastic MC3T3-E1 cells. *Am. J. Physiol.* 281, C1917–C1925.

44. Vander Molen, M. A., Donahue, H. J., Rubin, C. T. and McLeod, K. J. (2000). Osteoblastic networks with deficient coupling: differential effects of magnetic and electric field exposure. *Bone* 27, 227–231.
45. Palumbo, C., Palazzini, S. and Marotti, G. (1990). Morphological study of intercellular junctions during osteocyte differentiation. *Bone* 11, 401–406.
46. Doty, S. B. (1981). Morphological evidence of gap junctions between bone cells. *Calcif. Tissue Int.* 33, 509–512.
47. Turner, C. H., Forwood, M. R. and Otter, M. W. (1994). Mechanotransduction in bone: do bone cells act as sensors of fluid flow? *FASEB J.* 8, 875–878.
48. Duncan, R. L. and Turner, C. H. (1995). Mechanotransduction and the functional response of bone to mechanical strain [Review]. *Calcif. Tissue Int.* 57, 344–358.
49. Jeansonne, B. G., Feagin, F. F., McMinn, R. W., Shoemaker, R. L. and Rehm, W. S. (1979). Cell-to-cell communication of osteoblasts. *J. Dent. Res.* 58, 1415–1423.
50. Civitelli, R., Beyer, E. C., Warlow, P. M., Robertson, A. J., Geist, S. T. and Steinberg, T. H. (1993). Connexin43 mediates direct intercellular communication in human osteoblastic cell networks. *J. Clin. Invest.* 91, 1888–1896.
51. Schiller, P. C., Mehta, P. P., Roos, B. A. and Howard, G. A. (1992). Hormonal regulation of intercellular communication: parathyroid hormone increases connexin 43 gene expression and gap-junctional communication in osteoblastic cells. *Mol. Endocrinol.* 6, 1433–1440.
52. Yamaguchi, D. T., Ma, D., Lee, A., Huang, J. and Gruber, H. E. (1994). Isolation and characterization of gap junctions in the osteoblastic MC3T3-E1 cell line. *J. Bone and Miner. Res.* 9, 791–803.
53. Schirrmacher, K., Brummer, F., Dusing, R. and Bingmann, D. (1993). Dye and electric coupling between osteoblast-like cells in culture. *Calcif. Tissue Int.* 53, 53–60.
54. Donahue, H. J., McLeod, K. J., Rubin, C. T., Andersen, J., Grine, E. A., Hertzberg, E. L. and Brink, P. R. (1995). Cell-to-cell communication in osteoblastic networks: cell line-dependent hormonal regulation of gap junction function. *J. Bone and Miner. Res.* 10, 881–889.
55. Yellowley, C. E., Li, Z., Zhou, Z., Jacobs, C. R. and Donahue, H. J. (2000). Functional gap junctions between osteocytic and osteoblastic cells. *J. Bone and Miner. Res.* 15, 209–217.
56. Lo, C. W. (1996). The role of gap junction membrane channels in development. *J. Bioenerg. and Biomembr.* 28, 379–385.
57. Minkoff, R., Rundus, V. R., Parker, S. B., Hertzberg, E. L., Laing, J. G. and Beyer, E. C. (1994). Gap junction proteins exhibit early and specific expression during intramembranous bone formation in the developing chick mandible. *Anat. and Embryol.* 190, 231–241.
58. Minkoff, R., Bales, E. S., Kerr, C. A. and Struss, W. E. (1999). Antisense oligonucleotide blockade of connexin expression during embryonic bone formation: evidence of functional compensation within a multigene family. *Dev. Genet.* 24, 43–56.
59. Ralphs, J. R., Benjamin, M., Waggett, A. D., Russell, D. C., Messner, K. and Gao, J. (1998). Regional differences in cell shape and gap junction expression in rat Achilles tendon: relation to fibrocartilage differentiation. *J. Anat.* 193, 215–222.
60. Lecanda, F., Warlow, P. M., Sheikh, S., Furlan, F., Steinberg, T. H. and Civitelli, R. (2000). Connexin 43 deficiency causes delayed ossification, craniofacial abnormalities, and osteoblast dysfunction. *J. Cell Biol.* 151, 931–944.
61. Chiba, H., Sawada, N., Oyamada, M., Kojima, T., Nomura, S., Ishii, S. and Mori, M. (1993). Relationship between the expression of the gap junction protein and

osteoblast phenotype in a human osteoblastic cell line during cell proliferation. *Cell Struct. and Funct.* 18, 419–426.

62. Schiller, P. C., Roos, B. A. and Howard, G. A. (1997). Parathyroid hormone up-regulation of connexin 43 gene expression in osteoblasts depends on cell phenotype. *J. Bone and Miner. Res.* 12, 2005–2013.
63. Schiller, P. C., D'Ippolito, G., R., B., Roos, B. A. and Howard, G. A. (2001). Inhibition of gap-junctional communication induces the trans-differentiation of osteoblasts to an adipocytic phenotype *in vitro. J. Biol. Chem.* 276, 14133–14138.
64. Ratanatharathorn, V., Ayash, L., Lazarus, H. M., Fu, J. and Uberti, J. P. (2001). Chronic graft-versus-host disease: clinical manifestation and therapy. *Bone Marrow Transplant.* 28, 121–129.
65. Lecanda, F., Towler, D. A., Ziambaras, K., Cheng, S., Koval, M., Steinberg, T. H. and Civitelli, R. (1998). Gap junctional communication modulates gene expression in osteoblastic cells. *Mol. Biol. Cell* 9, 2249–2258.
66. Li, Z., Zhou, Z., Yellowley, C. E. and Donahue, H. J. (1999). Inhibiting gap junctional intercellular comunication alters expression of differentiation markers in osteoblastic cells. *Bone* 25, 661–666.
67. Furlan, F., Lecanda, F., Screen, J. and Civitelli, R. (2001). Proliferation, differentiation and apoptosis in connexin43-null osteoblasts. *Cell Commun. and Adhes.* 8, 367–371.
68. Grinnell, F. (1978). Cellular adhesiveness and extracellular substrata. *Int. Rev. Cytol.* 53, 65–144.
69. Anselme, K. (2000). Osteoblast adhesion on biomaterials. *Biomaterials* 21, 667–681.
70. Vogler, E. R. (1993). Interfacial chemistry in biomaterials science. in *Wettability,* Berg, J., Ed., Vol. 49, Marcel Dekker, New York, pp. 184–250.
71. Horbett, T. A. and Klumb, L. A. (1996). Cell culturing: surface aspects and considerations. in *Interfacial Phenomena and Bioproducts,* Brash, J. L. and Wojciechowski, W., Eds., Marcel Dekker, New York, pp. 351–445.
72. Vogler, E. R. (2001). Biological properties of water. in *Water in Biomaterials Surface Science,* Morra, M., Ed., John Wiley and Sons, New York, pp. 4–24.
73. Vogler, E. R. (2001). How water wets biomaterials. in *Water in Biomaterials Surface Science,* Morra, M., ed., John Wiley and Sons, New York, pp. 269–290.
74. Bowers, K. T., Keller, J. C., Randolph, B. A., Wick, D. G. and Michaels, C. M. (1992). Optimization of surface micromorphology for enhanced osteoblast responses *in vitro. Int. J. Oral and Maxillofac. Implants* 7, 302–310.
75. Lincks, J., Boyan, B. D., Blanchard, C. R., Lohmann, C. H., Liu, Y., Cochran, D. L., Dean, D. D. and Schwartz, Z. (1998). Response of MG63 osteoblast-like cells to titanium and titanium alloy is dependent on surface roughness and composition. *Biomaterials* 19, 2219–2232.
76. Martin, J. Y., Schwartz, Z., Hummert, T. W., Schraub, D. M., Simpson, J., Lankford, J., Jr., Dean, D. D., Cochran, D. L. and Boyan, B. D. (1995). Effect of titanium surface roughness on proliferation, differentiation, and protein synthesis of human osteoblast-like cells (MG63). *J. Biomed. Mater. Res.* 29, 389–401.
77. Boyan, B. D., Lohmann, C. H., Sisk, M., Liu, Y., Sylvia, V. L., Cochran, D. L., Dean, D. D. and Schwartz, Z. (2001). Both cyclooxygenase-1 and cyclooxygenase-2 mediate osteoblast response to titanium surface roughness. *J. Biomed. Mater. Res.* 55, 350–359.
78. Boyan, B. D., Batzer, R., Kieswetter, K., Liu, Y., Cochran, D. L., Szmuckler-Moncler, S., Dean, D. D. and Schwartz, Z. (1998). Titanium surface roughness alters responsiveness of MG63 osteoblast-like cells to 1 alpha,25-(OH)2D3. *J. Biomed. Mater. Res.* 39, 77–85.

79. Deligianni, D. D., Katsala, N. D., Koutsoukos, P. G. and Missirlis, Y. F. (2001). Effect of surface roughness of hydroxyapatite on human bone marrow cell adhesion, proliferation, differentiation and detachment strength. *Biomaterials* 22, 87–96.
80. Montanaro, L., Arciola, C. R., Campoccia, D. and Cervellati, M. (2002). *In vitro* effects on MG63 osteoblast-like cells following contact with two roughness-differing fluorohydroxyapatite-coated titanium alloys. *Biomaterials* 23, 3651–3659.
81. Laczka-Osyczka, A., Laczka, M., Kasugai, S. and Ohya, K. (1998). Behavior of bone marrow cells cultured on three different coatings of gel-derived bioactive glass-ceramics at early stages of cell differentiation. *J. Biomed. Mater. Res.* 42, 433–442.
82. Dalby, M. J., Yarwood, S. J., Riehle, M. O., Johnstone, H. J., Affrossman, S. and Curtis, A. S. (2002). Increasing fibroblast response to materials using nanotopography: morphological and genetic measurements of cell response to 13-nm-high polymer demixed islands. *Exp. Cell Res.* 276, 1–9.
83. Dalby, M. J., Riehle, M. O., Johnstone, H., Affrossman, S. and Curtis, A. S. (2002). *In vitro* reaction of endothelial cells to polymer demixed nanotopography. *Biomaterials* 23, 2945–2954.
84. Riehle, M. O., Dalby, M. J., Johnstone, H., MacIntosh, A. and Affrossman, S. (2003). Cell behaviour of rat calvaria bone cells on surfaces with random nanometric features. *Mater. Sci. Eng.* C, 337–340.
85. Hukkanen, M. V., Batten, J. J., Buttery, L. D., Hench, L. L., Polak, J. M. and Xynos, I. D. (2000). Bioglass 45S5 stimulates osteoblast turnover and enhances bone formation *In vitro*: implications and applications for bone tissue engineering. *Calcif. Tissue Int.* 67, 321–329.
86. Loty, C., Sautier, J. M., Tan, M. T., Oboeuf, M., Jallot, E., Boulekbache, H., Greenspan, D. and Forest, N. (2001). Bioactive glass stimulates *in vitro* osteoblast differentiation and creates a favorable template for bone tissue formation. *J. Bone Miner. Res.* 16, 231–239.
87. Xynos, I. D., Edgar, A. J., Buttery, L. D., Hench, L. L. and Polak, J. M. (2000). Ionic products of bioactive glass dissolution increase proliferation of human osteoblasts and induce insulin-like growth factor II mRNA expression and protein synthesis. *Biochem. and Biophys. Res. Commun.* 276, 461–465.
88. Hay, D. I. and Moreno, E. C. (1979). Differential adsorption and chemical affinities of proteins for apatitic surfaces. *J. Dental Res.* 58, 930–942.
89. Stephansson, S. N., Byers, B. A. and Garcia, A. J. (2002). Enhanced expression of the osteoblastic phenotype on substrates that modulate fibronectin conformation and integrin receptor binding. *Biomaterials* 23, 2527–2534.
90. Garcia, A. J., Vega, M. D. and Boettiger, D. (1999). Modulation of cell proliferation and differentiation through substrate-dependent changes in fibronectin conformation. *Mol. Biol. Cell* 10, 785–798.
91. Rezania, A. and Healy, K. E. (2000). The effect of peptide surface density on mineralization of a matrix deposited by osteogenic cells. *J. Biomed. Mater. Res.* 52, 595–600.
92. McFarland, C. D., Thomas, C. H., DeFilippis, C., Steele, J. G. and Healy, K. E. (2000). Protein adsorption and cell attachment to patterned surfaces. *J. Biomed. Mater. Res.* 49, 200–210.
93. Yadav, K. L. and Brown, P. W. (2003). Formation of hydroxyapatite in water, Hank's solution, and serum at physiological temperature. *J. Biomed. Mater. Res.* 65A, 158–163.
94. Durucan, C. and Brown, P. W. (2002). Kinetic model of tricalcium phosphate hydrolysis. *J. Am. Ceram. Soc.* 85, 2013–2017.
95. Brown, P. W. and Martin, R. I. (1999). An analysis of hydroxyapatite surface layer formation. *J. Phys. Chem. B* 103, 1671–1675.

96. Brown, P. W. and Fulmer, M. (1996). Effects of electrolytes on hydroxyapatite formation at 25° and 38°C. *J. Biomed. Mater. Res.* 31, 395–400.
97. Brown, P. W. (1992). Phase relationships in the ternary system $CaO–P_2O_5–H_2O$ at 25°C. *J. Am. Ceram. Soc.* 75, 17–22.
98. Flades, K., Lau, C., Mertig, M. and Pompe, W. (2001). Osteocalcin-controlled dissolution-reprecipitation of calcium phosphate under biomemetic conditions. *Chem. Mater.* 13, 3596–3602.
99. Vogler, E. A. (1996). On the biomedical relevance of surface spectroscopy. *J. Electron Spectrosc. Relat. Phenom.* 81, 237–247.
100. Vogler, E. A. (1998). Structure and reactivity of water at biomaterial surfaces. *Adv. Colloid Interface Sci.* 74, 69–117.
101. Vogler, E. A. (1999). Water and the acute biological response to surfaces. *J. Biomater. Sci. Polym. Edn.* 10, 1015–1045.
102. van Kooten, T. G., Schakenraad, J. M., van der Mei, H. C. and Busscher, H. J. (1992). Influence of substratum wettability on the strength of adhesion of human fibroblasts. *Biomaterials* 13, 897–904.
103. Meyer, R. A., Lampe, P. D., Malewicz, B., Baumann, W. J. and Johnson, R. G. (1991). Enhanced gap junction formation with LDL and apolipoprotein B. *Exp. Cell Res.* 196, 72–81.
104. Vogler, E. A. (2001). On the origins of water wetting terminology. In *Water in Biomaterials Surface Science* (Morra, M., ed.), pp. 150–182. John Wiley and Son, New York.
105. Vogler, E. A. (1989). A thermodynamic model of short-term cell adhesion *in vitro*. *Colloids Surfaces* 42, 233–254.
106. Dekker, A., Reitsma, K., Beugeling, T., Bantjes, A., Feijen, J. and van Aken, W. G. (1991). Adhesion of endothelial cells and adsorption of serum proteins on gas plasma-treated polytetrafluoroethylene. *Biomaterials* 12, 130–138.
107. Schakenraad, J. M., Busscher, H. J., Wildevuur, C. R. and Arends, J. (1988). Thermodynamic aspects of cell spreading on solid substrata. *Cell Biophys.* 13, 75–91.
108. Daw, R., Candan, S., Beck, A. J., Devlin, A. J., Brook, I. M., MacNeil, S., Dawson, R. A. and Short, R. D. (1998). Plasma copolymer surfaces of acrylic acid/1,7 octadiene: surface characterisation and the attachment of ROS 17/2.8 osteoblast-like cells. *Biomaterials* 19, 1717–1725.
109. Redey, S. A., Nardin, M., Bernache-Assolant, D., Rey, C., Delannoy, P., Sedel, L. and Marie, P. J. (2000). Behavior of human osteoblastic cells on stoichiometric hydroxyapatite and type A carbonate apatite: role of surface energy. *J. Biomed. Mater. Res.* 50, 353–364.
110. Cai, K., Yao, K., Cui, Y., Lin, S., Yang, Z., Li, X., Xie, H., Qing, T. and Luo, J. (2002). Surface modification of poly (D,L-lactic acid) with chitosan and its effects on the culture of osteoblasts *in vitro*. *J. Biomed. Mater. Res* 60, 398–404.
111. Lim, J. Y., Liu, X., Vogler, E. A. and Donahue, H. J. (2003). Systematic variation in osteoblast adhesion and phenotype with substratum surface characteristics. *J. Biomed. Mater. Res.,* 65A, 504–512.
112. Makowski, L., Caspar, D. L., Phillips, W. C. and Goodenough, D. A. (1977). Gap junction structures. II. Analysis of the x-ray diffraction data. *J. Cell Biol.* 74, 629–45.
113. Robin, C. H. (1849). Sur l'existence de deux espèces nouvelles d'éléments anatomiques qui se trouvent dans le canal medullaire des os. *CR Seances Soc. Biol. Paris* 1, 149.
114. Kölliker, A. (1873). *Die normale resorption des Knochengewehes und ihre bedentung für die eststehung der typischen Knochenformen*, Vogel, Leipzig.

115. Gay, C. V. and Mueller, W. J. (1974). Carbonic anhydrase and osteoclasts: localization by labeled inhibitor autoradiography. *Science* 183, 432.
116. Sly, W. S., Hewett-Emmett, D., Whyte, M.P., Yu, Y.S. and Tashian, R.E. (1983). Carbonic anhydrase II deficiency identified as the primary defect in the autosomal recessive syndrome of osteopetrosis with renal tubular acidosis and cerebral calcitonin. *Proc. Natl. Acad. Sci. USA* 80, 2752–2756.
117. Sly, W. S. (1991). Carbonic anhydrase II deficiency syndrome. in *The Carbonic Anhydrases,* Dodgson, S., Tashian, R. E., Gros, G. and Carter, N. D., Eds., Plenum Press, New York, pp. 183–193.
118. Blair, H. C., Teitelbaum, S. L., Ghiselli, R. and Gluck, S. (1989). Osteoclastic bone resorption by a polarized vacuolar proton pump. *Science* 245, 855–857.
119. Bekker, P. J. and Gay, C. V. (1990). Biochemical characterization of an electrogenic vacuolar proton pump in purified chicken osteoclast plasma membrane vesicles. *J. Bone Miner. Res.* 5, 569–579.
120. Väänänen, H. K., Karhukorpi, E. K., Sundquist, K., Wallmark, B., Roininen, I., Hentunen, T., Tuukkanen, J. and Lakkakorpi, P. (1990). Evidence for the presence of a proton pump of the vacuolar H(+)-ATPase type in the ruffled borders of osteoclasts. *J. Cell Biol.* 111, 1305–1311.
121. Minkin, C. and Jennings, J. M. (1972). Carbonic anhydrase and bone remodeling: sulfonamide inhibition of bone resorption in organ culture. *Science* 176, 1031–1033.
122. Gay, C. V., Kief, N. L. and Bekker, P. J. (1993). Effect of estrogen on acidification in osteoclasts. *Biochem. Biophys. Res. Commun.* 192, 1251–1259.
123. Hunter, S. J., Rosen, C. J. and Gay, C. V. (1991). *In vitro* resorptive activity of isolated chick osteoclasts: effects of carbonic anhydrase inhibition. *J. Bone Miner. Res.* 6, 61–66.
124. Scott, B. L. and Pease, D. D. (1956). Electron microscopy of the epiphyseal apparatus. *Anat. Rec.* 126, 465.
125. Holtrop, M. E. (1991). Light and electron microscopic structure of osteoclasts. in *Bone,* Hall, B. K., Ed., Vol. 1. CRC Press, Boca Raton, FL, pp. 1–39.
126. King, G. J. and Holtrop, M. E. (1975). Actin-like filaments in bone cells of cultured mouse calvaria as demonstrated by binding to heavy meromyosin. *J. Cell Biol.* 66, 445–451.
127. Ross, F. P., Chappel, J., Alvarez, J. I., Sander, D., Butler, W. T., Farach-Carson, M. C., Mintz, K. A., Robey, P. G., Teitelbaum, S. L. and Cheresh, D. A. (1993). Interactions between the bone matrix proteins osteopontin and bone sialoprotein and the osteoclast integrin alpha v beta 3 potentiate bone resorption. *J. Biol. Chem.* 268, 9901–9907.
128. Holtrop, M. E. and Raisz, L. G. (1979). Comparison of the effects of 1,25-dihydroxycholecalciferol, prostaglandin E2, and osteoclast-activating factor with parathyroid hormone on the ultrastructure of osteoclasts in cultured long bones of fetal rats. *Calcif. Tissue Int.* 29, 201–205.
129. Lucht, U. (1972). Osteoclasts and their relationship to bone as studied by electron microscopy. *Z. Zellforsch. Mikrosk. Anat.* 135, 211–228.
130. Fukushima, O., Bekker, P. J. and Gay, C. V. (1991). Characterization of the functional stages of osteoclasts by enzyme histochemistry and electron microscopy. *Anat. Rec.* 231, 298–315.
131. Kanehisa, J. and Heersche, J. N. (1988). Osteoclastic bone resorption: *in vitro* analysis of the rate of resorption and migration of individual osteoclasts. *Bone* 9, 73–79.
132. Miller, S. C. (1977). Osteoclast cell-surface changes during the egg-laying cycle in Japanese quail. *J. Cell Biol.* 75, 104–118.

133. Gay, C. V. (1992). Osteoclast ultrastructure and enzyme histochemistry: functional implications. in *Biology and Physiology of the Osteoclast,* Rifkin, B. R. and Gay, C. V., Eds., CRC Press, Boca Raton, FL, pp. 129–150.
134. Pierce, A. and Lindskog, S. (1988). Coated pits and vesicles in the osteoclast. *J. Submicrosc. Cytol. Pathol.* 20, 161–167.
135. Baron, R., Neff, L., Brown, W., Courtoy, P. J., Louvard, D. and Farquhar, M. G. (1988). Polarized secretion of lysosomal enzymes: co-distribution of cation-independent mannose-6-phosphate receptors and lysosomal enzymes along the osteoclast exocytic pathway. *J. Cell Biol.* 106, 1863–1872.
136. Lehninger, A. L. (1970). Mitochondria and calcium ion transport. *Biochem. J.* 119, 129–138.
137. Hohman, W. and Schraer, H. (1966). The intracellular distribution of calcium in the mucosa of the avian shell gland. *J. Cell Biol.* 30, 317–331.
138. Takahashi, N., Akatsu, T., Udagawa, N., Sasaki, T., Yamaguchi, A., Moseley, J. M., Martin, T. J. and Suda, T. (1988). Osteoblastic cells are involved in osteoclast formation. *Endocrinology* 123, 2600–2602.
139. Suda, T., Takahashi, N. and Martin, T. J. (1992). Modulation of osteoclast differentiation. *Endocr. Rev.* 13, 66–80.
140. Chambers, T. J., Owens, J. M., Hattersley, G., Jat, P. S. and Noble, M. D. (1993). Generation of osteoclast-inductive and osteoclastogenic cell lines from the H-$2K^b$tsA58 transgenic mouse. *Proc. Natl. Acad. Sci. USA* 90, 5578–5582.
141. Liu, B. Y., Guo, J., Lanske, B., Divieti, P., Kronenberg, H. M. and Bringhurst, F. R. (1998). Conditionally immortalized murine bone marrow stromal cells mediate parathyroid hormone-dependent osteoclastogenesis *in vitro*. *Endocrinology* 139, 1952–1964.
142. Sakuma, Y., Tanaka, K., Suda, M., Yasoda, A., Natsui, K., Tanaka, I., Ushikubi, F., Narumiya, S., Segi, E., Sugimoto, Y., Ichikawa, A. and Nakao, K. (2000). Crucial involvement of the EP4 subtype of prostaglandin E receptor in osteoclast formation by proinflammatory cytokines and lipopolysaccharide. *J. Bone Miner. Res.* 15, 218–227.
143. Takeda, S., Yoshizawa, T., Nagai, Y., Yamato, H., Fukumoto, S., Sekine, K., Kato, S., Matsumoto, T. and Fujita, T. (1999). Stimulation of osteoclast formation by 1,25-dihydroxyvitamin D requires its binding to vitamin D receptor (VDR) in osteoblastic cells: studies using VDR knockout mice. *Endocrinology* 140, 1005–1008.
144. Yoshida, H., Hayashi, S., Kunisada, T., Ogawa, M., Nishikawa, S., Okamura, H., Sudo, T. and Shultz, L. D. (1990). The murine mutation osteopetrosis is in the coding region of the macrophage colony stimulating factor gene. *Nature* 345, 442–444.
145. Niida, S., Kaku, M., Amano, H., Yoshida, H., Kataoka, H., Nishikawa, S., Tanne, K., Maeda, N. and Kodama, H. (1999). Vascular endothelial growth factor can substitute for macrophage colony-stimulating factor in the support of osteoclastic bone resorption. *J. Exp. Med.* 190, 293–298.
146. Takahashi, N., Udagawa, N., Takami, M. and Suda, T. (2002). Cells of bone: osteoclast generation. In *Principles of Bone Biology* (Bilezikian, J. P., Raisz, L. G. and Rodan, G. A., eds.), Vol. 1. Academic Press, San Diego, CA, pp. 109–126.
147. Jimi, E., Akiyama, S., Tsurukai, T., Okahashi, N., Kobayashi, K., Udagawa, N., Nishihara, T., Takahashi, N. and Suda, T. (1999). Osteoclast differentiation factor acts as a multifunctional regulator in murine osteoclast differentiation and function. *J. Immunol.* 163, 434–442.
148. The American Society for Bone and Mineral Research President's Committee on Nomenclature (2000). Proposed standard nomenclature for new tumor necrosis fac-

tor family members involved in the regulation of bone resorption. *J. Bone Miner. Research.* 15, 2293–2296.

149. Hsu, H., Lacey, D. L., Dunstan, C. R., Solovyev, I., Colombero, A., Timms, E., Tan, H. L., Elliott, G., Kelley, M. J., Sarosi, I., Wang, L., Xia, X. Z., Elliott, R., Chiu, L., Black, T., Scully, S., Capparelli, C., Morony, S., Shimamoto, G., Bass, M. B. and Boyle, W. J. (1999). Tumor necrosis factor receptor family member RANK mediates osteoclast differentiation and activation induced by osteoprotegerin ligand. *Proc. Natl. Acad. Sci. USA* 96, 3540–3545.
150. Nakagawa, N., Kinosaki, M., Yamaguchi, K., Shima, N., Yasuda, H., Yano, K., Morinaga, T. and Higashio, K. (1998). RANK is the essential signaling receptor for osteoclast differentiation factor in osteoclastogenesis. *Biochem. Biophys. Res. Commun.* 253, 395–400.
151. Kartsogiannis, V., Zhou, H., Horwood, N. J., Thomas, R. J., Hards, D. K., Quinn, J. M., Niforas, P., Ng, K. W., Martin, T. J. and Gillespie, M. T. (1999). Localization of RANKL (receptor activator of NF kappa B ligand) mRNA and protein in skeletal and extraskeletal tissues. *Bone* 25, 525–534.
152. Felix, R., Cecchini, M. G. and Fleisch, H. (1990). Macrophage colony stimulating factor restores *in vivo* bone resorption in the op/op osteopetrotic mouse. *Endocrinology* 127, 2592–2594.
153. Wood, G. W., Hausmann, E. and Choudhuri, R. (1997). Relative role of CSF-1, MCP-1/JE, and RANTES in macrophage recruitment during successful pregnancy. *Mol. Reprod. Dev.* 46, 62–9; discussion 69–70.
154. Itoh, K., Udagawa, N., Matsuzaki, K., Takami, M., Amano, H., Shinki, T., Ueno, Y., Takahashi, N. and Suda, T. (2000). Importance of membrane- or matrix-associated forms of M-CSF and RANKL/ODF in osteoclastogenesis supported by SaOS-4/3 cells expressing recombinant PTH/PTHrP receptors. *J. Bone Miner. Res.* 15, 1766–1775.
155. Onyia, J. E., Libermann, T. A., Bidwell, J., Arnold, D., Tu, Y., McClelland, P. and Hock, J. M. (1997). Parathyroid hormone (1-34)-mediated interleukin-6 induction. *J. Cell Biochem.* 67, 265–274.
156. Suzawa, T., Miyaura, C., Inada, M., Maruyama, T., Sugimoto, Y., Ushikubi, F., Ichikawa, A., Narumiya, S. and Suda, T. (2000). The role of prostaglandin E receptor subtypes (EP1, EP2, EP3, and EP4) in bone resorption: an analysis using specific agonists for the respective EPs. *Endocrinology* 141, 1554–1559.
157. O'Brien, C. A., Gubrij, I., Lin, S. C., Saylors, R. L. and Manolagas, S. C. (1999). STAT3 activation in stromal/osteoblastic cells is required for induction of the receptor activator of NF-kappaB ligand and stimulation of osteoclastogenesis by gp130-utilizing cytokines or interleukin-1 but not 1,25-dihydroxyvitamin D3 or parathyroid hormone. *J. Biol. Chem.* 274, 19301–19308.
158. Kitazawa, R., Kitazawa, S. and Maeda, S. (1999). Promoter structure of mouse RANKL/TRANCE/OPGL/ODF gene. *Biochim. Biophys. Acta* 1445, 134–141.
159. Takami, M., Woo, J. T., Takahashi, N., Suda, T. and Nagai, K. (1997). Ca^{2+}-ATPase inhibitors and Ca^{2+}-ionophore induce osteoclast-like cell formation in the cocultures of mouse bone marrow cells and calvarial cells. *Biochem. Biophys. Res. Commun.* 237, 111–115.
160. Takami, M., Takahashi, N., Udagawa, N., Miyaura, C., Suda, K., Woo, J. T., Martin, T. J., Nagai, K. and Suda, T. (2000). Intracellular calcium and protein kinase C mediate expression of receptor activator of nuclear factor-kappaB ligand and osteoprotegerin in osteoblasts. *Endocrinology* 141, 4711–4719.
161. Li, J., Sarosi, I., Yan, X. Q., Morony, S., Capparelli, C., Tan, H. L., McCabe, S., Elliott, R., Scully, S., Van, G., Kaufman, S., Juan, S. C., Sun, Y., Tarpley, J., Martin,

L., Christensen, K., McCabe, J., Kostenuik, P., Hsu, H., Fletcher, F., Dunstan, C. R., Lacey, D. L. and Boyle, W. J. (2000). RANK is the intrinsic hematopoietic cell surface receptor that controls osteoclastogenesis and regulation of bone mass and calcium metabolism. *Proc. Natl. Acad. Sci. USA* 97, 1566–1571.
162. Darnay, B. G., Haridas, V., Ni, J., Moore, P. A. and Aggarwal, B. B. (1998). Characterization of the intracellular domain of receptor activator of NF-kappa B (RANK). Interaction with tumor necrosis factor receptor-associated factors and activation of NF-kappa B and c-Jun N-terminal kinase. *J. Biol. Chem.* 273, 20551–20555.
163. Darnay, B. G., Ni, J., Moore, P. A. and Aggarwal, B. B. (1999). Activation of NF-kappa B by RANK requires tumor necrosis factor receptor-associated factor (TRAF) 6 and NF-kappa B-inducing kinase. Identification of a novel TRAF6 interaction motif. *J. Biol. Chem.* 274, 7724–7731.
164. Galibert, L., Tometsko, M. E., Anderson, D. M., Cosman, D. and Dougall, W. C. (1998). The involvement of multiple tumor necrosis factor receptor (TNFR)-associated factors in the signaling mechanisms of receptor activator of NF-kappaB, a member of the TNFR superfamily. *J. Biol. Chem.* 273, 34120–34127.
165. Kim, H. H., Lee, D. E., Shin, J. N., Lee, Y. S., Jeon, Y. M., Chung, C. H., Ni, J., Kwon, B. S. and Lee, Z. H. (1999). Receptor activator of NF-kappaB recruits multiple TRAF family adaptors and activates c-Jun N-terminal kinase. *FEBS Lett.* 443, 297–302.
166. Wong, B. R., Besser, D., Kim, N., Arron, J. R., Vologodskaia, M., Hanafusa, H. and Choi, Y. (1999). TRANCE, a TNF family member, activates Akt/PKB through a signaling complex involving TRAF6 and c-Src. *Mol. Cell* 4, 1041–1049.
167. Lomaga, M. A., Yeh, W. C., Sarosi, I., Duncan, G. S., Furlonger, C., Ho, A., Morony, S., Capparelli, C., Van, G., Kaufman, S., van der Heiden, A., Itie, A., Wakeham, A., Khoo, W., Sasaki, T., Cao, Z., Penninger, J. M., Paige, C. J., Lacey, D. L., Dunstan, C. R., Boyle, W. J., Goeddel, D. V. and Mak, T. W. (1999). TRAF6 deficiency results in osteopetrosis and defective interleukin-1, CD40, and LPS signaling. *Genes Dev.* 13, 1015–1024.
168. Takayanagi, H., Ogasawara, K., Hida, S., Chiba, T., Murata, S., Sato, K., Takaoka, A., Yokochi, T., Oda, H., Tanaka, K., Nakamura, K. and Taniguchi, T. (2000). T-cell-mediated regulation of osteoclastogenesis by signalling cross-talk between RANKL and IFN-gamma. *Nature* 408, 600–605.
169. Miyazaki, T., Katagiri, H., Kanegae, Y., Takayanagi, H., Sawada, Y., Yamamoto, A., Pando, M. P., Asano, T., Verma, I. M., Oda, H., Nakamura, K. and Tanaka, S. (2000). Reciprocal role of ERK and NF-kappaB pathways in survival and activation of osteoclasts. *J. Cell Biol.* 148, 333–342.
170. Matsuo, K., Owens, J. M., Tonko, M., Elliott, C., Chambers, T. J. and Wagner, E. F. (2000). Fosl1 is a transcriptional target of c-Fos during osteoclast differentiation. *Nat. Genet.* 24, 184–187.
171. Matsumoto, M., Sudo, T., Saito, T., Osada, H. and Tsujimoto, M. (2000). Involvement of p38 mitogen-activated protein kinase signaling pathway in osteoclastogenesis mediated by receptor activator of NF-kappa B ligand (RANKL). *J. Biol. Chem.* 275, 31155–31161.
172. Soriano, P., Montgomery, C., Geske, R. and Bradley, A. (1991). Targeted disruption of the c-src proto-oncogene leads to osteopetrosis in mice. *Cell* 64, 693–702.
173. Boyce, B. F., Yoneda, T., Lowe, C., Soriano, P. and Mundy, G. R. (1992). Requirement of pp60c-src expression for osteoclasts to form ruffled borders and resorb bone in mice. *J. Clin. Invest.* 90, 1622–1627.
174. Lowe, C., Yoneda, T., Boyce, B. F., Chen, H., Mundy, G. R. and Soriano, P. (1993). Osteopetrosis in Src-deficient mice is due to an autonomous defect of osteoclasts. *Proc. Natl. Acad. Sci. USA* 90, 4485–4489.

175. Kong, Y. Y., Feige, U., Sarosi, I., Bolon, B., Tafuri, A., Morony, S., Capparelli, C., Li, J., Elliott, R., McCabe, S., Wong, T., Campagnuolo, G., Moran, E., Bogoch, E. R., Van, G., Nguyen, L. T., Ohashi, P. S., Lacey, D. L., Fish, E., Boyle, W. J. and Penninger, J. M. (1999). Activated T cells regulate bone loss and joint destruction in adjuvant arthritis through osteoprotegerin ligand. *Nature* 402, 304–309.

176. Teng, Y. T., Nguyen, H., Gao, X., Kong, Y. Y., Gorczynski, R. M., Singh, B., Ellen, R. P. and Penninger, J. M. (2000). Functional human T-cell immunity and osteoprotegerin ligand control alveolar bone destruction in periodontal infection. *J. Clin. Invest.* 106, R59–R67.

177. Nesbitt, S., Nesbit, A., Helfrich, M. and Horton, M. (1993). Biochemical characterization of human osteoclast integrins. Osteoclasts express alpha v beta 3, alpha 2 beta 1, and alpha v beta 1 integrins. *J. Biol. Chem.* 268, 16737–16745.

178. Nakamura, I., Takahashi, N., Sasaki, T., Jimi, E., Kurokawa, T. and Suda, T. (1996). Chemical and physical properties of the extracellular matrix are required for the actin ring formation in osteoclasts. *J. Bone Miner. Res.* 11, 1873–1879.

179. Butler, W. T., Ridall, A. L. and McKee, M. D. (1996). Osteopontin. in *Principles of Bone Biology,* Bilezikian, J. P., Raisz, L. G. and Rodan, G. A., Eds.. Academic Press, San Diego, CA, pp. 167–181.

180. Horton, M. A., Townsend, P. and Nesbitt, S. (1996). Cell surface attachment molecules in bone. in *Principles of Bone Biology,* Bilezikian, J. P., Raisz, L. G. and Rodan, G. A., Eds.. Academic Press, San Diego, CA, pp. 217–230.

181. Duong, L. T., Lakkakorpi, P., Nakamura, I. and Rodan, G. A. (2000). Integrins and signaling in osteoclast function. *Matrix Biol.* 19, 97–105.

182. Nakamura, I., Pilkington, M. F., Lakkakorpi, P. T., Lipfert, L., Sims, S. M., Dixon, S. J., Rodan, G. A. and Duong, L. T. (1999). Role of alpha v beta 3 integrin in osteoclast migration and formation of the sealing zone. *J. Cell Sci.* 112 (Pt 22), 3985–3993.

183. Weir, E. C., Horowitz, M. C., Baron, R., Centrella, M., Kacinski, B. M. and Insogna, K. L. (1993). Macrophage colony-stimulating factor release and receptor expression in bone cells. *J. Bone. Miner. Res.* 8, 1507–1518.

184. Pilkington, M. F., Sims, S. M. and Dixon, S. J. (1998). Wortmannin inhibits spreading and chemotaxis of rat osteoclasts *in vitro. J. Bone Miner. Res.* 13, 688–694.

185. Grey, A., Chen, Y., Paliwal, I., Carlberg, K. and Insogna, K. (2000). Evidence for a functional association between phosphatidylinositol 3-kinase and c-src in the spreading response of osteoclasts to colony-stimulating factor-1. *Endocrinology* 141, 2129–2138.

186. Nakamura, I., Takahashi, N., Sasaki, T., Tanaka, S., Udagawa, N., Murakami, H., Kimura, K., Kabuyama, Y., Kurokawa, T., Suda, T. and Fukui, Y (1995). Wortmannin, a specific inhibitor of phosphatidylinositol-3 kinase, blocks osteoclastic bone resorption. *FEBS Lett.* 361, 79–84.

187. Hall, T. J., Jeker, H. and Schaueblin, M. (1995). Wortmannin, a potent inhibitor of phosphatidylinositol 3-kinase, inhibits osteoclastic bone resorption *in vitro. Calcif. Tissue Int.* 56, 336–338.

188. Chellaiah, M. and Hruska, K. (1996). Osteopontin stimulates gelsolin-associated phosphoinositide levels and phosphatidylinositol triphosphate-hydroxyl kinase. *Mol. Biol. Cell* 7, 743–753.

189. Duong, L. T., Sanjay, A., Horne, W. C., Baron, R. and Rodan, G. A. (2002). Integrin and calcitonin signaling in the regulation of the cytoskeleton and function of osteoclasts. In *Principles of Bone Biology,* 2nd ed., Bilezikian, J. P., Raisz, L. G. and Rodan, G. A., Eds., Academic Press, San Diego, CA, pp. 141–150.

190. Hall, T. J., Nyugen, H., Schaueblin, M. and Fournier, B. (1995). The bone-specific estrogen centchroman inhibits osteoclastic bone resorption *in vitro*. *Biochem. Biophys. Res. Commun.* 216, 662–668.
191. Laitala, T. and Väänänen, K. (1993). Proton channel part of vacuolar H(+)-ATPase and carbonic anhydrase II expression is stimulated in resorbing osteoclasts. *J. Bone Miner. Res.* 8, 119–126.
192. Kurihara, N., Gluck, S. and Roodman, G. D. (1990). Sequential expression of phenotype markers for osteoclasts during differentiation of precursors for multinucleated cells formed in long-term human marrow cultures. *Endocrinology* 127, 3215–3221.
193. Wang, Z. Q., Hemken, P., Menton, D. and Gluck, S. (1992). Expression of vacuolar H(+)-ATPase in mouse osteoclasts during *in vitro* differentiation. *Am. J. Physiol.* 263, F277–F283.
194. Gay, C. V. (1996). Role of microscopy in elucidating the mechanism and regulation of the osteoclast resorptive apparatus. *Microsc. Res. Tech.* 33, 165–170.
195. Väänänen, H. K. and Parvinen, E. K. (1983). High active isoenzyme of carbonic anhydrase in rat calvaria osteoclasts. Immunohistochemical study. *Histochemistry* 78, 481–485.
196. Anderson, R. E., Schraer, H. and Gay, C. V. (1982). Ultrastructural immunocytochemical localization of carbonic anhydrase in normal and calcitonin-treated chick osteoclasts. *Anat. Rec.* 204, 9–20.
197. Cao, H. and Gay, C. V. (1985). Effects of parathyroid hormone and calcitonin on carbonic anhydrase location in osteoclasts of cultured embryonic chick bone. *Experientia* 41, 1472–1474.
198. Waite, L. C., Volkert, W. A. and Kenny, A. D. (1970). Inhibition of bone resorption by acetazolamide in the rat. *Endocrinology* 87, 1129–1139.
199. Kenny, A. D. (1985). Role of carbonic anhydrase in bone: partial inhibition of disuse atrophy of bone by parenteral acetazolamide. *Calcif. Tissue Int.* 37, 126–133.
200. Hall, G. E. and Kenny, A. D. (1987). Role of carbonic anhydrase in bone resorption: effect of acetazolamide on basal and parathyroid hormone-induced bone metabolism. *Calcif. Tissue Int.* 40, 212–218.
201. Hunter, S. J., Schraer, H. and Gay, C. V. (1988). Characterization of isolated and cultured chick osteoclasts: the effects of acetazolamide, calcitonin, and parathyroid hormone on acid production. *J. Bone Miner. Res.* 3, 297–303.
202. Karhukorpi, E. K. (1991). Carbonic anhydrase II in rat acid secreting cells: comparison of osteoclasts with gastric parietal cells and kidney intercalated cells. *Acta Histochem.* 90, 11–20.
203. Raisz, L. G., Simmons, H. A., Thompson, W. J., Shepard, K. L., Anderson, P. S. and Rodan, G. A. (1988). Effects of a potent carbonic anhydrase inhibitor on bone resorption in organ culture. *Endocrinology* 122, 1083–1086.
204. Hott, M. and Marie, P. J. (1989). Carbonic anhydrase activity in fetal rat bone resorbing cells: inhibition by acetazolamide infusion. *J. Dev. Physiol.* 12, 277–281.
205. Hall, T. J., Higgins, W., Tardif, C. and Chambers, T. J. (1991). A comparison of the effects of inhibitors of carbonic anhydrase on osteoclastic bone resorption and purified carbonic anhydrase isozyme II. *Calcif. Tissue Int.* 49, 328–332.
206. Teti, A., Blair, H. C., Teitelbaum, S. L., Kahn, J. A., Carano, A., Grano, M., Santacroce, G., Schlesinger, P. and Zambonin Zallone, A. (1989). Cytoplasmic pH is regulated in isolated avian osteoclasts by a Cl^-/HCO_3^- exchanger. *Boll. Soc. Ital. Biol. Sper.* 65, 589–595.
207. Lehenkari, P., Hentunen, T. A., Laitala-Leinonen, T., Tuukkanen, J. and Väänänen, H. K. (1998). Carbonic anhydrase II plays a major role in osteoclast differentiation and

bone resorption by effecting the steady state intracellular pH and Ca^{2+}. *Exp. Cell Res.* 242, 128–137.

208. Laitala-Leinonen, T., Lowik, C., Papapoulos, S. and Väänänen, H. K. (1999). Inhibition of intravacuolar acidification by antisense RNA decreases osteoclast differentiation and bone resorption *in vitro*. *J. Cell Sci.* 112 (Pt 21), 3657–3666.
209. Quelo, I., Kahlen, J. P., Rascle, A., Jurdic, P. and Carlberg, C. (1994). Identification and characterization of a vitamin D_3 response element of chicken carbonic anhydrase-II. *DNA Cell Biol.* 13, 1181–1187.
210. Biskobing, D. M., Fan, D., Fan, X. and Rubin, J. (1997). Induction of carbonic anhydrase II expression in osteoclast progenitors requires physical contact with stromal cells. *Endocrinology* 138, 4852–4857.
211. Woods, C., Domenget, C., Solari, F., Gandrillon, O., Lazarides, E. and Jurdic, P. (1995). Antagonistic role of vitamin D_3 and retinoic acid on the differentiation of chicken hematopoietic macrophages into osteoclast precursor cells. *Endocrinology* 136, 85–95.
212. Quelo, I. and Jurdic, P. (2000). Differential regulation of the carbonic anhydrase II gene expression by hormonal nuclear receptors in monocytic cells: identification of the retinoic acid response element. *Biochem. Biophys. Res. Commun.* 271, 481–491.
213. Baron, R., Neff, L., Louvard, D. and Courtoy, P. J. (1985). Cell-mediated extracellular acidification and bone resorption: evidence for a low pH in resorbing lacunae and localization of a 100-kD lysosomal membrane protein at the osteoclast ruffled border. *J. Cell Biol.* 101, 2210–2222.
214. Francis, M. J., Lees, R. L., Trujillo, E., Martin-Vasallo, P., Heersche, J. N. and Mobasheri, A. (2002). ATPase pumps in osteoclasts and osteoblasts. *Int. J. Biochem. Cell. Biol.* 34, 459–476.
215. Akisaka, T. and Gay, C. V. (1986). Ultracytochemical evidence for a proton-pump adenosine triphosphatase in chick osteoclasts. *Cell Tissue Res.* 245, 507–512.
216. Lee, B. S., Gluck, S. L. and Holliday, L. S. (1999). Interaction between vacuolar H(+)-ATPase and microfilaments during osteoclast activation. *J. Biol. Chem.* 274, 29164–29171.
217. Sato, M. and Grasser, W. (1990). Myosin II antibodies inhibit the resorption activity of isolated rat osteoclasts. *Cell Motil. Cytoskeleton* 17, 250–263.
218. Laitala-Leinonen, T., Howell, M. L., Dean, G. E. and Väänänen, H. K. (1996). Resorption-cycle-dependent polarization of mRNAs for different subunits of V-ATPase in bone-resorbing osteoclasts. *Mol. Biol. Cell* 7, 129–142.
219. Nishi, T. and Forgac, M. (2000). Molecular cloning and expression of three isoforms of the 100-kDa a subunit of the mouse vacuolar proton-translocating ATPase. *J. Biol. Chem.* 275, 6824–6830.
220. Toyomura, T., Oka, T., Yamaguchi, C., Wada, Y. and Futai, M. (2000). Three subunit a isoforms of mouse vacuolar H(+)-ATPase. Preferential expression of the a3 isoform during osteoclast differentiation. *J. Biol. Chem.* 275, 8760–8765.
221. Kornak, U., Schulz, A., Friedrich, W., Uhlhaas, S., Kremens, B., Voit, T., Hasan, C., Bode, U., Jentsch, T. J. and Kubisch, C. (2000). Mutations in the a3 subunit of the vacuolar H(+)-ATPase cause infantile malignant osteopetrosis. *Hum. Mol. Genet.* 9, 2059–2063.
222. Li, Y. P., Chen, W., Liang, Y., Li, E. and Stashenko, P. (1999). Atp6i-deficient mice exhibit severe osteopetrosis due to loss of osteoclast-mediated extracellular acidification. *Nat. Genet.* 23, 447–451.
223. Scimeca, J. C., Franchi, A., Trojani, C., Parrinello, H., Grosgeorge, J., Robert, C., Jaillon, O., Poirier, C., Gaudray, P. and Carle, G. F. (2000). The gene encoding the

mouse homologue of the human osteoclast-specific 116-kDa V-ATPase subunit bears a deletion in osteosclerotic (oc/oc) mutants. *Bone* 26, 207–213.

224. Akisaka, T. and Gay, C. V. (1986). An ultracytochemical investigation of ouabain-sensitive p-nitrophenylphosphatase in chick osteoclasts. *Cell Tissue Res.* 244, 57–62.
225. Baron, R., Neff, L., Roy, C., Boisvert, A. and Caplan, M. (1986). Evidence for a high and specific concentration of (Na^+,K^+)ATPase in the plasma membrane of the osteoclast. *Cell* 46, 311–320.
226. Hunter, S. J., Gay, C. V., Osdoby, P. A. and Peters, L. L. (1998). Spectrin localization in osteoclasts: immunocytochemistry, cloning, and partial sequencing. *J. Cell. Biochem.* 71, 204–215.
227. Nakamura, I., Sasaki, T., Tanaka, S., Takahashi, N., Jimi, E., Kurokawa, T., Kita, Y., Ihara, S., Suda, T. and Fukui, Y. (1997). Phosphatidylinositol-3 kinase is involved in ruffled border formation in osteoclasts. *J. Cell. Physiol.* 172, 230–239.
228. Akisaka, T., Yamamoto, T. and Gay, C. V. (1988). Ultracytochemical investigation of calcium-activated adenosine triphosphatase (Ca^{++}-ATPase) in chick tibia. *J. Bone Miner. Res.* 3, 19–25.
229. Bekker, P. J. and Gay, C. V. (1992). Demonstration of calmodulin-sensitive calcium translocation by isolated osteoclast plasma membrane vesicles. *Calcif. Tissue Int.* 51, 312–316.
230. Yamakawa, K., Duncan, R. and Hruska, K. A. (1994). An Arg-Gly-Asp peptide stimulates Ca^{2+} efflux from osteoclast precursors through a novel mechanism. *Am. J. Physiol.* 266, F651–F657.
231. Silver, I. A., Murrills, R. J. and Etherington, D. J. (1988). Microelectrode studies on the acid microenvironment beneath adherent macrophages and osteoclasts. *Exp. Cell Res.* 175, 266–276.
232. Adebanjo, O. A., Moonga, B. S., Yamate, T., Sun, L., Minkin, C., Abe, E. and Zaidi, M. (1998). Mode of action of interleukin-6 on mature osteoclasts. Novel interactions with extracellular Ca^{2+} sensing in the regulation of osteoclastic bone resorption. *J. Cell Biol.* 142, 1347–1356.
233. Zaidi, M., Shankar, V. S., Tunwell, R., Adebanjo, O. A., Mackrill, J., Pazianas, M., O'Connell, D., Simon, B. J., Rifkin, B. R., Venkitaraman, A. R. and *et al.* (1995). A ryanodine receptor-like molecule expressed in the osteoclast plasma membrane functions in extracellular Ca^{2+} sensing. *J. Clin. Invest.* 96, 1582–1590.
234. Blair, H. C., Teitelbaum, S. L., Tan, H. L., Koziol, C. M. and Schlesinger, P. H. (1991). Passive chloride permeability charge coupled to H(+)-ATPase of avian osteoclast ruffled membrane. *Am. J. Physiol.* 260, C1315–C1324.
235. Kornak, U., Kasper, D., Bosl, M. R., Kaiser, E., Schweizer, M., Schulz, A., Friedrich, W., Delling, G. and Jentsch, T. J. (2001). Loss of the ClC-7 chloride channel leads to osteopetrosis in mice and man. *Cell* 104, 205–215.
236. Teti, A., Blair, H. C., Teitelbaum, S. L., Kahn, A. J., Koziol, C., Konsek, J., Zambonin-Zallone, A. and Schlesinger, P. H. (1989). Cytoplasmic pH regulation and chloride/bicarbonate exchange in avian osteoclasts. *J. Clin. Invest.* 83, 227–233.
237. Razzouk, S., Lieberherr, M. and Cournot, G. (1999). Rac-GTPase, osteoclast cytoskeleton and bone resorption. *Eur. J. Cell Biol.* 78, 249–255.
238. Brown, M. T. and Cooper, J. A. (1996). Regulation, substrates and functions of src. *Biochim. Biophys. Acta* 1287, 121–149.
239. Tezuka, K., Tezuka, Y., Maejima, A., Sato, T., Nemoto, K., Kamioka, H., Hakeda, Y. and Kumegawa, M. (1994). Molecular cloning of a possible cysteine proteinase predominantly expressed in osteoclasts. *J. Biol. Chem.* 269, 1106–1109.

240. Väänänen, H. K., Zhao, H., Mulari, M. and Halleen, J. M. (2000). The cell biology of osteoclast function. *J. Cell Sci.* 113 (Pt 3), 377–381.
241. Vu, T. H., Shipley, J. M., Bergers, G., Berger, J. E., Helms, J. A., Hanahan, D., Shapiro, S. D., Senior, R. M. and Werb, Z. (1998). MMP-9/gelatinase B is a key regulator of growth plate angiogenesis and apoptosis of hypertrophic chondrocytes. *Cell* 93, 411–422.
242. Yamashita, D. S. and Dodds, R. A. (2000). Cathepsin K and the design of inhibitors of cathepsin K. *Curr. Pharm. Des.* 6, 1–24.
243. Nesbitt, S. A. and Horton, M. A. (1997). Trafficking of matrix collagens through bone-resorbing osteoclasts. *Science* 276, 266–269.
244. Salo, J., Lehenkari, P., Mulari, M., Metsikko, K. and Väänänen, H. K. (1997). Removal of osteoclast bone resorption products by transcytosis. *Science* 276, 270–273.
245. Jimi, E., Nakamura, I., Duong, L. T., Ikebe, T., Takahashi, N., Rodan, G. A. and Suda, T. (1999). Interleukin 1 induces multinucleation and bone-resorbing activity of osteoclasts in the absence of osteoblasts/stromal cells. *Exp. Cell Res.* 247, 84–93.
246. Azuma, Y., Kaji, K., Katogi, R., Takeshita, S. and Kudo, A. (2000). Tumor necrosis factor-alpha induces differentiation of and bone resorption by osteoclasts. *J. Biol. Chem.* 275, 4858–4864.
247. Kobayashi, K., Takahashi, N., Jimi, E., Udagawa, N., Takami, M., Kotake, S., Nakagawa, N., Kinosaki, M., Yamaguchi, K., Shima, N., Yasuda, H., Morinaga, T., Higashio, K., Martin, T. J. and Suda, T. (2000). Tumor necrosis factor alpha stimulates osteoclast differentiation by a mechanism independent of the ODF/RANKL-RANK interaction. *J. Exp. Med.* 191, 275–286.
248. Langub, M. C., Monier-Faugere, M. C., Qi, Q., Geng, Z., Koszewski, N. J. and Malluche, H. H. (2001). Parathyroid hormone/parathyroid hormone-related peptide type 1 receptor in human bone. *J. Bone Miner. Res.* 16, 448–456.
249. Rao, L. G., Murray, T. M. and Heersche, J. N. (1983). Immunohistochemical demonstration of parathyroid hormone binding to specific cell types in fixed rat bone tissue. *Endocrinology* 113, 805–810.
250. Gay, C. V., Zheng, B. and Gilman, V. R. (2003). Co-detection of PTH/PTHrP receptor and tartrate resistant acid phosphatase in osteoclasts. *J. Cell. Biochem.* 89, 902–908.
251. Watson, P. H., Kisiel, M., Patterson, E. K., Hodsman, A. B., Sims, S. M., Dixon, S. J. and Fraher, L. J. (2002). Expression of type I PTH/PTHrP Receptors in the rat osteoclast and osteoclast-like RAW 264.7 cells. *J. Bone Miner. Res.* 17:S287.
252. Faucheux, C., Horton, M. A. and Price, J. S. (2002). Nuclear localization of type I parathyroid hormone/parathyroid hormone-related protein receptors in deer antler osteoclasts: evidence for parathyroid hormone-related protein and receptor activator of NF-kappaB-dependent effects on osteoclast formation in regenerating mammalian bone. *J. Bone Miner. Res.* 17, 455–464.
253. Gay, C. V. and Weber, J. A. (2000). Regulation of differentiated osteoclasts. *Crit. Rev. Eukaryot. Gene Expr.* 10, 213–230.
254. Amizuka, N., Karaplis, A. C., Henderson, J. E., Warshawsky, H., Lipman, M. L., Matsuki, Y., Ejiri, S., Tanaka, M., Izumi, N., Ozawa, H. and Goltzman, D. (1996). Haploinsufficiency of parathyroid hormone-related peptide (PTHrP) results in abnormal postnatal bone development. *Dev. Biol.* 175, 166–176.
255. Kartsogiannis, V., Udagawa, N., Ng, K. W., Martin, T. J., Moseley, J. M. and Zhou, H. (1998). Localization of parathyroid hormone-related protein in osteoclasts by *in situ* hybridization and immunohistochemistry. *Bone* 22, 189–194.
256. Tam, C. S., Heersche, J. N., Murray, T. M. and Parsons, J. A. (1982). Parathyroid hormone stimulates the bone apposition rate independently of its resorptive action: differential effects of intermittent and continuous administration. *Endocrinology* 110, 506–512.

257. Agarwala, N. and Gay, C. V. (1992). Specific binding of parathyroid hormone to living osteoclasts. *J. Bone Miner. Res.* 7, 531–539.
258. Niendorf, A., Dietel, M., Arps, H. and Childs, G. V. (1988). A novel method to demonstrate parathyroid hormone binding on unfixed living target cells in culture. *J. Histochem. Cytochem.* 36, 307–309.
259. Niendorf, A., Dietel, M., Arps, H., Lloyd, J. and Childs, G. V. (1986). Visualization of binding sites for bovine parathyroid hormone (PTH 1-84) on cultured kidney cells with a biotinyl-b-PTH (1-84) antagonist. *J. Histochem. Cytochem.* 34, 357–361.
260. Rickard, D., Harris, S. A., Turner, R., Khosla, S. and Spelsberg, T. (2002). Estrogens and progestins. In *Principles of Bone Biology* 2nd ed., Bilezikian, J. P., Raisz, L. G. and Rodan, G. A., Eds., Academic Press, San Diego, CA. pp. 655–675.
261. Sunyer, T., Lewis, J., Collin-Osdoby, P. and Osdoby, P. (1999). Estrogen's bone-protective effects may involve differential IL-1 receptor regulation in human osteoclast-like cells. *J. Clin. Invest.* 103, 1409–1418.
262. Oursler, M. J., Osdoby, P., Pyfferoen, J., Riggs, B. L. and Spelsberg, T. C. (1991). Avian osteoclasts as estrogen target cells. *Proc. Natl. Acad. Sci. USA* 88, 6613–6617.
263. Oursler, M. J., Pederson, L., Pyfferoen, J., Osdoby, P., Fitzpatrick, L. and Spelsberg, T. C. (1993). Estrogen modulation of avian osteoclast lysosomal gene expression. *Endocrinology* 132, 1373–1380.
264. Oursler, M. J., Pederson, L., Fitzpatrick, L., Riggs, B. L. and Spelsberg, T. (1994). Human giant cell tumors of the bone (osteoclastomas) are estrogen target cells. *Proc. Natl. Acad. Sci. USA* 91, 5227–5231.
265. Mano, H., Yuasa, T., Kameda, T., Miyazawa, K., Nakamaru, Y., Shiokawa, M., Mori, Y., Yamada, T., Miyata, K., Shindo, H., Azuma, H., Hakeda, Y. and Kumegawa, M. (1996). Mammalian mature osteoclasts as estrogen target cells. *Biochem. Biophys. Res. Commun.* 223, 637–642.
266. Hong, M. H., Williams, H., Jin, C. H. and Pike, J. W. (1994). 17 beta-estradiol suppresses mouse osteoclast differentiation and function in vitro via the estrogen receptor. *J. Bone Miner. Res.* 9, 8161.
267. Williams, J. P., Blair, H. C., McKenna, M. A., Jordan, S. E. and McDonald, J. M. (1996). Regulation of avian osteoclastic H^+ -ATPase and bone resorption by tamoxifen and calmodulin antagonists. Effects independent of steroid receptors. *J. Biol. Chem.* 271, 12488–12495.
268. Collier, F. M., Huang, W. H., Holloway, W. R., Hodge, J. M., Gillespie, M. T., Daniels, L. L., Zheng, M. H. and Nicholson, G. C. (1998). Osteoclasts from human giant cell tumors of bone lack estrogen receptors. *Endocrinology* 139, 1258–1267.
269. Pederson, L., Kremer, M., Foged, N. T., Winding, B., Ritchie, C., Fitzpatrick, L. A. and Oursler, M. J. (1997). Evidence of a correlation of estrogen receptor level and avian osteoclast estrogen responsiveness. *J. Bone Miner. Res.* 12, 742–752.
270. Sugiyama, T., Kusuhara, S. and Gay, C. V. (1999). Parathyroid hormone and estrogen effects on adhesion of chicken medullary bone osteoclasts. in *Calcium Metabolism: Comparative Endocrinology,* Danks, J., Dacke, C., Flik, G. and Gay, C. V., Eds. Bioscientifica Ltd., Bristol, pp. 107–111.
271. Kremer, M., Judd, J., Rifkin, B., Auszmann, J. and Oursler, M. J. (1995). Estrogen modulation of osteoclast lysosomal enzyme secretion. *J. Cell. Biochem.* 57, 271–279.
272. Brubaker, K. D. and Gay, C. V. (1994). Specific binding of estrogen to osteoclast surfaces. *Biochem. Biophys. Res. Commun.* 200, 899–907.
273. Brubaker, K. D. and Gay, C. V. (1999). Estrogen stimulates protein tyrosine phosphorylation and Src kinase activity in avian osteoclasts. *J. Cell. Biochem.* 76, 206–216.

274. Fiorelli, G., Gori, F., Frediani, U., Franceschelli, F., Tanini, A., Tosti-Guerra, C., Benvenuti, S., Gennari, L., Becherini, L. and Brandi, M. L. (1996). Membrane binding sites and non-genomic effects of estrogen in cultured human pre-osteoclastic cells. *J. Steroid Biochem. Mol. Biol.* 59, 233–240.
275. Brubaker, K. D. and Gay, C. V. (1999). Depolarization of osteoclast plasma membrane potential by 17beta-estradiol. *J. Bone Miner. Res.* 14, 1861–1866.
276. Berger, C. E., Horrocks, B. R. and Datta, H. K. (1999). Direct non-genomic effect of steroid hormones on superoxide anion generation in the bone resorbing osteoclasts. *Mol. Cell. Endocrinol.* 149, 53–59.
277. Hall, M. R., Kief, N. L., Gilman, V. R. and Gay, C. V. (1994). Surface binding and clearance of calcitonin by avian osteoclasts. *Comp. Biochem. Physiol. Comp. Physiol.* A 108, 59–63.
278. Hunter, S. J., Schraer, H. and Gay, C. V. (1989). Characterization of the cytoskeleton of isolated chick osteoclasts: effect of calcitonin. *J. Histochem. Cytochem.* 37, 1529–1537.
279. Datta, H. K., Manning, P., Rathod, H. and McNeil, C. J. (1995). Effect of calcitonin, elevated calcium and extracellular matrices on superoxide anion production by rat osteoclasts. *Exp. Physiol.* 80, 713–719.
280. Kaji, H., Sugimoto, T., Miyauchi, A., Fukase, M., Tezuka, K., Hakeda, Y., Kumegawa, M. and Chihara, K. (1994). Calcitonin inhibits osteopontin mRNA expression in isolated rabbit osteoclasts. *Endocrinology* 135, 484–487.
281. Yasuoka, T., Kawashima, M., Takahashi, T., Tatematsu, N. and Tanaka, K. (1998). Calcitonin receptor binding properties in bone and kidney of the chicken during the oviposition cycle. *J. Bone Miner. Res.* 13, 1412–1419.
282. Inoue, D., Shih, C., Galson, D. L., Goldring, S. R., Horne, W. C. and Baron, R. (1999). Calcitonin-dependent down-regulation of the mouse C1a calcitonin receptor in cells of the osteoclast lineage involves a transcriptional mechanism. *Endocrinology* 140, 1060–1068.
283. Liu, B. Y., Wang, J. T., Leu, J. S., Chiang, C. P., Hsieh, C. C. and Kwan, H. W. (2000). Effects of continuous calcitonin treatment on osteoclasts derived from cocultures of mouse marrow stromal and spleen cells. *J. Formos Med. Assoc.* 99, 140–150.
284. Dong, S. S., Williams, J. P., Jordan, S. E., Cornwell, T. and Blair, H. C. (1999). Nitric oxide regulation of cGMP production in osteoclasts. *J. Cell. Biochem.* 73, 478–487.
285. Ueno, M., Fukuda, K., Oh, M., Asada, S., Nishizaka, F., Hara, F. and Tanaka, S. (1998). Protein kinase C modulates the synthesis of nitric oxide by osteoblasts. *Calcif. Tissue Int.* 63, 22–26.
286. Silverton, S. F., Adebanjo, O. A., Moonga, B. S., Awumey, E. M., Malinski, T. and Zaidi, M. (1999). Direct microsensor measurement of nitric oxide production by the osteoclast. *Biochem. Biophys. Res. Commun.* 259, 73–77.
287. van't Hof, R. J. and Ralston, S. H. (1997). Cytokine-induced nitric oxide inhibits bone resorption by inducing apoptosis of osteoclast progenitors and suppressing osteoclast activity. *J. Bone Miner. Res.* 12, 1797–1804.
288. Lorget, F., Kamel, S., Mentaverri, R., Wattel, A., Naassila, M., Maamer, M. and Brazier, M. (2000). High extracellular calcium concentrations directly stimulate osteoclast apoptosis. *Biochem. Biophys. Res. Commun.* 268, 899–903.
289. Greenfield, E. M., Bi, Y. and Miyauchi, A. (1999). Regulation of osteoclast activity. *Life Sci.* 65, 1087–1102.
290. Sakai, K., Mohtai, M., Shida, J., Harimaya, K., Benvenuti, S., Brandi, M. L., Kukita, T. and Iwamoto, Y. (1999). Fluid shear stress increases interleukin-11 expression in human osteoblast-like cells: its role in osteoclast induction. *J. Bone Miner. Res.* 14, 2089–2098.

291. Ilvesaro, J., Väänänen, K. and Tuukkanen, J. (2000). Bone-resorbing osteoclasts contain gap-junctional connexin-43. *J. Bone Miner. Res.* 15, 919–926.
292. van't Hof, R. J., von Lindern, M., Nijweide, P. J. and Beug, H. (1997). Stem cell factor stimulates chicken osteoclast activity *in vitro*. *FASEB J.* 11, 287–293.
293. Abu-Amer, Y., Ross, F. P., Schlesinger, P., Tondravi, M. M. and Teitelbaum, S. L. (1997). Substrate recognition by osteoclast precursors induces C-src/microtubule association. *J. Cell Biol.* 137, 247–258.
294. Duong, L. T., Lakkakorpi, P. T., Nakamura, I., Machwate, M., Nagy, R. M. and Rodan, G. A. (1998). PYK2 in osteoclasts is an adhesion kinase, localized in the sealing zone, activated by ligation of alpha v beta 3 integrin, and phosphorylated by src kinase. *J. Clin. Invest.* 102, 881–892.
295. Chellaiah, M., Fitzgerald, C., Alvarez, U. and Hruska, K. (1998). c-Src is required for stimulation of gelsolin-associated phosphatidylinositol 3-kinase. *J. Biol. Chem.* 273, 11908–11916.
296. Huang, C., Ni, Y., Wang, T., Gao, Y., Haudenschild, C. C. and Zhan, X. (1997). Down-regulation of the filamentous actin cross-linking activity of cortactin by Src-mediated tyrosine phosphorylation. *J. Biol. Chem.* 272, 13911–13915.
297. Yamaji, Y., Tsuganezawa, H., Moe, O. W. and Alpern, R. J. (1997). Intracellular acidosis activates c-Src. *Am. J. Physiol.* 272, C886–C893.
298. Kerr, J. F., Wyllie, A. H. and Currie, A. R. (1972). Apoptosis: a basic biological phenomenon with wide-ranging implications in tissue kinetics. *Br. J. Cancer* 26, 239–257.
299. Wyllie, A. H., Kerr, J. F. and Currie, A. R. (1980). Cell death: the significance of apoptosis. *Int. Rev. Cytol.* 68, 251–306.
300. Fuller, K., Owens, J. M., Jagger, C. J., Wilson, A., Moss, R. and Chambers, T. J. (1993). Macrophage colony-stimulating factor stimulates survival and chemotactic behavior in isolated osteoclasts. *J. Exp. Med.* 178, 1733–1744.
301. Boyce, B. F., Wright, K., Reddy, S. V., Koop, B. A., Story, B., Devlin, R., Leach, R. J., Roodman, G. D. and Windle, J. J. (1995). Targeting simian virus 40 T antigen to the osteoclast in transgenic mice causes osteoclast tumors and transformation and apoptosis of osteoclasts. *Endocrinology* 136, 5751–5759.
302. Rani, S., Xing, L., Chen, H., Dai, A. and Boyce, B. F. (1997). Phosphatidylinositol 3-kinase activity is required for bone resorption and osteoclast survival in vitro. *J. Bone Miner. Res.* 12, S420.
303. Rodan, S. B. and Rodan, G. A. (1997). Integrin function in osteoclasts. *J. Endocrinol.* 154 Suppl, S47–S56.
304. Villanova, I., Townsend, P. A., Uhlmann, E., Knolle, J., Peyman, A., Amling, M., Baron, R., Horton, M. A. and Teti, A. (1999). Oligodeoxynucleotide targeted to the alphav gene inhibits alphav integrin synthesis, impairs osteoclast function, and activates intracellular signals to apoptosis. *J. Bone Miner. Res.* 14, 1867–1879.
305. Kameda, T., Miyazawa, K., Mori, Y., Yuasa, T., Shiokawa, M., Nakamaru, Y., Mano, H., Hakeda, Y., Kameda, A. and Kumegawa, M. (1996). Vitamin K_2 inhibits osteoclastic bone resorption by inducing osteoclast apoptosis. *Biochem. Biophys. Res. Commun.* 220, 515–519.
306. Sakagami, H., Satoh, K., Hakeda, Y. and Kumegawa, M. (2000). Apoptosis-inducing activity of vitamin C and vitamin K. *Cell. Mol. Biol. (Noisy-le-grand)* 46, 129–143.
307. Okahashi, N., Nakamura, I., Jimi, E., Koide, M., Suda, T. and Nishihara, T. (1997). Specific inhibitors of vacuolar H(+)-ATPase trigger apoptotic cell death of osteoclasts. *J. Bone Miner. Res.* 12, 1116–1123.
308. Boyce, B. F., Xing, L., Jilka, R. L., Bellido, T., Weinstein, R. S., Parfitt, A. M. and Manolagas, S. C. (2002). Apoptosis in bone cells. In *Principles of Bone Biology,*

Bilezikian, J. P., Raisz, L. G. and Rodan, G. A., Eds., Vol. 1. Academic Press, San Diego, CA, pp 151–168.
309. Ozaki, K., Takeda, H., Iwahashi, H., Kitano, S. and Hanazawa, S. (1997). NF-kappaB inhibitors stimulate apoptosis of rabbit mature osteoclasts and inhibit bone resorption by these cells. *FEBS Lett.* 410, 297–300.
310. Lacey, D. L., Tan, H. L., Lu, J., Kaufman, S., Van, G., Qiu, W., Rattan, A., Scully, S., Fletcher, F., Juan, T., Kelley, M., Burgess, T. L., Boyle, W. J. and Polverino, A. J. (2000). Osteoprotegerin ligand modulates murine osteoclast survival *in vitro* and *in vivo*. *Am. J. Pathol.* 157, 435–448.
311. Selander, K. S., Harkonen, P. L., Valve, E., Monkkonen, J., Hannuniemi, R. and Väänänen, H. K. (1996). Calcitonin promotes osteoclast survival *in vitro*. *Mol. Cell. Endocrinol.* 122, 119–129.
312. Kanaoka, K., Kobayashi, Y., Hashimoto, F., Nakashima, T., Shibata, M., Kobayashi, K., Kato, Y. and Sakai, H. (2000). A common downstream signaling activity of osteoclast survival factors that prevent nitric oxide-promoted osteoclast apoptosis. *Endocrinology* 141, 2995–3005.
313. Kallio, D. M., Garant, P. R. and Minkin, C. (1972). Ultrastructural effects of calcitonin on osteoclasts in tissue culture. *J. Ultrastruct. Res.* 39, 205–216.
314. Parfitt, A. M., Mundy, G. R., Roodman, G. D., Hughes, D. E. and Boyce, B. F. (1996). A new model for the regulation of bone resorption, with particular reference to the effects of bisphosphonates. *J. Bone Miner. Res.* 11, 150–159.

SECTION II

Basic Bone Biology and Scaffold Designs for Tissue

4 The Organic and Inorganic Matrices

Adele L Boskey

Contents

0-8493-1621-9/05/$0.00+$1.50
© 2005 by CRC Press LLC

INTRODUCTION

The structure and composition of bone varies with the tissue site and its origin (e.g., lamellar versus cortical, intramembranous versus endochondral) as well as with age, diet, and health status. Engineering bone requires mimicking of key aspects of bone structure and composition to achieve the optimal functional tissue. To appreciate these characteristics, this chapter will review current knowledge of the functions of the mineral and matrix constituents, and how these functions have been assessed by studies of age and disease variations in tissue composition. Some comment will also be made on the current state of bone tissue engineering from the mineral and matrix points of view. The functions of the bone cells and their importance in bone tissue engineering were discussed in chapter 3.

Tissue-engineered bone and other mineralized tissues will be important for the repair of lesions caused by cancer surgery, birth defects, and trauma. Since bone is unique in terms of being able to repair itself[1] (e.g., fracture healing in humans and animals, limb regeneration in lower species), lessons can not only be learned by examination of these processes, but there may be instances when tissue-engineered products may only be used to enhance the natural repair process.

BONE AS A COMPOSITE

Bone is a composite material consisting of mineral, matrix, cells, and water. Developmentally as the cartilaginous anlage of the bone shaft is replaced by a boney matrix, that matrix is predominantly matrix (osteoid) and calcified cartilage.[2] With age the mineral content increases, reaching a maximum value, approximated by "peak bone density," in male and female humans at different ages.[3]

In general, as the animal matures even further the total bone mineral content decreases.[4] Diseases such as osteoporosis are associated with a decrease in total bone density, but not necessarily with decreases in the proportion of mineral in any bone, while osteomalacia is defined as a loss of bone mineral and an increase in osteoid.[5] Conversely, osteopetrosis is an increase in both bone mineral and bone matrix, beyond that necessary for normal function.[6] Like many other composite materials, the integration of minerals within the organic matrix enables bone to have mechanical properties that are enhanced relative to the properties of the mineral (a brittle material) or the matrix (an elastic material) alone.

Bone has both mechanical and homeostatic functions, providing protection for the internal organs of the body and serving as a storage site and source of mineral ions.[7] In terms of evolution, the vertebrates developed calcium phosphate skeletons, but there are lower species with other mineralized exo- and endo-skeletons that contain calcium carbonates and other phases. The composition of those species' skeletons are reviewed elsewhere, along with discussions of how the study of these species can contribute to the development of tissue-engineered products.[8–12] Bone strength is determined by the shape of the bone, the structure (arrangement of its components), that is, its geometry, and by its so-called "material" properties.[7, 13] The stress–strain curve of the bone (Figure 4.1) has areas that have been shown to be determined predominantly by the mineral phase (elastic region) and principally by the organic phase (plastic region).[14] The slope of the stress–strain curve is the elastic (Young's) modulus, which describes the intrinsic stiffness of the bone. Hence, all other components being equal, a more highly mineralized material will have a greater elastic modulus. The area under the stress–strain curve is the amount of energy needed to cause a fracture; thus, bones with comparable Young's moduli that have stress–strain curves extended into the plastic region will be more resistant to load than those that do not.

There is a well-established correlation between bone mineral density (BMD) measured radiographically and bone strength, but BMD does not completely account for bone strength.[15] Rather, it is other properties of the composite tissue that are believed to explain this variation.

THE ORGANIC MATRIX

COLLAGEN

Type I collagen is the principal component of the organic matrix of bone, accounting for approximately 30% of the dry nondemineralized matrix. Type I collagen is a heteropolymer of two identical and one distinct chain, each having the primary structure $(Gly–X–Y)_n$, where X and Y are frequently proline or hydroxyproline. The collagen is post-translationally modified to contain hydroxylysine, hydroxyproline, and glycoslyated hydroxylysine.[16] In the extracellular matrix, the hydroxylysine residues are involved in the formation of stable collagen cross-links[17] (Figure 4.2).

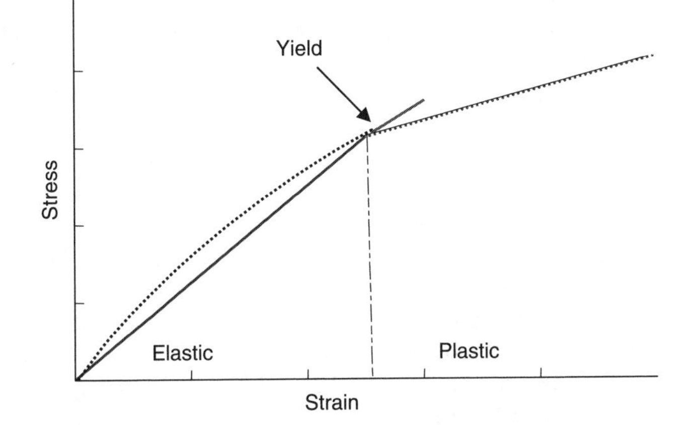

Figure 4.1 Theoretical stress–strain curve showing the elastic and plastic regions defined by the yield point. The dotted line represents typical data, and the straight lines represent the slopes used to calculate various parameters.

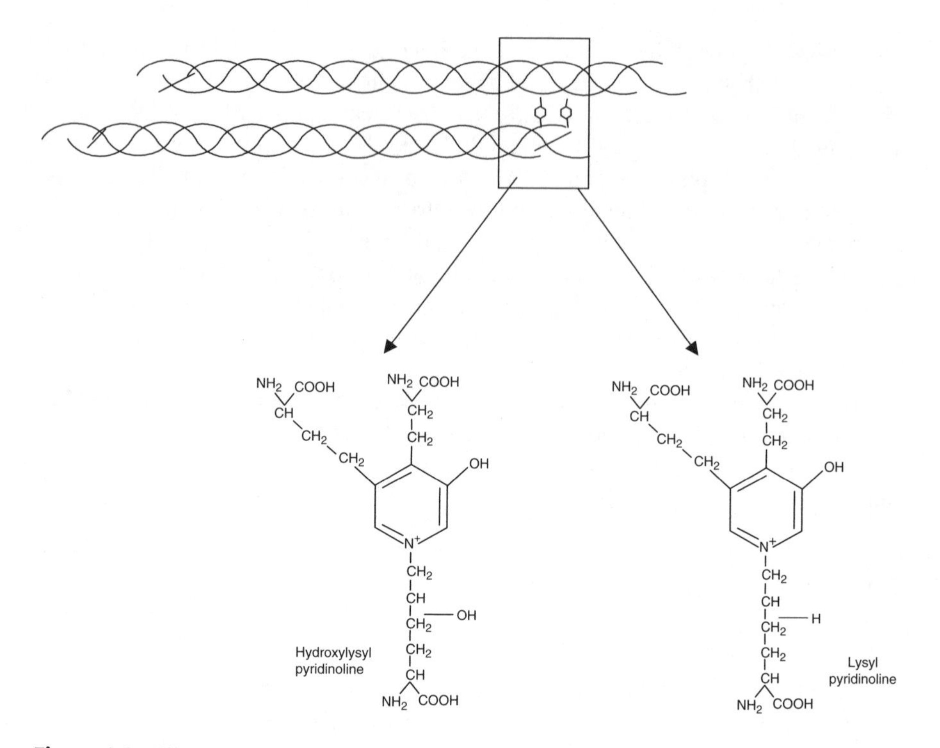

Figure 4.2 Illustration of trivalent collagen cross-links formed between collagen triple helical molecules (box). The structures of the two major stable cross-links, hydroxylysyl pyridinoline and lysyl pyridinoline, are illustrated.

In addition, bone contains small amounts of type III, type V, and type XII collagen. These collagens trim the type I fibrils, and may affect the properties of the tissue. Thus, in the *fro/fro* mouse,[18] a naturally occurring mutant with decreased type III collagen, the bones are deformed, undermineralized, and fragile, and the proportion of collagen fibrils with a large diameter is decreased (in the tendon). This mutant also has decreased levels of the noncollagenous protein, osteonectin (see below). Mice engineered with a type V collagen mutation have thinner fibrils, skeletal deformities, and abnormalities in all the type I-containing tissues.[19]

Type I collagen provides a backbone for the deposition of bone mineral. The bone mineral crystals are aligned with their long axis parallel to the collagen axis. Where the collagen molecule is altered, as in the case of the genetic abnormalities in osteogenesis imperfecta, otherwise known as brittle bone disease, mineral crystals are generally smaller in size than those in age-matched healthy bone[20] and the mineral may be found outside the collagen fibrils.[21] The altered bone and mineral properties in this collagen-based disease are the best evidence for proving the importance of the collagen for proper bone mineralization. Figure 4.3 shows the stress–strain curve for bones from *oim/oim* (osteogenesis imperfecta mouse) bones and those of age-matched wild-type animals, showing the increased brittle character of the animals, which have a type I-homotrimer collagen as opposed to the heteropolymer in

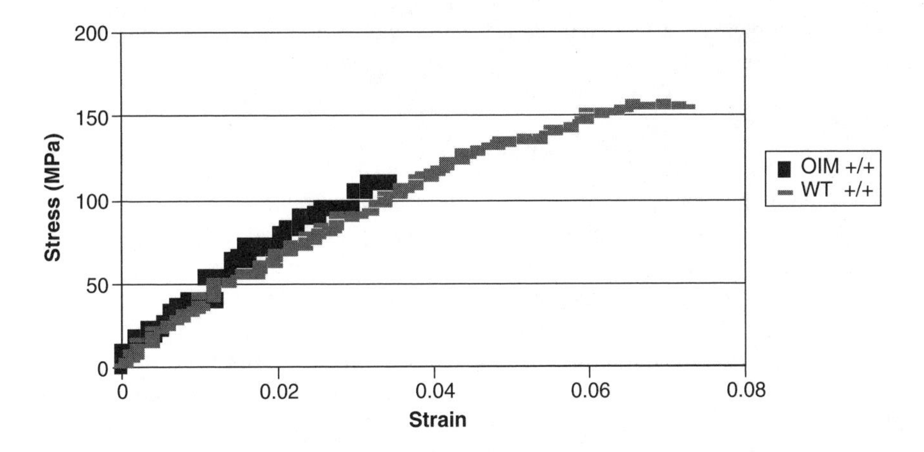

Figure 4.3 Measured stress–strain curves from 4-month-old osteogenesis imperfecta (*oim/oim*) mice and the age-matched controls. Note that the *oim/oim* mice (darker symbols) have no plastic region, and a yield point lower than the wildtype (lighter symbols), and thus have a decreased ability to bear load, and are thus more brittle. Courtesy of Dr. Nancy Camacho.

the wild-type animals. The collagen gives bone its elastic properties, and there have been several studies demonstrating that the connective tissues, as well as the bone, in animal models of osteogenesis imperfecta are brittle (i.e., absorbs little energy before failure), and fails more easily when load is applied.[22,23]

But collagen itself is not the bone matrix component that facilitates mineral formation (the process of nucleation by which the initial mineral forms will be discussed in section IV of this chapter). This was demonstrated many years ago when it was noted that when decalcified bone collagen was implanted *in situ*, it took many months for mineral deposition to commence.[24] Termine et al. later demonstrated that it was the other proteins associated with the bone collagen matrix that were facilitating bone mineralization.[25] Similarly, Glimcher and Endo showed that the ability to mineralize collagen *in vitro* was directly related to the extent to which its associated proteins were phosphorylated.[26] Thus, it is apparent that bone collagen re-mineralization requires, at a minimum, phosphorylated noncollagenous proteins.

Bone Matrix Proteins

The noncollagenous proteins account for approximately 5% of the dry bone matrix; yet, it is these proteins that appear to regulate the organization, turnover, and mineralization of the bone matrix. Most of these proteins appear to have more than one function. However, because their functions in humans have been obtained from investigations *in vitro*, or in transgenic animals, the only proof of function comes from the limited patient data available. These proteins fall into a series of families, and their functions appear to be redundant. This redundancy is crucial, as the presence of a mineralized skeleton is a key feature of a viable vertebrate, and were the functional proteins not redundant in activity, it would be impossible to maintain a functional tissue.

SIBLINGs

The SIBLING proteins (Small Integrin Binding LIgand with N-Glycosylation) are all located on human chromosome 4, all have one or more arginine–glycine–aspartate (RGD) sequences, and all have some N-glycosylated residues. Thus named by Fisher et al.[27] because of their shared ability to activate complement factor H,[28] the SIBLING proteins include osteopontin, bone sialoprotein (BSP), dentin matrix protein-1, dentin sialoprotein, and matrix extracellular glyco-phosphoprotein (MEPE). The genes for each of these proteins are expressed in bone, and the proteins have been isolated from bone. Of those studied in more detail, each seem to have more than one function, either indicating that they are multifunctional, or that their true function *in situ* has not yet been identified. All the SIBLING proteins are located on human chromosome 4q, and of those investigated all seem to have a high affinity for bone mineral[29] when they are either phosphorylated[30] or post-transitionally modified, but not when these anionic groups are removed.

Osteopontin

Osteopontin is the major phosphorylated glycoprotein of bone.[31] Its molecular weight ranges from ~42 to 70 kDa depending on the method of analysis and its post-translational modifications.[32] It contains several sialic acid residues, one RGD sequence, and has 29 possible phosphorylation sites.[32] Preliminary primary structure analysis in my laboratory has indicated that in solution, in the absence of calcium, osteopontin is mainly a random coil, obtaining more beta-sheet structure in the presence of physiologic concentrations of calcium.

Osteopontin is found in almost every tissue in the body, including the skin, brain, body fluids, cartilage, and a variety of tumors.[32] Its expression is increased in response to increasing phosphate concentrations in soft tissues,[33] suggesting that it is a key regulator of mineralization.

Osteopontin may be sulfated and it is variably phosphorylated. The extent of phosphorylation differs in various tissues, but also appears to vary with bone tissue age. *In vitro*, the phosphorylated osteopontin isolated from bone is an effective inhibitor of mineralization.[34] Dephosphorylation of the protein eliminates this ability. Osteopontin's kidney analog, uropontin, inhibits calcium oxalate formation and growth.[35] The osteopontin present in milk, which has 27 of its 29 residues phosphorylated, in contrast, is an *in vitro* promoter of bone mineral formation, although preliminary data from our laboratory indicate that its true function in milk may be to help retain the calcium phosphate, casein-micelles, in a noncrystalline (amorphous) form. A more phosphorylated variant of osteopontin isolated from bone, was also shown to have some ability to act as a nucleator, losing this ability when dephosphorylated. *In vitro*, osteopontin is also chemotactic for osteoclasts.[36]

The osteopontin knockout mouse[37,38] has a bone phenotype that is related to both these *in vitro* functions. The bones are thickened and resistant to ovariectomy-stimulated or PTH-induced remodeling (Figure 4.4),[39] and the mineral crystals in the bones of the knockout are more abundant and of a larger crystal size.[40]

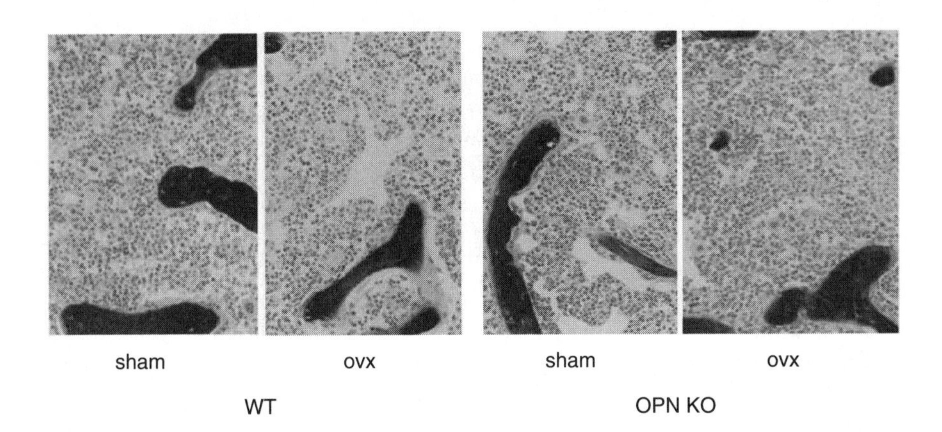

Figure 4.4 Micro CT images of tibia of ovariectomized osteopontin knockout and wild-type animals, from reference 60, figure 2. Wild-type (WT) (A and B) or $OPN^{-/-}$ (KO) (C and D) mice were either ovariectomized (OVX, A and C) or sham-operated (SHAM, B and D). Images in the midsaggital plane were taken 4 weeks postoperatively. Note the loss of trabecular bone in the WT after ovariectomy and the lack of such a change in the ovariectomized KO. Reprinted with permission from Proc. Natl. Acad. Sci. USA.

Bone sialoprotein

Bone sialoprotein, BSP, also called integrin-binding bone sialoprotein (IBSP) accounts for approximately 15% of the human noncollagenous bone proteins.[41] The protein has a molecular mass of approximately 50–70 kDa, the range being due to phosphorylation and post-translational glycation. It has both N- and O-based glycosylations, is sulfated in some species, and has fewer phosphorylation sites than osteopontin. The sulfation is lost prior to mineralization in culture.[42] It also has two poly-glutamate repeat regions, one of which is essential for mineral interactions.[43] Analysis of the BSP circular dichroism spectra revealed that in solution the structure consisted of ~5% alpha-helix, 32% beta-sheet, 17% beta-turn, and 46% random coil;[41] however, in contrast to the case of osteopontin, no conformational changes were observed in the presence of calcium and phosphate solutions.[41]

Electron microscopic localization of the bone matrix proteins has demonstrated that both BSP and osteopontin are concentrated at the same sites in mineralized bone: areas of active remodeling, over the cement lines, and on nascent mineralization foci.[44] However, BSP tends to be concentrated where new mineralization is occurring, while osteopontin is concentrated in the already mineralized tissues.

BSP is expressed by osteoblast cells after they have started matrix production,[45] and, based on analysis of the BSP gene promoter, it has recently been suggested that BSP and type I collagen production is coordinately linked.[46] BSP binds to collagen[47] and promotes osteoblast binding *in vitro*; this binding is, in part, RGD dependent.[48]

In vitro BSP promotes mineral formation,[49] functioning as a mineral nucleator (see section IV below). However, it can also regulate bone mineral crystal growth when present at higher concentrations.[50] The BSP knockout animals showed minimal skeletal changes, but detailed analyses have not been reported.

Dentin matrix protein-1

The first dentin protein to be cloned,[51] dentin matrix protein-1 (DMP1), is a highly hydrophilic phosphorylated glycoprotein with both poly-aspartic acid and poly-glutamic acid repeats. It shows some homology to osteopontin, BSP, and, as discussed below, BAG-75, but does not cross-react with antibodies against those proteins.[52] Developmentally, DMP1 expression is first noted in hypertrophic chondrocytes, then in osteoblasts, and then in osteocytes. Hence, it is considered to be a differentiation marker for osteoblasts.[53] There appears to be as much DMP1 in bone as there is in dentin; however, in bone the DMP1 is fragmented, while in dentin it is intact.[54]

Dentin matrix protein-1 is an ~150 kDa protein. It has 55 possible phosphorylation sites, but few of these are occupied. It has one RGD domain that is active in cell attachment,[55] and the N terminus has some sequence homology to the other SIBLINGs. DMP1 is ~30% by weight sialic acid. Like the other SIBLINGs, it binds to factor H and activates complement.[28] It is found localized around the nonmineralized areas of odontoblasts (dentin-forming cells) and osteoblasts (in the lamina limitans).[55]

In vitro, recombinant, nonphosphorylated DMP1 is a bone mineral (apatite) nucleator, which when partially enzymatically phosphorylated has no effect on apatite formation. A human DMP1 expressed in bovine marrow stromal cells, and therefore, probably phosphorylated similar to the native protein, is an effective inhibitor of mineral formation and proliferation in solution.[55a]

Overexpression of DMP1 leads to the differentiation of mesenchymal cells into osteoblasts[56] and enhanced mineralization. Thus, in culture, DMP1 appears to act as a signaling molecule, activating osteoblasts. The knockout animal, developed by Feng, has an interesting phenotype. The cortices in the knockout animals' bones are thinner than those of the wildtype; but the epiphyses are excessively calcified [Jian Feng, personal communication]. Since DMP1 expression is greatest in hypertrophic chondrocytes,[53] this supports the view that the native protein is an inhibitor of mineralization. However, the thinned cortical bone agrees with its importance for determining the osteoblast patterning, indicating that DMP1 like the other SIBLING proteins is multifunctional.

Dentin sialoprotein

Another dentin protein that is also found in bone[57] DSP (MW ~53 kDa) is only slightly phosphorylated (less than one of its two phosphorylation sites is occupied); but it too, like the other SIBLING family members, has an RGD sequence and is N- and O-glycosylated. DSP is expressed as part of the dspp gene;[58] the other product is dentin phosphophoryn (DPP), a highly phosphorylated protein that is unique to dentin. However, DSP is also found in bone, and although initial studies suggested that DPP was not in bone, it does raise some questions.

Structural studies indicate that DSP is in a random coil conformation in solution, but like the other SIBLINGs it binds to apatite, although not with as high an affinity as the other family members.[59] *In vitro*, DSP can inhibit mineral formation and growth, but its activity is nowhere as strong as that of osteopontin.[60] DPP, in contrast, can act both as an apatite nucleator and an inhibitor of mineralization.[61] It has been postulated that the intact gene product of dspp may be a potent regulator of mineralization in both bones and teeth;[62] however, the intact protein has not yet been investigated.

There are no human bone diseases associated with dspp gene alterations; however, improper dentin formation associated with opalescent teeth and dentin attrition (dentinogenesis imperfecta type II) has been linked to mutations in this protein.[63] Patients with this dental disorder sometimes have a high-frequency hearing loss,[64] which might suggest a bone phenotype, as hearing loss is a feature of osteogenesis imperfecta,[65] and demineralization of the cochlear capsule in the ear is associated with hearing loss.[66] Further analyses will obviously be required to discern the importance of both DSP and dssp for bone mineralization.

MEPE

Matrix extracellular phosphoglycoprotein was first cloned[67] from tumors of patients with oncogenic hypophosphatemic osteomalacia. Similar to the other SIBLINGs on human chromosome 4q, this 525-residue protein (431 residues in the mouse) has one RGD domain, 2 N-glycosylation motifs, and a short N-terminal signal peptide.[67] Low levels of expression occur in the bone marrow, while high levels are associated with these tumors. The gene and the protein, also named "osteoregulin," are also expressed in osteoblast cultures, associated with matrix mineralization.[68] Patients with x-linked hypophosphatemic rickets, and mice that model this disease associated with an inactivating mutation in Phex (an endopeptidase with no known substrate that regulates tissue phosphate levels[69] have elevated levels of MEPE. Further, Phex inhibits hydrolysis of MEPE, but does not interact with it, indicating that this pathway is complicated.[70]

Other non-SIBLING phosphorylated glycoproteins

Bone acidic glycoprotein -75. BAG-75 shares extensive structural homology with dentin matrix protein -1 and with the other SIBLING molecules, but is considered separately, as its chromosomal localization has not yet been identified. In tissue sections, it does not colocalize with DMP1,[52] and hence BAG-75 is a unique molecule with distinctive features, which may be quite important for its function in bone, and may explain why it has not been as extensively studied as the other SIBLING proteins.

First identified by Gorski in 1988,[71] BAG-75 is an ~75,000 MW phosphorylated glycoprotein with 44 phosphoryl groups per molecule, ~7% sialic acid, and 30 mole percent acidic amino acids. It tends to form macromolecular aggregates in solution,[72] a feature unique to this among the other SIBLING proteins, and a complication that has limited *in vitro* studies of its function. Among its unique functional properties is its ability to inhibit osteoclast activity.[73] The mechanism of this inhibition is via interaction with a newly described osteoclast membrane receptor, Galectin-3.[74] Similar to the other phosphorylated glycoproteins, BAG-75 has a high affinity for calcium, collagen, and bone mineral. Its affinity for calcium is 1.7 times higher (mole/mole) than that of BSP and 2.8 times higher than that of osteopontin. Additionally, it undergoes Ca-dependent changes in its conformation, similar to that noted above for osteopontin.[75] BAG-75 has been localized to newly forming mineralized nodules in both normal and pathologic calcifications.[76] Thus, this protein may be important both for the initiation of mineralization on the collagenous matrix and for protecting the newly formed mineral from being removed too rapidly [Gorski, Jeffrey, personal communication].

Osteonectin. Osteonectin[25] is another widely distributed glycosylated phosphoprotein, found in endothelial cells, thrombocytes, and placenta, as well as in the mineralized tissues. It is also called secreted protein, acidic, cysteine-rich (SPARC), and BM-40. Not being on human chromosome q4,[77] osteonectin is not really in the SIBLING family, although it is often discussed with the SIBLINGs. Osteonectin, molecular mass ~32 kDa, has 1 RGD domain, a calcium-binding domain,[78] and multiple potential phosphorylation sites. In bone, it accounts for ~23% of the total noncollagenous protein. It has a calcium-binding EF-hand structure,[79] not seen in the SIBLINGs.

In vitro, osteonectin binds to collagen; the binding occurs through the Ca-binding domain as shown by a recent crystal structure determination of the isolated domains' interactions with both type IV and type I collagen.[80] Like many of the other matrix proteins, it has been reported to enhance mineral deposition,[25] and to be an inhibitor of mineral crystal growth.[81] As will be discussed below, these two functions are not contradictory in terms of calcification mechanisms.

Osteonectin has also been implicated *in vitro* as having a role in regulating the activities of the metalloproteases that control the degradation of the extracellular matrix,[82] although osteonectin itself is modulated by these proteases.[83] These activities are consistent with some of the observations in the knockout mice. The first osteonectin-deficient mice generated by Gilmour and Norose[84] were found to develop cataracts early during development, but not to have any early skeletal phenotype. They were also later shown to have impaired wound healing.[85] At 11 weeks, the cortices of their bones were thinned, and their trabecular spacing had increased (Figure 4.5).

Older osteonectin-deficient mice had more fragile bones at 17 weeks of age than the wildtype ones, and they had altered rates of bone formation and bone remodeling.[86] With age, the trabecular spacing in these animals increased significantly and few trabeculae were visible by 35 weeks of age. Using infrared imaging, we characterized the bones of the knockout and wild-type animals at 11, 17, and 35 weeks, and found that while bone shape was not different, the osteonectin-deficient animals had a higher mineral content and larger/more perfect crystalline mineral than age-matched controls. Most importantly, the collagen maturity (an index of the ratio of nonreducible to reducible cross-links) was much higher throughout the bones of the knockout animals.[87]

There are no human diseases to which osteonectin mutations have been linked by genetic analyses; however, osteoblasts from some fibrous dysplasia patients express elevated levels of osteonectin in addition to other proteins,[88] and there are markedly reduced levels of osteonectin in bone from patients with osteogenesis imperfecta.[89] In this light, it is of interest to note that the *fro/fro* mouse, a naturally occurring mutant mouse with brittle bones reminiscent of osteogenesis imperfecta, has a 50% reduction in osteonectin expression,[18] all demonstrating that osteonectin has a marked, albeit as yet undefined effect on bone.

Vitronectin and tetranectin. These proteins are glycoproteins that are expressed during mineralization, but direct roles for vitronectin in the mineralization process have not been established.

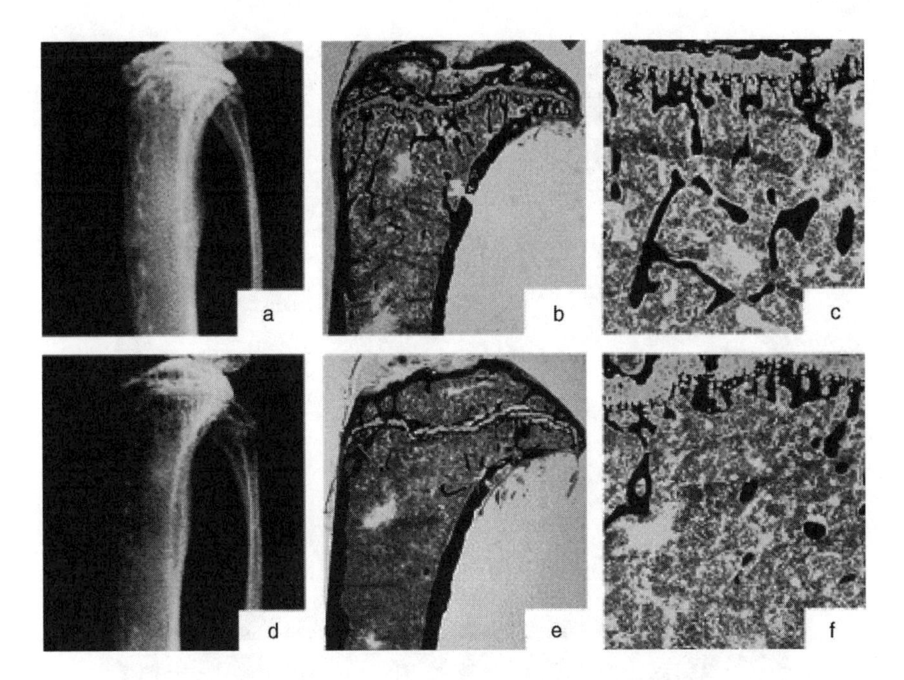

Figure 4.5 The bones of osteonectin-deficient animals show trabecular thinning and decreased bone formation. Radiographs and von Kiss stained histological sections from the tibia of the wild-type (a,b,c) and osteonectin knockout animal (d, e, f) at 11 weeks. (From Delaney, A. et al., Osteopenia and decreased bone formation in osteonectin-deficient mice, *J. Clin. Invest.*, 105(7), 915–923, 2000. With permission.)

Tetranectin is immunolocalized in developing woven bone and is expressed by osteoblastic cultures during matrix mineralization.[90] Overexpression of tetranectin by tumor cells implanted into nude mice caused an increase in matrix mineralization,[90] while downregulation by vitamin A prevents tetranectin expression and blocks mineralization,[91] suggesting that tetranectin may play a role in mineral deposition. The tetranectin knockout mouse has an interesting skeletal phenotype, with severe spinal deformities (Figure 4.6), but no other visible bone abnormalities.[92]

Gla-proteins

Proteins containing gamma-carboxyglutamic acid (gla) residues are found in bone and cartilage, as well as in proteins associated with blood clotting. Gla is formed by a vitamin K-dependent post-translational modification,[93] explaining both why vitamin K supplements have been proposed as therapies for improving bone quality,[94] and why abnormal bone forms when warfin, an inhibitor of the vitamin K-dependent carboxylase, is ingested.[95]

The two vitamin K-dependent GLA proteins in bone are osteocalcin (also known as bone gla protein) and matrix gla protein.[96] Bone gla protein is the most abundant noncollagenous protein in most species' bones, while matrix gla protein is more abundant in soft tissues, such as cartilage and blood vessels. In humans, these proteins contain 2 or 3, and 5 gla residues, respectively.[97] There is a good deal of

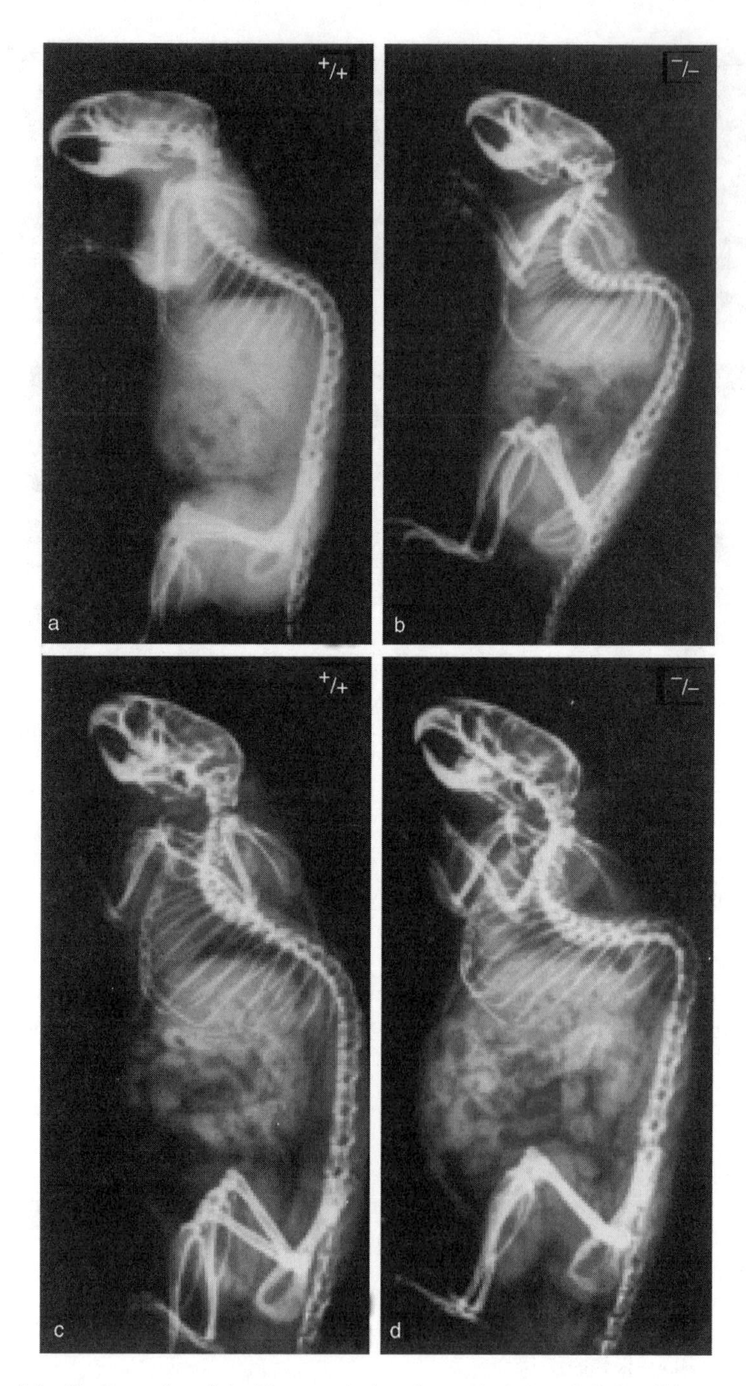

Figure 4.6 Radiographs of the Tetranectin knockouts (−/−) and their wild-type controls (+/+) show severe spinal deformities at 6 (A,B) and 12(C,D) months. (Reproduced with permission from Iba, K. et al., *Molecular and Cellular Biology*, 21:7817, 2001. © 2001 American Society for Microbiology.)

sequence homology between the osteocalcin and MGP genes, and no obvious indication of why one should be so insoluble, except for the single S–S bond in the MGP, which may alter its conformation.

Matrix gla protein (MGP)

Matrix gla protein, MGP, an 84 residue protein, is extremely hydrophobic. The insolubility of matrix gla protein for a long period of time prevented analysis of its function. However, recently a method was developed for synthesizing the protein *de novo*,[98] and this should lead to more studies of the protein's mechanism of action.

The protein is found in cartilage, smooth muscle cells, and the matrix adjacent to sites of vascular calcification.[99] Knockout animals lacking the matrix gla protein gene die prematurely of excessive calcification of their blood vessels and trachea.[100] Their growth and articular cartilage as well as their bones were also excessively mineralized,[101] suggesting that matrix gla protein is a mineralization inhibitor, which normally prevents the soft tissues in which it is expressed from becoming calcified. In cell culture systems, ablating of the matrix gla protein gene similarly caused excessive calcification, while overexpression of this gene led to a lack of calcification in this system.[102] Patients with Keutel's syndrome, a disease characterized by excessive cartilage calcification, have mutations in the MGP gene.[103] Similarly, mutations in the MGP gene are weakly associated with vascular calcification.[104] All these data support the view that MGP is a calcification inhibitor.

Osteocalcin

Osteocalcin,[105] also known as bone gla protein (BGP), accounts for ~15% of the noncollagenous protein pool. In contrast to MGP, osteocalcin is water-soluble. In fact, serum osteocalcin levels are used frequently as a clinical measure of bone formation. Osteocalcin is a relatively small protein of molecular weight 5.3 kDa. It has one intramolecular disulfide bond and three to five residues of γ-carboxy glutamic acid depending on the species.

Osteocalcin in a cell-free solution inhibits bone mineral crystal formation and crystal growth.[81] Its action is dependent on the presence of three γ-carboxy-glutamate residues, as decarboxylated osteocalcin has no ability to bind to mineral crystals,[106] and the variant of osteocalcin with only two γ-carboxy residues did not bind calcium.

Solution-based structural studies predicted that osteocalcin would bind to specific faces on the surfaces of the bone mineral crystal,[107] such that the C-terminal residues were unattached. This C-terminal domain was hypothesized to be chemotactic for osteoclasts,[108] a hypothesis that was verified by a series of cell culture and implantation studies.[109] It thus appeared from *in vitro* data that osteocalcin was both a regulator of bone mineral formation and growth, and by virtue of its interaction with bone mineral crystals it could regulate the remodeling of the bone mineral.

The knockout animal, which lacked all four osteocalcin genes,[110] however, did not show the expected characteristics, nor did it have a severe skeletal phenotype. There was increased bone formation, and osteoclast-mediated resorption was not impaired, even in animals that were ovariectomized to stimulate osteoclastic remodeling. However, in the rodent, osteoclastic resorption is not extensive; thus, the small change in osteoclast number, although not significant, may have indicated a trend.

The most striking feature of the knockout animals was the increased width of their bones, a feature interpreted as an indication of osteocalcin-regulated bone formation.[110] Detailed mineral analyses of the bones of the osteocalcin knockout mice and their age- and sex-matched controls showed that the bone mineral crystals, although more abundant, were not maturing,[111] consistent with the hypothesized role for osteocalcin in recruiting osteoclasts to remodel bone. Since the greatest changes were observed in the cortices of these bones and in the center of their trabeculae, it was argued that there was impairment of bone remodeling. During the remodeling cycle, osteoclasts cause the local release of acid, which dissolves the bone and enzymes, such as MMP-13 (collagenase 3), that degrade the matrix. Were osteoclasts not functioning properly because they could not bind to the mineral (due to the absence of a chemotactic factor), then the mineral would not be remodeled, and therefore would not mature. This situation is similar to what is observed in some forms of osteopetrosis in humans and other animals, although in osteopetrosis it is the osteoclast that is not functional. These data then agreed with the data derived from the *in vitro* studies, suggesting that by binding to the mineral surface, osteocalcin was acting as a chemotactic factor for osteoclasts, and that this function was more important in the living animal than its ability to coat the crystal and retard mineral growth and formation.

Human mutations of osteocalcin gene associated with bone diseases have not been reported. From the above data it appears that osteocalcin, although a useful marker of bone formation and an extremely useful marker when manipulating osteoblastic-specific protein expression (see below), may not be critical for bone formation.

Proteoglycans

Proteoglycans are proteins with glycosaminoglycan (GAG) side chains attached covalently. Generally, the GAGs are long chains of repeating disaccharides that are frequently sulfated. The proteoglycan family members that are found in bone can be divided into large aggregating proteoglycans, the small leucine-rich proteoglycans (SLRPs), the heparan sulfate proteoglycans, and some other small proteoglycans. Proteoglycan core proteins are post-translationally modified in several different ways. The core proteins are glycosylated by the addition of N- and O-linked oligosaccharides. In addition to sulfation of the GAG chains, the oligosaccharides can be phosphorylated and/or sulfated.

Large aggregating proteoglycans

The large aggregating proteoglycans have molecular masses exceeding 400 kDa.[112] The core proteins often have 40 or more GAG chains attached, and they aggregate by noncovalent interactions with hyaluronan, forming higher-molecular-weight species (proteoglycan aggregates) that are stabilized by link protein. The major noncollagenous protein of articular cartilage, aggrecan, with chondroitin and keratin sulfate GAG chains persist in bone.[113] However, the major aggregating proteoglycan in bone is versican, which is a chondroitin sulfate proteoglycan whose core protein structure is distinct from that of aggrecan, and has fewer (~12–15 GAG chains) in contrast to the ~100 in aggrecan.[114] Versican from soft tissues and from dental pulp is capable of forming aggregates, but that from bone does not form aggregates *in vitro*.[115] Versican seems to be important for matrix organization, and for promoting cell motility and growth. In culture, it is localized to mineralizing bone nodules.

The aggregating proteoglycans from cartilage are effective inhibitors of mineral deposition, both because of their ability to sequester calcium, and the steric hindrance that they impose.[115] The effects of versican on mineral deposition *in vitro* have not yet been reported. However it is important to note that with age, in normal bone, the amount of bone proteoglycans decreases and the relative proportion of versican decreases. In contrast, in patients with osteogenesis imperfecta, independent of patient age, the proteoglycan distribution remains similar to that in fetal bone, but the proteoglycan content is reduced to that of older individuals.[116] Thus, it may be suggested that proper bone development initially requires a higher proportion of aggregating proteoglycans.

SLRPs

The small leucine-rich proteoglycans (SLRPs) consist of a protein core, tandem leucine repeats bounded by cysteine residues,[117] and a few N-linked glycosaminoglycan chains. The SLRPs all bind collagen, growth factors, and other matrix molecules. *In vitro*, as a class, they regulate collagen fibrillogenesis, as well as matrix organization. In bone, they include decorin, biglycan, lumican, fibromodulin, and osteoadherin.[115]

Knockout mice deficient in decorin, biglycan, lumican, and fibromodulin were generated by Marion Young.[118] All have altered collagen fibrils, indicating the role of these small proteoglycans in determining the shape and the mechanical properties of the matrix. In bone, biglycan is the most abundant of the SLRPs, and knockout mice lacking this protein have osteopenia and defective bone growth,[119] while the double decorin biglycan knockout shows a complete loss of collagen fibril integrity.[120] The lumican decorin double knockout shows structural and mechanical abnormalities in the tendon collagen fibril, and ectopic calcification.[121] Further characterization is in progress.

In addition to the effects on collagen fibrillogenesis and growth factor release, the SLRPs can affect bone mineral formation *in vitro*. In bone and dentin, biglycan and decorin are chondroitin sulfate proteoglycans, whereas in cartilage they are dermatan sulfate proteoglycans. Lumican and fibromodulin are keratan sulfate proteoglycans in both tissues. The keratan sulfate proteoglycans tend to accumulate where mineralization commences.[122] In contrast, the mineralized bone matrix has decreased amounts of decorin relative to the unmineralized matrix.[123] This is in agreement with early reports[124] showing proteoglycans disappearing from the site of initial mineral formation in collagen fibrils as mineralization commenced. Differences in the expression and distribution profiles of decorin and fibromodulin suggest that fibromodulin, which is deposited closer to cells than decorin, may have a primary role in collagen fibrillogenesis, and decorin may have a role in the maintenance of fibril structures.[125] Similarly, the distribution of lumican and fibromodulin in the tooth suggested that these SLRPs have a role in the regulation of mineralization.[126]

In vitro, in the absence of cells, biglycan can function as a bone mineral nucleator in the absence of cells, while decorin is an inhibitor.[127] There are no reports on the effects of the other SLRPs on cell-free mineralization. In cell culture, increased lumican expression is associated with maturation of the bone matrix, but not directly with mineralization,[128] whereas calcification is associated with increased turnover of biglycan and decorin.[129] Interestingly, immunohistochemistry in another mineralizing tissue, dentin, shows a gradient of proteoglycans with decorin increasing toward the mineralization front and biglycan decreasing.[126]

Osteoadherin is predominantly a product of osteoblasts,[130] and is expressed at the same sites as the bone sialoproteins.[131] It is a cell-binding protein that is associated with mineralization that binds cells more efficiently than fibronectin.[130] Osteoadherin has a molecular weight of 85,000, is rich in aspartic and glutamic acids, and has a high affinity for bone mineral,[130] but its effects on bone formation in cell culture or on mineralization in the absence of cells have not yet been described.

The human diseases associated with SLRP mutations are usually related to abnormal collagen formation, for example, ulcer formation in Ehler's Danlos syndrome,[132] progeroid syndromes,[133] Kleinfelter's syndrome (associated with overexpression of biglycan), and x-linked Turner syndrome (underexpression of biglycan).[134] Human data suggest that the cell culture predictions that the SLRPs are most important for regulated collagen fibrillogenesis, and hence, indirectly, collagen-based mineralization, are probably valid.

Perlecan and the heparan sulfate proteoglycans

The heparin sulfate proteoglycans are found associated with almost all cell membranes. In bone, there are three heparan sulfate proteoglycans present on osteoblasts. The most predominant type is similar to the syndecan family.[135] The intact syndecan molecule has a molecular weight of ~400 kDa, with an 89 kDa core protein and several heparin sulfate side chains. Betaglycan, a helper receptor for the type I and type II TGF β receptors, is also present in bone cells,[136] and interestingly, the number of these receptors, but not their activity, is increased in patients with osteogenesis imperfecta.[136] Thus, these receptors are responsive to changes in collagen structure and/or mineralization.

Perlecan, the large basement membrane heparan sulfate proteoglycan (HSPG2) is expressed mainly in cartilage, but when absent there are visible bone abnormalities because of disruption of the epiphysial growth plate.[137] Mutations in the perlecan core protein gene[138] are associated with dyssegmental dysplasia, a disease characterized by neonatal short-limbed dwarfism, which is lethal in its more severe forms. Perlecan knockout mice have a similar phenotype.[139] So named because of its "beads on a string" appearance when observed by rotary shadowing under the electron microscope,[140] perlecan has a 467 kDa core protein, with five distinct domains.[141] Domain one has three GAG chains associated with it, which account for 50% of the molecular mass. It binds growth factors, mediates cell attachment, and interacts with other molecules. Its precise function in bone has not yet been addressed.

Glycoproteins

Glycoproteins have covalentently linked sugars attached to the protein via asparaginyl or seryl residues. Distinct from proteoglycans where the majority of the protein mass is sugars, the glycoproteins are predominantly protein. The SIBLINGs, tetranectin and osteonectin, discussed above are glycoproteins. Alkaline phosphatase, an enzyme crucial to the mineralization process, will not be considered here, as it is not truly a matrix molecule. More details on the role of alkaline phosphatase in mineralization can be found in a recent review[115] and in studies of the tissue nonspecific alkaline phosphatase knockout mouse.[142] As noted above, the glycoproteins may be further post-translationally modified by sulfation and phosphorylation.

Thrombospondin

Thrombospondin is a homotrimeric disulfide-linked glycoprotein (MW ~450 kDa) found in the extracellular matrix of many tissues. Its main function is thought to be the regulation of angiogenesis. Thrombospondin-1 knockout mice do not show abnormal angiogenesis, but at birth they do show severe spinal deformities along with impaired lung development.[143] Overexpression of thrombospondin-1 was associated with prevention of tumor growth.[144] Both of these findings could be associated with the regulation of matrix metalloproteinase. The thrombospondin-2 knockout mice[145] have a distinct bone phenotype with increased cortical bone thickness. Their skin has abnormal collagen fibrils[146] and their marrow stromal cells do not differentiate normally.[147] Nethier the detailed mineral analysis of these bones nor the effect of either of these thrombospondins on bone formation *in vitro* has been reported.

Fibronectin

Fibronectin is one of the most abundant extracellular matrix proteins. All connective cells produce it. Fibronectin is one of the first noncollagenous proteins to be expressed during bone development,[115] where it appears to be important for organizing the matrix.[148] Fibronectin is a large dimeric protein (molecular weight 400 kDa) that is composed of two homologous subunits, held together by two disulfide bonds near the carboxy termini. Each subunit has multiple domains that bind to fibrin, heparin, collagen, and cell surfaces. No human bone diseases have been associated with fibronectin mutations, and the tissue protein has not been knocked out, although the serum protein has.

In solution, fibronectin inhibits bone mineral (hydroxyapatite) crystal growth.[149] However, it promoted mineralization when hydroxyapatite crystals were present.[149] Since fibronectin has a high affinity for hydroxyapatite,[150] these effects, as will be discussed in section IV, are probably due to the interaction with the mineral crystals.

Fibrillin

Fibrillin is the principal component of the microfibrillar aggregates in the skeleton. It is a cell-binding (RGD-containing), calcium-binding, cysteine-rich protein with a cysteine-poor carboxyl terminus. Defects in fibrillin lead to Marfan's syndrome, a disease characterized by skeletal, ocular, and cardiovascular abnormalities.[151] It has not been shown to have a direct effect on bone formation or mineralization, but seems to be important in patterning of the skeleton. Analysis of the enzyme resistance of fibrillin in Marfan's patients demonstrated that it was more easily degraded, providing an explanation for the musculoskeletal tissue phenotype.[152]

LIPIDS

The bone matrix contains less than 2–4% lipid,[153] much of which are protein associated. However, these lipids are extremely important for bone metabolism, and for bone mineralization. The majority of these lipids are derived from cells. The cell membrane lipids are important for bone cell function, as they determine fluxes of components into and out of the cell.[153] They are also the "sea" in which the receptors that regulate cell function are floating.[154] The most important of these are the caveolae, or lipid rafts, which modulate signaling and membrane function.[155] In addition, there are lipids within the extracellular matrix that appear to be associated with the collagen,[156] and the membrane-bound extracellular matrix vesicles (ECMVs). ECMVs are the site of initial calcification in cartilage, mantle dentin, and some bone.[157,158]

Wuthier initially identified a phospholipid complex, consisting of acidic phospholipids, calcium, and inorganic phosphate in mineralizing tumors.[159] It was subsequently shown by Boskey et al.[160] that these complexed acidic phospholipids were found in actively mineralizing tissues, and were present to lesser extents in tissues that were already mineralized. These lipids are not found in tissues that do not undergo physiologic calcification[161] however, they are present in mineralizing tissues prior to the onset of mineralization.[162] This includes tissues that normally mineralize (bone and hypertrophic cartilage[115]) and those that undergo dystrophic calcification.[163]

In vitro[164,165]and when implanted in a muscle pouch *in vivo*,[166] the complexed acidic phospholipids induce hydroxyapatite (bone mineral) formation in a dose-dependent manner, which can be altered by addition of foreign ions or other proteins. Wuthier's group first identified these complexed acidic phospholipids as components of ECMVs in 1977.[167] More detailed structural work by Wuthier's group has identified the complexed acidic phospholipids as components of the "nucleational core" of these matrix vesicles.[168] Wuthier has shown that the "nucleational core" consists of complexed acidic phospholipids and the proteins annexin and type II collagen.[169] However, both the complexed acidic phospholipids and the "nucleational core" with proteins are capable of nucleating mineral formation. They can also regulate the growth of preformed mineral crystals.[170]

There are some diseases where complexed acidic phospholipids formation and/or matrix vesicle formation are impaired. These include avian tibial dyschondroplasia,[171] a condition of impaired mineralization, and decreased ECMVs. The lpr/lpr lupus mouse, which has decreased apoptosis, consequently decreased ECMVs, and moderately impaired bone development [Gokhale Jashree, unpublished data]. In contrast, for example, rachitic animals[172] have elevated contents because their tissues are primed for mineralization, but they do not have sufficient extracellular calcium (and/or phosphate) to support mineralization.

In addition to acidic phospholipids, other lipid constituents of bone seem to be important for bone cell function. Perhaps this is why leptin, originally thought to be only a fat cell component, may be so important for the stimulation of osteoblast (bone formation) function and the suppression of osteoclast (bone resorbing) activity.[173] The ability of fat cells to produce anabolic signals and their ability to respond to osteogenic signals may explain why fat cells (adipocytes) have been used as osteoblast precursors by tissue engineers.[174]

Recently, with better analytical techniques, it has been recognized that many microorganisms are capable of inducing mineralization.[175] Most probably, this involves their membranes and membrane-associated proteins, and while not necessarily relevant for engineering bone, it is important to realize that bacteria as well as degraded cell membranes can cause mineralization — either wanted or unwanted.

ENZYMES AND CYTOKINES

The enzymes and growth factors that are associated with bone formation are reviewed in detail elsewhere in this book, both in Chapters 3 and 5. Because these are protein components found both in the bone cells and in the extracellular matrix, they are mentioned here for completion, but readers are referred to elsewhere in the text for more details.

APPLICATIONS OF BONE MATRIX PROTEINS IN TISSUE ENGINEERING

It should be apparent from the above sections that most of the functions of the bone matrix constituents are redundant, and that in many cases the true functions are yet to be elucidated. The presence of collagen or a collagen-like template is obviously essential for "engineering" a bone, but which of the other constituents are essential, and which could be omitted in a mimetic bone substitute? One way of avoiding this question is to use a scaffold that will stimulate bone cells to make a functional tissue. The other way is to consider the importance of the mineral, and perhaps use that mineral to stimulate bone formation. Thus, it is essential to understand the nature of the mineral, how it forms, and what controls and modulates its formation.

There have been several reports of the use of the SIBLING proteins or peptides derived from them to enhance cell binding in association with a variety of different scaffolds in tissue engineering applications.[176–179] An interesting chimeric protein consisting of decorin and domains known to interact with minerals in BSP has been used to direct new mineralization toward collagen fibrils, via the decorin interaction with collagen and BSP's nucleating ability.[179] Most of these used peptides and proteins with integrin-binding RGD sequences to enhance cell binding to scaffolds, but an interesting study by Wang et al [Wang, Jinxi, personal communication] used BSP without any carrier to cure critical-sized calvarial defects. Others used cytokines bound to matrices to increase cell recruitment, the most frequently used growth factor being the BMPs, either by inclusion of the cytokine on the implanted scaffold[180] or by gene delivery of the growth factor.[181] There has also been some attempt to use other matrix constituents, for example, lipids, fibronectin, and carbohydrates in bone tissue engineering.[182,183]

Several investigators have used information about the structure of the matrix proteins to derive artificial matrices. For example, taking advantage of the polyanionic character of the matrix proteins. Murphy and Mooney[184] prepared a biomimetic alpha-hydroxy ester polymer that induced mineral deposition in an *in vitro* system. One of the most exciting examples of a biomimetic structure that induces apatite formation emerged from Sam Stupp's group. Using a self-assembly approach, they used a disulfide-linked nano-fibril with associated acidic phosphate groups modeled after the dentin matrix protein, phosphophoryn, to induce bone-mineral-like deposition *in situ*[185] (Figure 4.7). The matrix formed was fibrillar; the mineral crystals that formed were oriented along the axis of the fibrils.

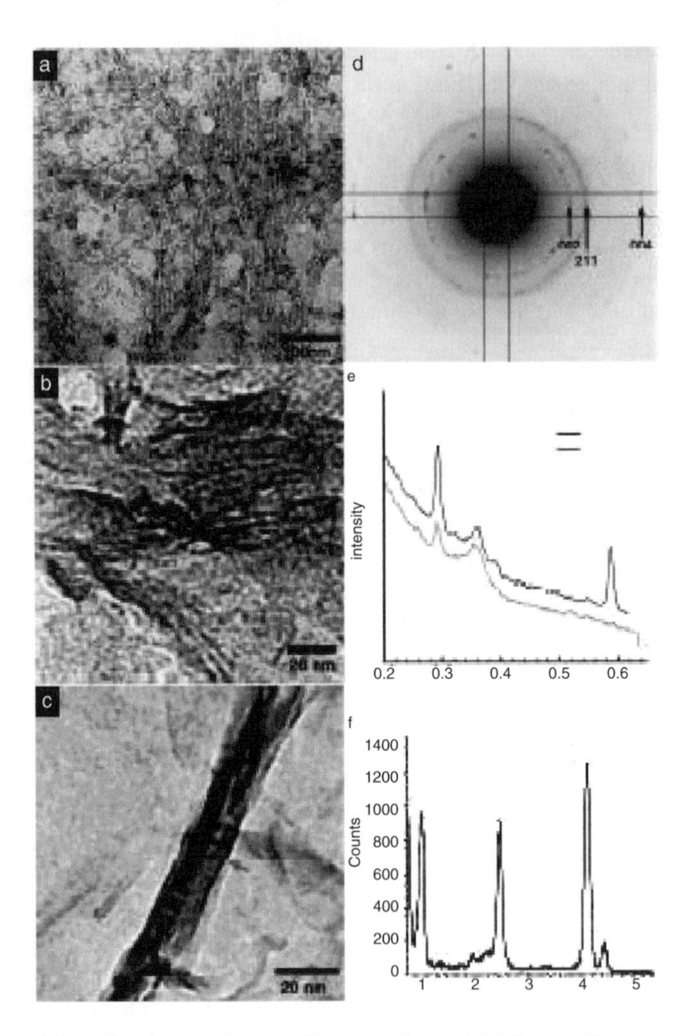

Figure 4.7 Mineralization of self-assembled peptide-amphiphile nanofibers, from reference 293, figure 3, shows electron micrographs of the fibers incubated in a mineralizing solution for 10 min (a), 20 min (b), and 30 min (c), showing mineral crystals (arrows) forming along the fibers. The 30-min samples were examined by electron diffraction (d and e), and by EDS (f) to demonstrate the apatitic nature of the mineral (d and f), and the preferred orientation of the mineral crystals along the fiber axis (e). (Reprinted with permission from Self-assembly and mineralization of peptide-amphile nanofibers. Hargerink, Beniash, and Stupp, *Science* 294:1684, 2001, Copyright 2001, American Association for the Advancement of Science.)

All of these tissue-engineered constructs, whether they are scaffold and matrix based or scaffold and cell based, have as an ultimate goal the formation of a functional bone, which resembles that found in healthy individuals. This requires that the load-bearing properties are retained. This in turn requires appropriate matrix mineralization. Hence, a detailed understanding of the mineralized matrix and how it is formed is essential.

THE MINERAL COMPONENT OF BONE

The Chemical Nature of Bone Mineral

Bone mineral, as reviewed in detail elsewhere, was recognized in the 1920s as being an analog of the naturally occurring geologic mineral, hydroxyapatite.[115] The unit cell of crystalline hydroxyapatite has the chemical formula $Ca_{10}(PO_4)_6(OH)_2$; however, analysis of bone mineral shows a Ca:P molar ratio ranging from 1.3:1 to 1.9:1. This, in part, is due to the contribution of the organic phosphate in the bone matrix to this ratio, but is also related to the nature of the bone mineral itself. It is now recognized that bone mineral is a hydroxyl-deficient, calcium-deficient, carbonated apatite. The crystals of bone mineral are small, and combined with their small size these crystal imperfections make the x-ray diffraction pattern of bone broader than that of synthetic apatites.

Mechanism of Bone Mineralization

The process of cell-mediated biomineralization has been extensively reviewed elsewhere.[115] As seen from these reviews of mineral formation in a wide range of species, the common features of biomineralization in each species, ranging from the newly described copper silicates[186] to the calcium carbonates and phosphates that are much more common in the exo- and endo-skeleton, the process is cell mediated. The mineral deposits in an oriented fashion on a template either inside (lower species) or outside the cell. Matrix proteins, generally anionic, as were discussed above, act as nucleators and/or regulators of the mineralization process.

Nucleation is an energy-requiring physico-chemical process through which the first stable crystalline material forms. Ions or ion clusters colliding in the solution form metastable structures that persist into the solution until the "critical" size or a stable "critical nucleus" is formed. The energy required for critical nucleus far

Table 4.1
Connective Tissue Matrix Components That Both Nucleate and Retard Apatite Crystal Formation in a Cell-Free Solution

SIBLINGs	
	Osteopontin [193]
	Bone Sialoprotein [49]
	Dentin Matrix Protein-1 [55a]
PROTEOGLYCANS	
	Biglycan [127]
GLYCOPROTEINS	
	Phosphophoryn [61]
	Osteonectin [25, 81]
	Fibronectin [149]
LIPIDS	
	Complexed Acidic Phospholipids [170]

exceeds that for crystal growth, which occurs as ions or ion clusters are added to the critical nucleus. Nucleation may occur *de novo* due to an increase in the ion concentration, an increase in temperature, or a change in solution composition. In fact, nucleation is favored by increases in concentration, which may be modulated by cellular efflux of calcium and phosphate ions or local increases in pH (increases in hydroxide ion concentration). This leads to the accumulation of larger ion clusters.[187] Such "homogeneous nucleation" is quite rare — most nucleation studied in the laboratory takes place on dust particles, scratches on the walls of containers, burette tips, etc.[188]

In contrast, heterogeneous nucleation occurs on already formed surfaces, and a subclass of this epitaxial nucleation occurs when the foreign surface matches one of the surfaces of the forming crystal.[189] After stable crystal nuclei are formed, they can grow in dimension by the addition of more ions or ion clusters, by secondary nucleation, in which new growth sites are formed on the surface of the existing crystals, or by agglomeration. Proteins and lipids participate in these processes as "heteroge-

Table 4.2
Summary of Functions of Bone Matrix Proteins Validated in Culture, Cell-Free Solutions, Transgenic Animals, and Human Diseases

Protein	Function in Vitro	Function in Transgenic or Mutant Animals	Human diseases
Collagen I	Template for mineral deposition; tissue elasticity	Regulates deposition of mineral & tissue mechanical properties	Osteogenesis Imperfecta
Osteopontin	Osteoclast recruitment; Mineralization inhibitor	Osteoclast recruitment; Mineralization inhibitor	?
BSP	Mineral nucleators; Cell signal	?	Elevated in end stage OA
DMP-1	Signaling molecule; Mineralization regulator	Signaling molecule; Mineralization regulator	?
Osteonectin	Metalloprotease regulator; Mineralization regulator	Metalloprotease regulator; Mineralization regulator	?
DSP	Regulator of Mineralization	?	?
Osteocalcin	Recruit osteoclasts; regulate mineralization	Increase bone diameter; Prevent mineral maturation	?
MGP	Prevents mineralization	Prevents mineralization	Keutel's Syndrome; Vascular calcification
Biglycan	Binds growth factors; mineral nucleator	Regulates collagen diameter; mineral nucleator	Kleinfelter's syndrome: Turner's syndrome
Decorin	Regulate collagen fibrillogenesis	Regulates collagen fibrillogenesis	?
Aggrecan	Matrix organization; inhibit calcification	?	?
Tetranectin	Regulation of mineral deposition	Regulation of mineral deposition	?

See text for full description of these proteins. ? = data not available.

neous nucleators," as chelators that control the accumulation of ions, and by forming protected environments in which mineral ions can accumulate protected from inhibitors of mineralization in solution.[190] The protected environment has been proposed as the mechanism of extracellular matrix vesicles in initial calcium phosphate mineral deposition and of the vesicles in iron oxide-forming species.[10]

During bone formation, while ECMVs have been suggested to be the site of initial mineral accumulation, collagen-based mineralization appears to occur concurrently.[191] The vesicles are thought to provide protected sites for mineral formation, and to provide enzymes that assist in the modification of inhibitors in the extravesicular matrix.[192] Initial mineralization occurs at many sites along the collagen matrix. It is the anionic noncollagenous matrix proteins that are believed to be acting as nucleators of initial apatite formation. Once the first mineral crystals are formed at these discrete sites, they grow by a process of lengthening, and agglomeration. Since the sizes of the mineral crystals in bone occur over a fairly narrow range, it is believed that both the spacing between the collagen fibrils and the matrix proteins that act as inhibitors regulate the size and shape of these mineral crystals. The same matrix proteins that bind to and stabilize the first formed nuclei, making the process of mineral proliferation energetically more favorable, can also coat these crystals and determine the shape the crystals take, and the extent to which they can grow. Table 4.1 lists those proteins that have been shown to be capable of BOTH acting as nucleators (initiators) of biomineralization and regulators (inhibitors) of mineral crystal proliferation. Proof of the participation of these proteins in both processes has come from both *in vitro* studies and the characterization of the mineral in transgenic and naturally occurring mutant animals, as delineated in Table 4.2.

CONCLUSIONS

Deposition of mineral in bone during development and remodeling is a complex process that involves the cell, the organic extracellular matrix, and physicochemical processes. Bone mineral optimally has a broad range of compositions and sizes, but the mineral crystals are always associated with the collagen matrix. Mimicking the physiologic mineralization process during tissue engineering will require the determination of the following: how cells interact with the matrix; how matrix molecules interact with one another; which matrix proteins and lipids are mandatory for the formation of the optimal mineralized matrix; and which of these proteins and lipids can be replaced with an alternative natural or mimetic material for the development of a functional bone.

ACKNOWLEDGEMENTS

Dr. Boskey's work as reported in this review was supported by NIH grants DE04141, AR037661, and AR041325. Dr. Boskey is grateful to all her collaborators for providing the proteins and knockout animals used for analysis of the effects of these proteins on the biomineralization process.

References

1. Reddi, A.H., Morphogenesis and tissue engineering of bone and cartilage: inductive signals, stem cells, and biomimetic biomaterials. *Tissue Eng*. 2000. 6(4): 351–359.
2. Mazhuga, P.M., Mechanisms of cartilage precursor replacement by bone in the mammalian skeleton. *Acta Biol. Hung*. 1984. 35(2-4): 219–225.
3. Deng, H.W., et al., Genetic determination of variation and covariation of peak bone mass at the hip and spine. *J. Clin. Densitom*. 1999. 2(3): 251–263.
4. Riggs, B.L., S. Khosla, and L.J. Melton 3rd, Sex steroids and the construction and conservation of the adult skeleton. *Endocr. Rev*. 2002. 23(3): 279–302.
5. Lane, J.M., et al., Overview of geriatric osteopenic syndromes. Part I: definition and pathophysiology. *Orthop. Rev*. 1988. 17(11): 1131-2, 1135-6, 1138–1139.
6. Raisz, L.G., Physiology and pathophysiology of bone remodeling. *Clin. Chem*. 1999. 45(8 Pt 2): 1353–1358.
7. Boskey, A., Bone mineralization. in *Bone Biomechanics*, 3rd ed., Vol. 5.1, S.C. Cowin, Ed., 2001, Boca Raton, FL: CRC Press.
8. Sarikaya, M. and I.A. Aksay, Nacre of abalone shell: a natural multifunctional nanolaminated ceramic–polymer composite material. *Results Probl. Cell Differ*. 1992. 19: 1–26.
9. Boskey, A.L., Biomineralization: conflicts, challenges, and opportunities. *J. Cell. Biochem. Suppl* 1998. 30-31: 83–91.
10. Mann, S., The chemistry of form. *Angew Chem. Int. Ed. Engl*. 2000. 39(19): 3392–3406.
11. Wilt, F.H., Biomineralization of the spicules of sea urchin embryos. *Zool. Sci*. 2002. 19(3): 253–261.
12. Green, D., et al., The potential of biomimesis in bone tissue engineering: lessons from the design and synthesis of invertebrate skeletons. *Bone* 2002. 30(6): 810–815.
13. van der Meulen, M.C., K.J. Jepsen, and B. Mikic, Understanding bone strength: size isn't everything. *Bone* 2001. 29(2): 101–114.
14. Burstein, A.H., et al., Contribution of collagen and mineral to the elastic–plastic properties of bone. *J. Bone Joint Surg. Am*. 1975. 57(7): 956–961.
15. Nielsen, S.P., The fallacy of BMD: a critical review of the diagnostic use of dual X-ray absorptiometry. *Clin. Rheumatol*. 2000. 19(3): 174–183.
16. Burgeson, R.E. and M.E. Nimni, Collagen types. Molecular structure and tissue distribution. *Clin. Orthop*. 1992(282): 250–272.
17. Knott, L. and A.J. Bailey, Collagen cross-links in mineralizing tissues: a review of their chemistry, function, and clinical relevance. *Bone* 1998. 22(3): 181–187.
18. Muriel, M.P., et al., Morphological and biochemical studies of a mouse mutant (fro/fro) with bone fragility. *Bone* 1991. 12(4): 241–248.
19. Andrikopoulos, K., et al., Targeted mutation in the col5a2 gene reveals a regulatory role for type V collagen during matrix assembly. *Nat. Genet*. 1995. 9(1): 31–36.
20. Vetter, U., et al., Changes in apatite crystal size in bones of patients with osteogenesis imperfecta. *Calcif. Tissue Int*. 1991. 49(4): 248–250.
21. Traub, W., et al., Ultrastructural studies of bones from patients with osteogenesis imperfecta. *Matrix Biol*. 1994. 14(4): 337–345.
22. Grabner, B., et al., Age- and genotype-dependence of bone material properties in the osteogenesis imperfecta murine model (oim). *Bone* 2001. 29(5): 453–457.
23. Camacho, N.P., et al., The material basis for reduced mechanical properties in oim mice bones. *J. Bone Miner. Res*. 1999. 14(2): 264–272.
24. Mergenhagen, S.E., et al., Calcification *in vivo* of implanted collagens. *Biochim. Biophys. Acta* 1960. 43: 563.

25. Termine, J.D., et al., Osteonectin, a bone-specific protein linking mineral to collagen. *Cell* 1981. 26(1 Pt 1): 99–105.
26. Endo, A. and M.J. Glimcher, The effect of complexing phosphoproteins to decalcified collagen on *in vitro* calcification. *Connect. Tissue Res.* 1989. 21(1-4): 179–90; discussion 191–196.
27. Fisher, L.W., et al., Flexible structures of SIBLING proteins, bone sialoprotein, and osteopontin. *Biochem. Biophys. Res. Commun.* 2001. 280(2): 460–465.
28. Jain, A., et al., Three SIBLINGs (small integrin-binding ligand, N-linked glycoproteins) enhance factor H's cofactor activity enabling MCP-like cellular evasion of complement-mediated attack. *J. Biol. Chem.* 2002. 277(16): 13700–13708.
29. Kasugai, S., T. Nagata, and J. Sodek, Temporal studies on the tissue compartmentalization of bone sialoprotein (BSP), osteopontin (OPN), and SPARC protein during bone formation *in vitro*. *J. Cell. Physiol.* 1992. 152(3): 467–477.
30. Benaziz, L., et al., Adsorption of O-phospho-L-serine and L-serine onto poorly crystalline apatite. *J. Colloid Interface Sci.* 2001. 238(1): 48–53.
31. Butler, W.T., The nature and significance of osteopontin. *Connect. Tissue Res.* 1989. 23(2-3): 123–136.
32. Denhardt, D.T., et al., Osteopontin as a means to cope with environmental insults: regulation of inflammation, tissue remodeling, and cell survival. *J. Clin. Invest.* 2001. 107(9): 1055–1061.
33. Beck, G.R., Jr., B. Zerler, and E. Moran, Phosphate is a specific signal for induction of osteopontin gene expression. *Proc. Natl. Acad. Sci. USA*, 2000. 97(15): 8352–8357.
34. Hunter, G.K., C.L. Kyle, and H.A. Goldberg, Modulation of crystal formation by bone phosphoproteins: structural specificity of the osteopontin-mediated inhibition of hydroxyapatite formation. *Biochem. J.* 1994. 300 (Pt 3): 723–728.
35. Shiraga, H., et al., Inhibition of calcium oxalate crystal growth *in vitro* by uropontin: another member of the aspartic acid-rich protein superfamily. *Proc. Natl. Acad. Sci. USA*, 1992. 89(1): 426–430.
36. Reinholt, F.P., et al., Osteopontin—a possible anchor of osteoclasts to bone. *Proc. Natl. Acad. Sci. USA*, 1990. 87(12): 4473–4475.
37. Liaw, L., et al., Altered wound healing in mice lacking a functional osteopontin gene (spp1). *J. Clin. Invest.* 1998. 101(7): 1468–1478.
38. Rittling, S.R., et al., Mice lacking osteopontin show normal development and bone structure but display altered osteoclast formation *in vitro*. *J. Bone Miner. Res.* 1998. 13(7): 1101–1111.
39. Yoshitake, H., et al., Osteopontin-deficient mice are resistant to ovariectomy-induced bone resorption. *Proc. Natl. Acad. Sci. USA*, 1999. 96(14): 8156–8160.
40. Boskey, A.L., et al., Osteopontin deficiency increases mineral content and mineral crystallinity in mouse bone. *Calcif. Tissue Int.* 2002. 71(2): 145–154.
41. Wuttke, M., et al., Structural characterization of human recombinant and bone-derived bone sialoprotein. Functional implications for cell attachment and hydroxyapatite binding. *J. Biol. Chem.* 2001. 276(39): 36839–36848.
42. Zhu, X.L., et al., Synthesis and processing of bone sialoproteins during de novo bone formation in vitro. *Biochem. Cell Biol.* 2001. 79(6): 737–746.
43. Harris, N.L., et al., Functional analysis of bone sialoprotein: identification of the hydroxyapatite-nucleating and cell-binding domains by recombinant peptide expression and site-directed mutagenesis. *Bone* 2000. 27(6): 795–802.
44. Riminucci, M., et al., The anatomy of bone sialoprotein immunoreactive sites in bone as revealed by combined ultrastructural histochemistry and immunohistochemistry. *Calcif. Tissue Int.* 1995. 57(4): 277–284.

45. Chen, J., et al., Bone sialoprotein mRNA expression and ultrastructural localization in fetal porcine calvarial bone: comparisons with osteopontin. *Histochem. J.* 1994. 26(1): 67–78.
46. Fujisawa, R., et al., Attachment of osteoblastic cells to hydroxyapatite crystals by a synthetic peptide (Glu7-Pro-Arg-Gly-Asp-Thr) containing two functional sequences of bone sialoprotein. *Matrix Biol.* 1997. 16(1): 21–28.
47. Fujisawa, R., Y. Nodasaka, and Y. Kuboki, Further characterization of interaction between bone sialoprotein (BSP) and collagen. *Calcif. Tissue Int.* 1995. 56(2): 140–144.
48. Grzesik, W.J. and P.G. Robey, Bone matrix RGD glycoproteins: immunolocalization and interaction with human primary osteoblastic bone cells *in vitro*. *J. Bone Miner. Res.* 1994. 9(4): 487–496.
49. Hunter, G.K. and H.A. Goldberg, Nucleation of hydroxyapatite by bone sialoprotein. *Proc. Natl. Acad. Sci. USA*, 1993. 90(18): 8562–8565.
50. Boskey, A.L., Osteopontin and related phosphorylated sialoproteins: effects on mineralization. *Ann. N Y Acad. Sci.* 1995. 760: 249–256.
51. George, A., et al., Characterization of a novel dentin matrix acidic phosphoprotein. Implications for induction of biomineralization. *J. Biol. Chem.* 1993. 268(17): 12624–12630.
52. Srinivasan, R., et al., Recombinant expression and characterization of dentin matrix protein 1. *Connect. Tissue Res.* 1999. 40(4): 251–258.
53. Feng, J.Q., et al., Dentin matrix protein 1, a target molecule for Cbfa1 in bone, is a unique bone marker gene. *J. Bone Miner. Res.* 2002. 17(10): 1822–1831.
54. Qin, C., et al., A comparative study of sialic acid-rich proteins in rat bone and dentin. *Eur. J. Oral Sci.* 2001. 109(2): 133–141.
55. Kulkarni, G.V., et al., Promotion of selective cell attachment by the RGD sequence in dentine matrix protein 1. *Arch. Oral Biol.* 2000. 45(6): 475–484.
55a. Tartaix, P.H., et. al., *In vitro* effects of dentin matrix proteins-1 on hydroxyapatite formation provides insights into *in vivo* functions *J. Biol. Chem.* 2004. 279(18): 18115-18120.
56. Narayanan, K., et al., Differentiation of embryonic mesenchymal cells to odontoblast-like cells by overexpression of dentin matrix protein 1. *Proc. Natl. Acad. Sci. USA*, 2001. 98(8): 4516–4521.
57. Butler, W.T., et al., Isolation, characterization and immunolocalization of a 53-kDal dentin sialoprotein (DSP). *Matrix* 1992. 12(5): 343–351.
58. MacDougall, M., et al., Dentin phosphoprotein and dentin sialoprotein are cleavage products expressed from a single transcript coded by a gene on human chromosome 4. Dentin phosphoprotein DNA sequence determination. *J. Biol. Chem.* 1997. 272(2): 835–842.
59. Butler, W.T., H.H. Ritchie, and A.L. Bronckers, Extracellular matrix proteins of dentine. Ciba Found Symp, 1997. 205: p. 107-15; discussion 115–117.
60. Boskey, A., et al., Dentin sialoprotein (DSP) has limited effects on *in vitro* apatite formation and growth. *Calcif. Tissue Int.* 2000. 67(6): 472–478.
61. Saito, T., et al., *In vitro* apatite induction by phosphophoryn immobilized on modified collagen fibrils. *J. Bone Miner. Res.* 2000. 15(8): 1615–1619.
62. MacDougall, M., et al., Developmental regulation of dentin sialophosphoprotein during ameloblast differentiation: a potential enamel matrix nucleator. *Connect. Tissue Res.* 1998. 39(1-3): 25-37; discussion 63–67.
63. Zhang, X., et al., DSPP mutation in dentinogenesis imperfecta Shields type II. *Nat. Genet.* 2001. 27(2): 151–152.
64. Xiao, S., et al., Dentinogenesis imperfecta 1 with or without progressive hearing loss is associated with distinct mutations in DSPP. *Nat. Genet.* 2001. 27(2): 201–204.

65. Ross, U.H., et al., Osteogenesis imperfecta: clinical symptoms and update findings in computed tomography and tympano-cochlear scintigraphy. *Acta Otolaryngol.* 1993. 113(5): 620–624.
66. Clark, K., et al., Age-related hearing loss and bone mass in a population of rural women aged 60 to 85 years. *Ann. Epidemiol.* 1995. 5(1): 8–14.
67. Rowe, P.S., et al., MEPE, a new gene expressed in bone marrow and tumors causing osteomalacia. *Genomics* 2000. 67(1): 54–68.
68. Argiro, L., et al., Mepe, the gene encoding a tumor-secreted protein in oncogenic hypophosphatemic osteomalacia, is expressed in bone. *Genomics* 2001. 74(3): 342–351.
69. Drezner, M.K., PHEX gene and hypophosphatemia. *Kidney Int.* 2000. 57(1): 9–18.
70. Guo, R., et al., Inhibition of MEPE cleavage by Phex. *Biochem. Biophys. Res. Commun.* 2002. 297(1): 38–45.
71. Gorski, J.P. and K. Shimizu, Isolation of new phosphorylated glycoprotein from mineralized phase of bone that exhibits limited homology to adhesive protein osteopontin. *J. Biol. Chem.* 1988. 263(31): 15938–15945.
72. Gorski, J.P., et al., Bone acidic glycoprotein-75 self-associates to form macromolecular complexes *in vitro* and *in vivo* with the potential to sequester phosphate ions. *J. Cell. Biochem.* 1997. 64(4): 547–564.
73. Sato, M., et al., Bone acidic glycoprotein 75 inhibits resorption activity of isolated rat and chicken osteoclasts. *FASEB J.* 1992. 6(11): 2966–2976.
74. Gorski, J.P., et al., New alternatively spliced form of galectin-3, a member of the beta-galactoside-binding animal lectin family, contains a predicted transmembrane-spanning domain and a leucine zipper motif. *J. Biol. Chem.* 2002. 277(21): 18840–18848.
75. Chen, Y., B.S. Bal, and J.P. Gorski, Calcium and collagen binding properties of osteopontin, bone sialoprotein, and bone acidic glycoprotein-75 from bone. *J. Biol. Chem.* 1992. 267(34): 24871–24878.
76. Gorski, J.P., Is all bone the same? Distinctive distributions and properties of non-collagenous matrix proteins in lamellar vs. woven bone imply the existence of different underlying osteogenic mechanisms. *Crit. Rev. Oral Biol. Med.* 1998. 9(2): 201–223.
77. Swaroop, A., B.L. Hogan, and U. Francke, Molecular analysis of the cDNA for human SPARC/osteonectin/BM-40: sequence, expression, and localization of the gene to chromosome 5q31–q33. *Genomics* 1988. 2(1): 37–47.
78. Yost, J.C. and E.H. Sage, Specific interaction of SPARC with endothelial cells is mediated through a carboxyl-terminal sequence containing a calcium-binding EF hand. *J. Biol. Chem.* 1993. 268(34): 25790–25796.
79. Sage, E.H., et al., Inhibition of endothelial cell proliferation by SPARC is mediated through a Ca(2+)-binding EF-hand sequence. *J. Cell. Biochem.* 1995. 57(1): 127–140.
80. Sasaki, T., et al., Crystal structure and mapping by site-directed mutagenesis of the collagen-binding epitope of an activated form of BM-40/SPARC/osteonectin. *EMBO J.* 1998. 17(6): 1625–1634.
81. Romberg, R.W., et al., Inhibition of hydroxyapatite crystal growth by bone-specific and other calcium-binding proteins. *Biochemistry* 1986. 25(5): 1176–1180.
82. Tremble, P.M., et al., SPARC, a secreted protein associated with morphogenesis and tissue remodeling, induces expression of metalloproteinases in fibroblasts through a novel extracellular matrix-dependent pathway. *J. Cell Biol.* 1993. 121(6): 1433–1444.
83. Tyree, B., The partial degradation of osteonectin by a bone-derived metalloprotease enhances binding to type I collagen. *J. Bone Miner. Res.* 1989. 4(6): 877–883.

84. Gilmour, D.T., et al., Mice deficient for the secreted glycoprotein SPARC/osteonectin/BM40 develop normally but show severe age-onset cataract formation and disruption of the lens. *EMBO J.* 1998. 17(7): 1860–1870.
85. Basu, A., et al., Impaired wound healing in mice deficient in a matricellular protein SPARC (osteonectin, BM-40). *BMC Cell Biol.* 2001. 2(1): 15.
86. Delany, A.M., et al., Osteopenia and decreased bone formation in osteonectin-deficient mice. *J. Clin. Invest.* 2000. 105(9): 1325.
87. Boskey, A.L., et al., Infrared analysis of the mineral and matrix in bones of osteonectin-null mice and their wildtype controls. *J. Bone Miner. Res.* 2003. 18(6): 1005–1011.
88. Marie, P., [Cellular and molecular biology of fibrous dysplasia]. *Ann. Pathol.* 2001. 21(6): 489–498.
89. Fedarko, N.S., P.G. Robey, and U.K. Vetter, Extracellular matrix stoichiometry in osteoblasts from patients with osteogenesis imperfecta. *J. Bone Miner. Res.* 1995. 10(7): 1122–1129.
90. Iba, K., et al., Transforming growth factor-beta 1 downregulates dexamethasone-induced tetranectin gene expression during the *in vitro* mineralization of the human osteoblastic cell line SV-HFO. *FEBS Lett.* 1995. 373(1): 1–4.
91. Iba, K., et al., Phase-independent inhibition by retinoic acid of mineralization correlated with loss of tetranectin expression in a human osteoblastic cell line. *Cell Struct. Funct.* 2001. 26(4): 227–233.
92. Iba, K., et al., Mice with a targeted deletion of the tetranectin gene exhibit a spinal deformity. *Mol. Cell. Biol.* 2001. 21(22): 7817–7825.
93. Berkner, K.L., The vitamin K-dependent carboxylase. *J. Nutr.* 2000. **130**(8): 1877–1880.
94. Wolf, J. and C. Vermeer, Potential effect of vitamin K on microgravity-induced bone loss. *J. Gravit. Physiol.* 1996. 3(2): 29–32.
95. Feteih, R., M.S. Tassinari, and J.B. Lian, Effect of sodium warfarin on vitamin K-dependent proteins and skeletal development in the rat fetus. *J. Bone Miner. Res.* 1990. 5(8): 885–894.
96. Price, P.A., Gla-containing proteins of bone. *Connect. Tissue Res.* 1989. 21(1-4): 51-7; discussion 57–60.
97. Price, P.A. and M.K. Williamson, Primary structure of bovine matrix Gla protein, a new vitamin K-dependent bone protein. *J. Biol. Chem.* 1985. 260(28): 14971–14975.
98. Hackeng, T.M., et al., Total chemical synthesis of human matrix Gla protein. *Protein Sci.* 2001. 10(4): 864–870.
99. Spronk, H.M., et al., Matrix Gla protein accumulates at the border of regions of calcification and normal tissue in the media of the arterial vessel wall. *Biochem. Biophys. Res. Commun.* 2001. 289(2): 485–490.
100. Luo, G., et al., Spontaneous calcification of arteries and cartilage in mice lacking matrix GLA protein. *Nature* 1997. 386(6620): 78–81.
101. Boskey, A.L., Karsenty, G. and McKee, M.D., Chemistry and Biology of Mineralized Tissues. in *Mineral Characterizeation of Bones and Soft Tissues in Matrix gla Protein Deficient Mice*, M. Goldberg, Boskey, A., Robinson, C., Eds., 2000, Chicago: Am Acad Orthopaed Surgeons.
102. Yagami, K., et al., Matrix GLA protein is a developmental regulator of chondrocyte mineralization and, when constitutively expressed, blocks endochondral and intramembranous ossification in the limb. *J. Cell Biol.* 1999. 147(5): 1097–1108.
103. Munroe, P.B., et al., Mutations in the gene encoding the human matrix Gla protein cause Keutel syndrome. *Nat. Genet.* 1999. 21(1): 142–144.

104. Farzaneh-Far, A., et al., A polymorphism of the human matrix gamma-carboxyglutamic acid protein promoter alters binding of an activating protein-1 complex and is associated with altered transcription and serum levels. *J. Biol. Chem.* 2001. 276(35): 32466–32473.
105. Lian, J.B. and C.M. Gundberg, Osteocalcin. Biochemical considerations and clinical applications. *Clin. Orthop.* 1988. 226, 267–291.
106. Pan, L.C. and P.A. Price, The propeptide of rat bone gamma-carboxyglutamic acid protein shares homology with other vitamin K-dependent protein precursors. *Proc. Natl. Acad. Sci. USA*, 1985. 82(18): 6109–6113.
107. Hauschka, P.V. and S.A. Carr, Calcium-dependent alpha-helical structure in osteocalcin. *Biochemistry* 1982. 21(10): 2538–2547.
108. Chenu, C., et al., Osteocalcin induces chemotaxis, secretion of matrix proteins, and calcium-mediated intracellular signaling in human osteoclast-like cells. *J. Cell Biol.* 1994. 127(4): 1149–1158.
109. Liggett, W.H., Jr., et al., Osteocalcin promotes differentiation of osteoclast progenitors from murine long-term bone marrow cultures. *J. Cell. Biochem.* 1994. 55(2): 190–199.
110. Ducy, P., et al., Increased bone formation in osteocalcin-deficient mice. *Nature* 1996. 382(6590): 448–452.
111. Boskey, A.L., et al., Fourier transform infrared microspectroscopic analysis of bones of osteocalcin-deficient mice provides insight into the function of osteocalcin. *Bone* 1998. 23(3): 187–196.
112. Roughley, P.J. and E.R. Lee, Cartilage proteoglycans: structure and potential functions. *Microsc. Res. Tech.* 1994. 28(5): 385–397.
113. Poole, A.R., et al., Cartilage macromolecules and the calcification of cartilage matrix. *Anat. Rec.* 1989. 224(2): 167–179.
114. Lee, I., et al., Immunocytochemical localization and biochemical characterization of large proteoglycans in developing rat bone. *J. Oral Sci.* 1998. 40(2): 77–87.
115. Gokhale, J., Robey, P.G., and Boskey, A.L., Osteoporosis. in Second ed. The *Biochemistry of Bone*, 2nd ed., R. Marcus, D. Feldman, J. Kelsey, Eds., 2001, San Diego: Academic Press.
116. Grzesik, W.J., et al., Age related changes in human bone proteoglycan structure: impact of osteogenesis imperfecta. *J. Biol. Chem.* 2002. 277(46): 43638–43647.
117. Hocking, A.M., T. Shinomura, and D.J. McQuillan, Leucine-rich repeat glycoproteins of the extracellular matrix. *Matrix Biol.* 1998. 17(1): 1–19.
118. Ameye, L. and M.F. Young, Mice deficient in small leucine-rich proteoglycans: novel *in vivo* models for osteoporosis, osteoarthritis, Ehlers-Danlos syndrome, muscular dystrophy, and corneal diseases. *Glycobiology* 2002. 12(9): 107R–116R.
119. Xu, T., et al., Targeted disruption of the biglycan gene leads to an osteoporosis-like phenotype in mice. *Nat. Genet.* 1998. 20(1): 78–82.
120. Corsi, A., et al., Phenotypic effects of biglycan deficiency are linked to collagen fibril abnormalities, are synergized by decorin deficiency, and mimic Ehlers-Danlos-like changes in bone and other connective tissues. *J. Bone Miner. Res.* 2002. 17(7): 1180–1189.
121. Ameye, L., et al., Abnormal collagen fibrils in tendons of biglycan/fibromodulin-deficient mice lead to gait impairment, ectopic ossification, and osteoarthritis. *FASEB J.* 2002. 16(7): 673–680.
122. Nakamura, H., et al., Immunolocalization of keratan sulfate proteoglycan in rat calvaria. *Arch. Histol. Cytol.* 2001. 64(1): 109–118.
123. Kamiya, N., K. Shigemasa, and M. Takagi, Gene expression and immunohistochemical localization of decorin and biglycan in association with early bone formation in the developing mandible. *J. Oral Sci.* 2001. 43(3): 179–188.

124. Scott, J.E. and M. Haigh, Proteoglycan-type I collagen fibril interactions in bone and non-calcifying connective tissues. *Biosci. Rep.* 1985. 5(1): 71–81.
125. Saamanen, A.M., et al., Murine fibromodulin: cDNA and genomic structure, and age-related expression and distribution in the knee joint. *Biochem. J.* 2001. 355(Pt 3): 577–585.
126. Embery, G., et al., Proteoglycans in dentinogenesis. *Crit. Rev. Oral Biol. Med.* 2001. 12(4): 331–349.
127. Boskey, A.L., et al., Effects of bone CS-proteoglycans, DS-decorin, and DS-biglycan on hydroxyapatite formation in a gelatin gel. *Calcif. Tissue Int.* 1997. 61(4): 298–305.
128. Raouf, A., et al., Lumican is a major proteoglycan component of the bone matrix. *Matrix Biol.* 2002. 21(4): 361–367.
129. Inoue, A., et al., Correlation between induction of expression of biglycan and mineralization by C-type natriuretic peptide in osteoblastic cells. *J. Biochem.* (Tokyo), 1999. 125(1): 103–108.
130. Wendel, M., Y. Sommarin, and D. Heinegard, Bone matrix proteins: isolation and characterization of a novel cell-binding keratan sulfate proteoglycan (osteoadherin) from bovine bone. *J. Cell Biol.* 1998. 141(3): 839–847.
131. Shen, Z., et al., Tissue distribution of a novel cell binding protein, osteoadherin, in the rat. *Matrix Biol.* 1999. 18(6): 533–542.
132. Wu, J., et al., Deficiency of the decorin core protein in the variant form of Ehlers-Danlos syndrome with chronic skin ulcer. *J. Dermatol. Sci.* 2001. 27(2): 95–103.
133. Beavan, L.A., et al., Deficient expression of decorin in infantile progeroid patients. *J. Biol. Chem.* 1993. 268(13): 9856–9862.
134. Geerkens, C., et al., The X-chromosomal human biglycan gene BGN is subject to X inactivation but is transcribed like an X-Y homologous gene. *Hum. Genet.* 1995. 96(1): 44–52.
135. Couchman, J.R., L. Chen, and A. Woods, Syndecans and cell adhesion. *Int. Rev. Cytol.* 2001. 207: 113–150.
136. Nakayama, H., et al., Dexamethasone enhancement of betaglycan (TGF-beta type III receptor) gene expression in osteoblast-like cells. *Exp. Cell Res.* 1994. 211(2): 301–306.
137. Forsberg, E. and L. Kjellen, Heparan sulfate: lessons from knockout mice. *J. Clin. Invest.* 2001. 108(2): 175–180.
138. Arikawa-Hirasawa, E., W.R. Wilcox, and Y. Yamada, Dyssegmental dysplasia, Silverman-Handmaker type: unexpected role of perlecan in cartilage development. *Am. J. Med. Genet.* 2001. 106(4): 254–257.
139. Arikawa-Hirasawa, E., et al., Perlecan is essential for cartilage and cephalic development. *Nat. Genet.* 1999. 23(3): 354–358.
140. Paulsson, M., et al., Structure of low density heparan sulfate proteoglycan isolated from a mouse tumor basement membrane. *J. Mol. Biol.* 1987. 197(2): 297–313.
141. Groffen, A.J., et al., Recent insights into the structure and functions of heparan sulfate proteoglycans in the human glomerular basement membrane. *Nephrol. Dial. Transplant.* 1999. 14(9): 2119–2129.
142. Fedde, K.N., et al., Alkaline phosphatase knock-out mice recapitulate the metabolic and skeletal defects of infantile hypophosphatasia. *J. Bone Miner. Res.* 1999. 14(12): 2015–2026.
143. Lawler, J., et al., Thrombospondin-1 is required for normal murine pulmonary homeostasis and its absence causes pneumonia. *J. Clin. Invest.* 1998. 101(5): 982–992.
144. Rodriguez-Manzaneque, J.C., et al., Thrombospondin-1 suppresses spontaneous tumor growth and inhibits activation of matrix metalloproteinase-9 and mobilization

Carnegie Mellon Bookstore

662 CASH-1 9420 0001 103

978084931621 NEW
HOLLING/BONE TISSU MDS 1N 189.95
978084932282 NEW
GUELCHE/INTRODUCTI MDS 1N 99.95
TOTAL 289.90

APPROVAL: 584535 SEQ NUM 8265
TICKET# 072280002886
ACCOUNT NUMBER XXXXXXXXXXXX0833 XX/XX
Visa/MasterCard 289.90

Last day for textbook returns is 8/31

8/16/07 12:11 PM

of vascular endothelial growth factor. *Proc. Natl. Acad. Sci. USA*, 2001. 98(22): 12485–12490.
145. Kyriakides, T.R., et al., Mice that lack thrombospondin 2 display connective tissue abnormalities that are associated with disordered collagen fibrillogenesis, an increased vascular density, and a bleeding diathesis. *J. Cell Biol.* 1998. 140(2): 419–430.
146. Bornstein, P., et al., Thrombospondin 2 modulates collagen fibrillogenesis and angiogenesis. *J. Investig. Dermatol. Symp. Proc.* 2000. 5(1): 61–66.
147. Hankenson, K.D. and P. Bornstein, The secreted protein thrombospondin 2 is an autocrine inhibitor of marrow stromal cell proliferation. *J. Bone Miner. Res.* 2002. 17(3): 415–425.
148. Nordahl, J., et al., Ultrastructural immunolocalization of fibronectin in epiphyseal and metaphyseal bone of young rats. *Calcif. Tissue Int.* 1995. 57(6): 442–449.
149. Couchourel, D., et al., Effects of fibronectin on hydroxyapatite formation. *J. Inorg. Biochem.* 1999. 73(3): 129–136.
150. Sammons, R.L., J. Sharpe, and P.M. Marquis, Use of enhanced chemiluminescence to quantify protein adsorption to calcium phosphate materials and microcarrier beads. *Biomaterials* 1994. 15(10): 842–847.
151. Hayward, C., M.E. Porteous, and D.J. Brock, Mutation screening of all 65 exons of the fibrillin-1 gene in 60 patients with Marfan syndrome: report of 12 novel mutations. *Hum. Mutat.* 1997. 10(4): 280–289.
152. Reinhardt, D.P., et al., Mutations in calcium-binding epidermal growth factor modules render fibrillin-1 susceptible to proteolysis. A potential disease-causing mechanism in Marfan syndrome. *J. Biol. Chem.* 2000. 275(16): 12339–12345.
153. Goldberg, M. and A.L. Boskey, Lipids and biomineralizations. *Prog. Histochem. Cytochem.* 1996. 31(2): 1–187.
154. McGuire, R.F. and R. Barber, Hormone receptor mobility and catecholamine binding in membranes. A theoretical model. *J. Supramol. Struct.* 1976. 4(2): 259–269.
155. Razani, B., S.E. Woodman, and M.P. Lisanti, Caveolae: from cell biology to animal physiology. *Pharmacol. Rev.* 2002. 54(3): 431–467.
156. Ennever, J., L.J. Riggan, and J.J. Vogel, Proteolipid and collagen calcification, *in vitro*. *Cytobios* 1984. 39(155-156): 151–157.
157. Anderson, H.C., Molecular biology of matrix vesicles. *Clin. Orthop.* 1995(314): 266–280.
158. Wuthier, R.E., Involvement of cellular metabolism of calcium and phosphate in calcification of avian growth plate cartilage. *J. Nutr.* 1993. 123(2 Suppl): 301–309.
159. Cotmore, J.M., G. Nichols, Jr., and R.E. Wuthier, Phospholipid-calcium phosphate complex: enhanced calcium migration in the presence of phosphate. *Science* 1971. 172(990): 1339–1341.
160. Boskey, A.L. and A.S. Posner, Extraction of a calcium-phospholipid-phosphate complex from bone. *Calcif. Tissue Res.* 1976. 19(4): 273–283.
161. Timchak, D.M., A.L. Boskey, and V. Vigorita, Lack of lipid involvement in nonosseous tissue repair. *Proc. Soc. Exp. Biol. Med.* 1983. 174(1): 59–64.
162. Boskey, A.L. and A.H. Reddi, Changes in lipids during matrix: induced endochondral bone formation. *Calcif. Tissue Int.* 1983. 35(4-5): 549–554.
163. Boskey, A.L., et al., Calcium-acidic phospholipid-phosphate complexes in human hydroxyapatite-containing pathologic deposits. *Am. J. Pathol.* 1988. 133(1): 22–29.
164. Boskey, A.L. and A.S. Posner, *In vitro* nucleation of hydroxyapatite by a bone calcium-phospholipid-phosphate complex. *Calcif. Tissue Res.* 1977. 22 Suppl: 197–201.

165. Boskey, A.L., F.H. Wians, Jr., and P.V. Hauschka, The effect of osteocalcin on *in vitro* lipid-induced hydroxyapatite formation and seeded hydroxyapatite growth. *Calcif. Tissue Int*. 1985. 37(1): 57–62.
166. Raggio, C.L., B.D. Boyan, and A.L. Boskey, *In vivo* hydroxyapatite formation induced by lipids. *J. Bone Miner. Res*. 1986. 1(5): 409–415.
167. Wuthier, R.E. and S.T. Gore, Partition of inorganic ions and phospholipids in isolated cell, membrane and matrix vesicle fractions: evidence for Ca-Pi-acidic phospholipid complexes. *Calcif. Tissue Res*. 1977. 24(2): 163–171.
168. Wuthier, R.E., et al., Mechanism of matrix vesicle calcification: characterization of ion channels and the nucleational core of growth plate vesicles. *Bone Miner*. 1992. 17(2): 290–295.
169. Kirsch, T., et al., Roles of the nucleational core complex and collagens (types II and X) in calcification of growth plate cartilage matrix vesicles. *J. Biol. Chem*. 1994. 269(31): 20103–20109.
170. Boskey, A.L., et al., Persistence of complexed acidic phospholipids in rapidly mineralizing tissues is due to affinity for mineral and resistance to hydrolytic attack: *in vitro* data. *Calcif. Tissue Int*. 1996. 58(1): 45–51.
171. Nie, D., et al., Defect in formation of functional matrix vesicles by growth plate chondrocytes in avian tibial dyschondroplasia: evidence of defective tissue vascularization. *J. Bone Miner. Res*. 1995. 10(11): 1625–1634.
172. Boskey, A.L., et al., Phospholipid changes in the bones of the hypophosphatemic mouse. *Bone* 1991. 12(5): 345–351.
173. Whitfield, J.F., Leptin: brains and bones. *Expert Opin. Investig. Drugs* 2001. 10(9): 1617–1622.
174. Zuk, P.A., et al., Multilineage cells from human adipose tissue: implications for cell-based therapies. *Tissue Eng*. 2001. 7(2): 211–228.
175. Geesey, G.G., et al., A review of spectroscopic methods for characterizing microbial transformations of minerals. *J. Microbiol. Methods* 2002. 51(2): 125–139.
176. Goldberg, M., et al., [Mineralization of the dental pulp: contributions of tissue engineering to tomorrow's therapeutics in odontology]. *Pathol. Biol*. (Paris), 2002. 50(3): 194–203.
177. Anderson, H.C., et al., Selective synthesis of bone morphogenetic proteins-1, -3, -4 and bone sialoprotein may be important for osteoinduction by Saos-2 cells. *J. Bone Miner. Metab*. 2002. 20(2): 73–82.
178. Dee, K.C., T.T. Andersen, and R. Bizios, Design and function of novel osteoblast-adhesive peptides for chemical modification of biomaterials. *J. Biomed. Mater. Res*. 1998. 40(3): 371–377.
179. Hunter, G.K., et al., Induction of collagen mineralization by a bone sialoprotein — decorin chimeric protein. *J. Biomed. Mater. Res*. 2001. 55(4): 496–502.
180. Teixeira, J.O. and M.R. Urist, Bone morphogenetic protein induced repair of compartmentalized segmental diaphyseal defects. *Arch. Orthop. Trauma Surg*. 1998. 117(1-2): 27–34.
181. Giannobile W.V., et al., Platelet-derived growth factor (PDGF) gene delivery for application in periodontal tissue engineering. *J. Periodontol*. 2001. 72(6): 815–823.
182. Daculsi, G., et al., Role of fibronectin during biological apatite crystal nucleation: ultrastructural characterization. *J. Biomed. Mater. Res*. 1999. 47(2): 228–233.
183. Lahiji, A., et al., Chitosan supports the expression of extracellular matrix proteins in human osteoblasts and chondrocytes. *J. Biomed. Mater. Res*. 2000. 51(4): 586–595.
184. Murphy, W.L. and D.J. Mooney, Bioinspired growth of crystalline carbonate apatite on biodegradable polymer substrata. *J. Am. Chem. Soc*. 2002. 124(9): 1910–1917.

185. Hartgerink, J.D., E. Beniash, and S.I. Stupp, Self-assembly and mineralization of peptide-amphiphile nanofibers. *Science* 2001. 294(5547): 1684–1688.
186. Wetherbee, R., Biomineralization. The diatom glasshouse. *Science* 2002. 298(5593): 547.
187. Nichols, G., et al., A review of the terms agglomerate and aggregate with a recommendation for nomenclature used in powder and particle characterization. *J. Pharm. Sci.* 2002. 91(10): 2103–2109.
188. Veloski, C.A., R.A. McCann, and J.W. Snow, An analytical model for the phase behavior of cholesteryl esters in intracellular inclusions. *Biochim. Biophys. Acta* 1994. 1213(2): 183–192.
189. Mitchell, C.A., L. Yu, and M.D. Ward, Selective nucleation and discovery of organic polymorphs through epitaxy with single crystal substrates. *J. Am. Chem. Soc.* 2001. 123(44): 10830–10839.
190. Falini, G., S. Fermani, and A. Ripamonti, Oriented crystallization of octacalcium phosphate into beta-chitin scaffold. *J. Inorg. Biochem.* 2001. 84(3-4): 255–258.
191. Landis, W.J., et al., Extracellular vesicles of calcifying turkey leg tendon characterized by immunocytochemistry and high voltage electron microscopic tomography and 3-D graphic image reconstruction. *Bone Miner.* 1992. 17(2): 237–241.
192. Boskey, A.L., B.D. Boyan, and Z. Schwartz, Matrix vesicles promote mineralization in a gelatin gel. *Calcif. Tissue Int.* 1997. 60(3): 309–315.
193. Boskey, A.L., et al., Osteopontin-hydroxyapatite interactions *in vitro*: inhibition of hydroxyapatite formation and growth in a gelatin-gel. *Bone Miner.* 1993. 22(2): 147–159.

5 Signaling Molecules for Tissue Engineering

C. Sfeir, J. Jadlowiec, H. Koch and P. Campbell

Contents

GROWTH FACTORS AS USEFUL COMPONENTS OF TISSUE ENGINEERING

A tissue-engineered implant designed to restore or modify the function of a tissue or an organ is usually composed of a combination of biocompatible materials and biological components of the tissue.[1] Biocompatible matrices have several key roles, which include facilitating cellular migration or invasion into the implanted material, guiding wound healing and tissue regeneration, and providing specific cues through cell/matrix interactions and tissue responses to the material. Regulated growth factor/hormone release from matrices or transplanted cells can create a refined and controlled approach to tissue regeneration. Immobilized bioactive ligands on or within biomaterials control single and multiple cellular morphologies and functions via receptor-mediated processes and are called biomimetic materials.[2] Inclusion of growth factors within tissue-engineered therapies mimics the natural tissue microenvironment and will presumably improve healing.

0-8493-1621-9/05/$0.00+$1.50
© 2005 by CRC Press LLC

This chapter describes representative growth factors used in tissue engineering (bone tissue in particular). The rationale for growth factor use in tissue engineering will be addressed, including basic biological and physiological principles. A discussion of the biological mechanisms is also included for the reader to further understand the logic of growth factor incorporation in tissue-engineered therapies.

GROWTH FACTOR INTRODUCTION

Growth factors are hormones that regulate cellular activity. Growth factor effects on cells are best characterized as activity modulators. These actions can either stimulate or inhibit cellular proliferation, differentiation, migration, adhesion, apoptosis, and gene expression.[3]

Growth factors are secreted proteins that exert their effects by interacting with specific receptors on the cell surface. This interaction could take place in the vicinity of the cell by either affecting a neighboring cell (paracrine) or the growth factor-producing cell itself (autocrine). Furthermore, there are additional short-range interactions termed juxtacrine (the growth factor/receptor complex interact with neighboring cells) and intracrine (the growth factor/receptor complex is internalized). If the growth factor binds to a receptor at a distant site, this is termed endocrine or more correctly hemocrine activity. In addition, multiple cell types can produce the same growth factors that can act on multiple cell types (pleiotropism) with similar or various effects. Different growth factors have been known to elicit the same biological effect (redundancy). Growth factor effects on cells are usually concentration-dependent and can up- or downregulate the number of cell surface receptors. The secretion and action of other growth factors in either an antagonistic or synergistic manner are usually influenced by one growth factor. In general, growth factor synthesis is a self-limited event and initiated by new gene transcription and translation. The produced messenger RNA (mRNA) is unstable, leading to transient synthesis usually followed by a rapid release. Post-translational processing may occur to render the polypeptide active from an inactive precursor. Once secreted, the growth factor may bind to matrix molecules or soluble carrier molecules or binding proteins for activity and stabilization. This interaction regulates growth factor activity on cells. Furthermore, growth factor availability, location, and temporal expression are important effectors of cellular response.

Using bone regeneration as our paradigm, it can be considered a special case of recapitulation of embryogenesis. Both processes involve complex spatial and temporal signaling arrays that regulate mitogenesis, cell shape, movement, differentiation, protein secretion, and apoptosis.[4,5] Insulin-like growth factors (IGFs), bone morphogenetic proteins (BMPs), fibroblast growth factors (FGFs), and vascular endothelial growth factors (VEGFs) are examples of secreted soluble growth factors that are important during wound healing for chemoattraction, mitogenesis, and differentiation.[4,5] Spatial localization of growth factors[6] drives the temporal sequence of wound healing.

The diverse biological effects of growth factors occur at picomolar to nanomolar concentrations.[7] The ability of an individual growth factor to have multiple effects dictates its receptors and/or its isoforms are distributed throughout a tissue. A degree of control may be produced by changes in the overall interstitial growth factor concentration or possibly total receptor population. However, in order to insure temporal and spatial accessibility to specific receptor subpopulations, the body does not release bioactive growth factors, that is, a molecular configuration capable of immediate binding to their receptors.[8]

The mechanism controlling receptor accessibility or bioavailabilty is by ***sequestering growth factors*** (also known as *latency*) within the interstitium or in the circulation.[8] For example, IGF-I and IGF-II,[9] transforming growth factor- beta (TGF-β),[8,10] and bone morphogenetic proteins (BMPs)[11] are associated with specific binding proteins that limit direct access to their receptors by maintaining both soluble and insoluble growth factor-binding protein complexes. Hepatocyte growth factor (HGF) is produced in a latent precursor molecular form.[12] Platelet-derived growth factor (PDGF),[13] fibroblast growth factor (FGF),[14] and vascular endothelial growth factor (VEGF)[15] can become immobilized by binding to specific extracellular matrix molecules. Lastly, epidermal growth factor (EGF),[16] stem cell factor,[17] and transforming growth factor alpha (TGFα)[18] are expressed in cells as transmembrane proteins.

Growth factor sequestration directly affects temporal and spatial functionality. Using IGF-I as an example, it is largely not in a bioactive state, but greater than 99% are bound to IGF binding proteins (IGFBPs) in both fluid and solidphases.[19] IGFBPs extend IGFs' half-life or residency within the circulation or a tissue compartment.[19] IGFs, TGF-β, PDGFs, FGFs, and endothelial growth factor (EGF) circulate in blood associated within platelets.[20] Platelet sequestration of these growth factors provides for a ~5 day circulating lifespan[21] compared to the short circulating half-lives (typically less than 15 min) of these growth factors when infused in a purified bioactive state. Sequestration also presents growth factors within specific locations in the extracellular matrix or on the cell surface.[19]

The conversion of sequestered growth factors to a bioactive state requires an activation event. Protease activation represents the best-understood mechanism of activation with the plasminogen–plasminogen activator system being involved in numerous growth factor systems, including IGFs,[22] TGF-β,[8] FGF-2,[23] VEGF,[24] and HGF.[25]

Therefore, controlling the physical placement, concentration, and sequestration of a growth factor is essential for tissue engineering applications

Spatial control in tissue engineering has been addressed through controlled growth factor release technologies including microencapsulation, nonspecific adsorption to native and synthetic degradable matrices, entrapment and release from multifunctional polymeric systems, and microfluidics.[26] These technologies release diffusible molecular cues into the *liquid phase*. Consequently, over time, a transient, nonpersistent, diffusion gradient results.[26] Local delivery of pharmacological doses of growth factors can also far exceed naturally occurring concentrations[26] with unclear or possibly undesirable side-effects.[27] To begin to address persistence and dosing issues, growth factors have been immobilized to engineered matrices to localize delivery and encourage wound healing. However, spatial patterning has not been accomplished.[28] A major obstacle for tissue engineering of complex structures, such as bone, remains the inability to *control spatial gradients of wound healing growth factors.*[26]

To design an effective growth factor delivery system for a tissue engineering application, it is important to understand the biological rationale for the growth factor of choice. It is also prudent to have an appreciation for the mechanism by which the growth factor elicits the signal to cells and the intracellular events that transpire therefrom.

GROWTH FACTORS IN BONE

In this section, we describe the importance of growth factors in tissue development and regeneration. Table 5.1 shows several growth factors under investigation for tissue engineering. For the remainder of this chapter, we will focus on the endeavors made using these signaling molecules for bone tissue engineering and their biological mechanisms that elicit cellular changes toward tissue regeneration.

Table 5.1 Representative Growth Factors with Potential for Bone Tissue Engineering Application

Growth factor	Putative function(s)	Tissue engineering applications	Delivery systems	References
BMPs	Bone, liver development, embryonic development	Spinal fusion, fracture healing, dental and craniofacial reconstruction	Adenoviral, recombinant protein, cell-mediated scaffold-mediated	[29–31]
TGF-β	Bone formation and resorption, growth arrest, metastasis, chondrocyte differentiation	Intervertebral disc regeneration, arthritis	Local adenoviral (percultaneous injection), plasmid, systemic, scaffold-mediated (collagen sponge)	[29, 32, 33]
IGFs	Embryonic and neonatal growth, bone matrix mineralization, cartilage development and homeostasis	Cartilage, bone, tendon	Adenoviral, systemic, scaffold-mediated	[34–36]
FGFs	Embryonic development, wound healing, bone and cartilage formation, enhancement of blood vessels	Bone, blood vessels	Recombinant protein, scaffold-mediated (gels), systemic	[37]
VEGFs	Angiogenesis, vessel remodeling and repair, vasodilatation, bone formation	Bone, blood vessel	Coated scaffolds, in conjunction with IGF-I	[38]
PDGF	Bone formation, osteoblast chemotaxis	Ligament and tendon, bone, periodontal	Retroviral, cell-mediated gene delivery, systemic	[28, 39]

BONE-SPECIFIC SIGNALING MOLECULES IN TISSUE ENGINEERING

The physiology of bone development, repair, and regeneration should be appreciated as a multi-component biological system in which an array of components contribute equally. The overall goal of tissue engineering is to establish an integrated organ. To reach this goal, two complementary elements are fundamental:

- firstly, bone-specific signaling molecules that ideally promote cell recruitment, mitogenesis, and differentiation of progenitor cells into osteoblasts, and renewal; and
- secondly, *in vivo* factors that initiate and enhance neovascularization to overcome the lack of solutions for sufficient nutrition and oxygenation.

Two pathways of tissue neovascularization are known: vasculogenesis, the *in situ* assembly of capillaries from undifferentiated endothelial cells, and angiogenesis, the sprouting of capillaries from preexisting blood vessels. Several growth factors serve as stimuli for endothelial cell proliferation and migration as well as the formation of new blood vessels.[40] The purification and cloning of BMPs and growth factors such as PDGFs, TGF-β, and IGF will allow the design of an optimal combination of signals to initiate and promote the development of mesenchymal stem cells into cartilage and bone.[41] The following classes of extracellular signaling molecules are currently being considered to be beneficial for bone tissue engineering:

- Transforming growth factor beta (TGF-β)
- Bone morphogenetic proteins (BMPs)
- Fibroblast growth factors (FGFs)
- Insulin-like growth factors (IGFs)
- Platelet-derived growth factor (PDGF)
- Vascular endothelial growth factors (VEGFs)

TRANSFORMING GROWTH FACTOR BETA

The TGF-β superfamily consists of TGF-β1 through TGF-β5, bone morphogenetic proteins, growth and differentiation factors (GDF), activins, inhibins, and Müllerian Inhibitory substance.[33]

TGF-β is mainly found in bone, platelets, and cartilage and triggers growth, differentiation, and extracellular matrix synthesis.[42] Since TGF-β receptors are found in increased quantity on chondrocytes[43,44] and osteoblasts,[45] it has been hypothesized that these growth factors participate in bone development and repair process at all stages.[42] It has been reported that after passage under serum-free conditions, TGF-β1 stimulation of osteogenic cells immediately resulted in the formation of three-dimensional cellular condensations within 24–48 h *in vitro*.[46] During days 3–7, there was an upregulation of alkaline phosphatase, type I collagen, and osteonectin, three important proteins in bone development.[46] Several studies examining TGF-β1 for experimental fracture healing have been published.[43,47] Initially, based on periosteal injections in rats, TGF-β1 was thought to induce bone formation *in vivo*.[48] Further studies indicated that TGF-β1 enhances mineralization of human osteoblasts on implant materials.[49] However, conflicting reports suggested that TGF-β1 might inhibit the normal development of peripheral callus in response to axial interfragmentary motion in post-fracture recovery in rabbits.[50] Although some *in vitro* experiments have suggested that TGF-β1 inhibits the activity of BMP-2,[51] *in vivo* experiments suggested that it may enhance the activity of BMP.[52] However, the use of different isoforms, doses, and animal models makes it difficult to compare directly these studies and reach stringent conclusions. Superphysiological doses or repeated dosing may be necessary to enhance bone generation.[47] It is also likely that promising results were achieved because of superphysiological doses of TGF-β1.[53] Although these parameters may be available for an *in vitro* tissue engineering approach, enhanced cell proliferation among a variety of different phenotypes by TGF-β1 should be considered as a possible source of unwanted side effects *in vivo*. To date, the function and effects of TGF-β1 are not entirely understood and TGF-β1 likely does not function alone as a bioactive factor for bone tissue engineering.[54] TGF-β1 binds primarily to two transmembrane receptors: type I and type II. A review on TGF-β1 and its receptors was published by Shi and Massague (2003).[55] TGF-β1 initiates signaling similar to that of BMPs, which are members of the TGF-β superfamily.

Bone Morphogenetic Proteins

As members of the TGF-β superfamily, 15 individual BMPs have been identified to date.[56]

Fundamental work by Marshall R. Urist in 1965 led to what he called the *bone induction principle*. Urist postulated that there is a substance in bone that can induce new bone formation. Urist implanted demineralized bone matrix in a muscle pouch in a rat, which resulted in the formation of a new ossicle.[57] Later, Urist identified a protein responsible for this phenomenon and named it *bone morphogenetic protein*. The protein was sequenced and a number of isoforms were identified.[58] Today, BMP-2, 4, and BMP-7 are established as being osteoinductive.[56] These BMPs stimulate differentiation of mesenchymal stem cells to an osteochondroblastic lineage.[59,60]

BMPs are also crucial for cell growth and bone formation.[56] BMP-2 promoted apoptosis in primary human calvarial osteoblasts and in immortalized human neonatal calvaria osteoblasts.[61] BMP-4 is a key morphogen for embryonic lung development and is expressed in high levels in the peripheral epithelium.[62] BMP-4 may contribute to eye development by promoting cell proliferation and programmed cell death.[63] Mice deficient in BMP-5 have short-ear deformities, and lack of BMP-7 has been associated with hind limb polydactyl and renal agenesis.[64] Furthermore, it has been suggested that BMP-3, the most abundant BMP in demineralized bone, may play an essential role as a modulator of the activity of osteogenic BMPs *in vivo*.[65] Han and co-workers engineered a recombinant BMP-3 protein to include an auxiliary collagen-targeting domain derived from von Willebrand coagulation factor (vWF). Local injection of rhBMP3-C infused in a collagen suspension induced new bone formation on the periosteal surface of rat calvaria, and in a rat cranial defect model. The authors concluded that engineering and manufacturing of targeted-BMPs may be useful in bone tissue engineering.[66]

A rabbit model of mandibular distraction osteogenesis was developed, and the outcome was that BMP-2 and -4 were highly expressed in osteoblastic cells during the distraction period.[67] The authors suggested that BMPs participate in the transformation of mechanical stimuli into a biological response.[67] Injectable bone morphogenetic proteins (BMPs) in a polymeric delivery system have also been used as an osteoinductive material.[68] In critical-sized cranial defects in rabbits with periosteal cells that produced BMP-7 after retroviral gene transfer, an indication of enhanced defect repair was observed.[69] The group of Gazit and Turgeman published promising results on transgenic BMP-2 produced by human mesenchymal stem cells (hMSCs) and their beneficial effect in bone regeneration.[70] Recombinant human BMP-2 was successfully used for the treatment of critical-sized defects in rat,[71] rabbit,[72] sheep,[73] and dog[74] models. However, initial enthusiasm over the compelling results did not hold with studies on phylogenetically higher species. Boden and co-workers did laparoscopic anterior spinal arthrodesis in five adult rhesus monkeys with a titanium interbody cage containing a collagen sponge with doses of rhBMP-2 of either 0.75 or 1.5 mg/mL.[75] Although a solid fusion was achieved with both doses and no fusion in the control animals, superphysiological doses of rhBMP-2 raise serious concerns.[76] The administration of milligrams of BMPs to a patient may unleash sinister, unpredictable, or unexpected sequalae.[56] Doses of BMP required in the milligram range increase the ease of manufacture and cost issues.[56] Moreover, several authors reported voids in the fusion mass,[74] which may have not affected biomechanical outcome but certainly need to be further assessed, although the recent FDA approval of INFUSE Bone Graft (Medtronics), an absorbable collagen sponge, offers promise. Two recent human clinical trials using INFUSE loaded with rhBMP-2 demonstrates feasibility in the spine model.[77, 78] However, the success of BMP-2 in the clinic requires controlled delivery and full understanding of its bioactivity in bone and other tissues.

Fibroblast Growth Factors

The growing family of fibroblast growth factors consists of mitogenic polypeptides that demonstrate affinity to heparan sulfate glycosaminoglycan-binding sites on cells. FGFs are important for embryonic development, regeneration, wound healing, angiogenesis, and mesenchymal cell mitogenesis.[79] FGF-1 and 2 are most abundant in adult tissue and were formerly known as acidic and basic FGF, respectively. Craniosynostosis is one disorder associated with the syndromes Apert, Crouzon, Pfeiffer, and Saethre-Chotzen and has been linked to mutations in the FGF receptors 1, 2, and 3.[80] Craniosynostosis results from precocious osteoblast differentiation and premature fusion of the sutures. Studies in calvaria osteoblasts have shown that FGF-2, FGFR-2, and BMP-2 are essential factors in proliferation, differentiation, and apopotosis.[81] The role of FGFs and their receptors in proper bone formation is just beginning to be understood.

FGF-2 appears to be more potent than FGF-1 and is expressed by osteoblasts,[82] while FGF-1 has been associated with chondrocyte proliferation.[83] Increased activity of FGF-1 and 2 is detected during early stages of fracture healing. Therefore, these growth factors may enhance bone repair[84] and angiogenesis.[85–87] It has been suggested that FGF-2 plays an important role in bone formation as a regulator of Runx2, an osteogenesis-related transcription factor.[88]

rhFGF-2 was shown to stimulate periodontal regeneration in bone defects in beagle dogs.[89] Also, there is an indication that local delivery of FGF-2 may be useful in revision of total joint replacements in treating early osteolytic lesions, and facilitating osseointegration.[90] Furthermore, local injection of rhFGF-4 may stimulate bone formation around titanium implants in bone.[91] Dunstan and co-workers studied the systemic administration of FGF-1 in ovariectomized rats and concluded that both local and systemic FGF-1 increased new bone formation and bone density.[92] Systemic FGF-1 also appeared to restore bone microarchitecture bone loss by estrogen—withdrawal was thwarted.[92] *In vivo* results suggest that FGF-2 treatment upregulates IGF-I gene expression in aged ovariectomized rats.[93] The authors concluded that this may mediate, at least in part, the increased gene expression for bone matrix proteins and the bone anabolic effects of FGF-2. Pandit and co-workers created a rabbit ulcer model to determine the effective angiogenic stimulatory dose of FGF-1 delivered via a modified fibrin matrix. An increase in the angiogenic and fibroblastic responses to FGF-1 as well as an increase in the epithelialization rate were observed.[94] Montero and co-workers examined mice with a disruption of the FGF-2 gene and observed significantly decreased trabecular bone volume, mineral apposition, and bone formation rates.[95] In addition, there was a profound decrease in mineralization of bone marrow stromal cultures isolated from fgf-2(-/-) mice.[95] From these findings, the authors concluded that FGF-2 helps determine bone mass as well as bone formation.[95]

Today, FGFs appear to be promising candidates to enhance bone tissue engineering since they promote bone formation and enhance the development of blood vessels.

Insulin-like Growth Factors

Insulin-like growth factors-I and -II (IGF) are produced by bone cells, stored in bone matrix, and stimulate bone cell DNA synthesis and type I collagen production. The IGFs have been shown to stimulate osteoblast proliferation and differentiation *in vitro* and *in vivo*.[96, 97] IGFs increase collagen production in osteoblasts and inhibit collagen degradation by decreasing collagenase synthesis.[98] IGF-II is the most abundant growth factor in bone;[99] however, it is less potent than IGF-I, which has been localized in healing fractures in rats and humans.[100]

IGFs transduce signals via the IGF-I and IGF-II receptors.[101] Receptors for IGF-I have been located on both osteoblasts and osteoclasts.[102] IGF-I has complex effects on both bone formation and bone resorption.[103] It has been suggested that IGF-I supports the formation and activation of osteoclasts,[104] and growth hormone-stimulatory effects on osteoclastic resorption are partly mediated by IGF-I.[105] It is generally accepted that systemic and locally synthesized IGF-I is important in longitudinal bone growth and the maintenance of bone mass,[106] and that IGF-I is involved in cell proliferation or differentiation in mesenchymal stem cells, periosteal cells, osteoblasts, and chondrocytes [107]. Mice overexpressing IGF-I demonstrated increased muscle mass, which was associated with larger bones.[108] However, only pure cortical bone increased in both area and mineral content in these animals.[108] Locally, IGF-I synthesis by osteoblasts appears to be crucial for bone remodeling. Both circulating parathyroid hormone (PTH) and locally synthesized prostaglandin E_2 (PGE_2) directly stimulate IGF-I gene induction, which may explain their anabolic effects on bone formation.[109] Further, studies by Bikle and co-workers indicate that IGF-I is required for the anabolic actions of PTH on bone formation.[110] Animal studies suggested bone formation,[111] an increase in intramembranous bone defect repair,[112] and even beneficial effects on age-related osteopenia[113] by IGF-I. IGF-I-transduced mesenchymal cells appeared to migrate to and repopulate the bone marrow after systemic injection.[114] Interestingly, these cells localized preferentially to a fracture site and appeared to accelerate fracture healing.[114] However, growth plate injuries in rabbits showed the supportive effect of IGF-I on physeal chondrocytes, while BMP-2 stimulates osteogenic activity in injured growth plates.[115] Enhancement of chondrogenesis by spatially defined overexpression of human IGF-I has also been reported.[116] IGF-I-mediated acceleration of functional recovery from Achilles tendon injury was achieved in a rat model.[36] Tanaka and co-workers focused on IGF-I and/or PDGF as possible therapeutic agents for the age-related decline in bone formation activity.[117] They found that IGF-I and, to a much lesser extent, PDGF may partially re-establish the expression of osteoblast markers in old bones.[117]

The effects of IGFs on bipotential hMSCs have also been investigated; no change in the expression of type I collagen or cbfa1/runx2 was detected, suggesting that IGFs do not participate in early commitment to the osteoblast lineage.[118] However, mature human osteoblast-like cells with IGF-I showed a significant increase in type I collagen synthesis. The authors conclude that whereas the IGFs exert mitogenic effects on human bone marrow stromal cells, their effects on differentiation are dependent on the stage of maturation of the cells.[118] Hence, IGFs appear to be proliferative and permissive factors that may synergistically enhance the effect of differentiative factors such as dexamethasone. These data support an important role for IGF-I in bone formation and suggest that IGF-I therapy might be useful in diseases related to an insufficiency of osteoprogenitor cells such as cancer (ablation by radiation).[118]

Platelet-Derived Growth Factor

In vitro studies showed that platelet-derived growth factor (PDGF), which is secreted by platelets during the early phase of fracture healing, is mitogenic for osteoblasts[82] and has been localized at fracture sites in both mice[119] and humans.[120] Gene delivery for PDGF and recombinant PDGF stimulated cementoblast activity in periodontal tissue engineering.[121] Primary human mesenchymal progenitor cells and osteoprogenitor cells that had differentiated into osteoblasts were examined for chemotactic responses to FGF-2, TGF-β1, PDGF-BB, BMP-2, and BMP-4. Migration of primary human progenitor cells was stimulated by BMP-2, BMP-4, and PDGF-bb in a dose-dependent manner, whereas TGF-β1 and FGF-2 did not stimulate cell migration.[122] Since chemotactic migration of osteoprogenitor cells is crucial for bone development, remodeling, and repair, the authors concluded that these growth factors have functional roles during bone development and remodeling, as well as fracture healing.[122]

A factor that inhibits differentiation of mouse osteoblast precursor-like cell line MC3T3-E1 to osteoblasts induced by bone morphogenetic protein 4 (BMP-4) was isolated and identified as platelet-derived growth factor BB (PDGF BB) homodimer.[123] The authors concluded that osteoclasts may regulate osteoblasts directly, and suggest that PDGF BB is a key factor in bone remodeling.[123]

It was demonstrated that rhPDGF-BB transcriptionally increases osteoblasts VEGF mRNA expression *in vitro;* similar mechanisms may occur *in vivo*, at a site of skeletal injury, to induce neoangiogenesis and promote fracture repair.[124] Studies on spontaneously diabetic BB Wistar rats indicated that in these animals, compared to nondiabetic controls, the early phase of fracture healing was affected by inhibiting cell proliferation through decreasing the expression of PDGF.[125] Experiments with PDGF-BB releasing porous poly(epsilon-caprolactone) (PCL)-chitosan matrices suggested its use with bone-regenerative efficacy.[126] PDGF may partially induce upregulation of osteoblast marker genes in old bones, and the combination of PDGF with IGF-I may improve the outcome by stimulating osteocalcin expression.[117] Although there are some indications that PDGF might be useful in bone regeneration and bone tissue engineering, its therapeutic impact is yet to be proven.

Vascular Endothelial Growth Factors

Vascular endothelial growth factors (VEGF) are vascular cytokines.[127,128] Currently, there are at least six isoforms known:[128] VEGF-A to E and placental growth factor (PLGF). VEGFs are endogenously produced and promote angiogenesis, vasodilatation, and increased microvascular permeability *in vivo*.[128] VEGF induces endothelial cell proliferation, promotes cell migration, and inhibits apoptosis. VEGFs are endothelial specific and result in increased permeability of endothelial cells in culture.[128] *In vivo,* VEGF induces angiogenesis as well as permeabilization of blood vessels, and plays a central role in the regulation of vasculogenesis. VEGF has a direct and singular action on endothelial cells,[129] which uniquely distinguishes this growth factor from other factors such as epidermal growth factor, TGF-α and TGF-β1, FGFs, tumor necrosis factor, angiogenin, and PGE_2.[87] VEGF binds two related receptor tyrosine kinases (RTKs): VEGFR-1 and -2. For a review, the reader is referred to Ferrara *et al.*[130] Signaling via VEGFR-1 has been associated with induction of matrix metalloproteinase-9 in lung epithelial cells and the progression of lung metastases. VEGFR-2 is encoded by the gene Flk-1. The key role of this receptor was in angiogenesis, which was identified by a lack of vasculogenesis and organized blood vessels in Flk1-null mice. Hence, it is thought that VEGFR-2 is a major regulator of the mitogenic and angiogenic effects of VEGF. In addition to the RTKs, VEGF also interacts with a family of co-receptors, called the neutropilins (NRP-1 and -2). NRP-1 enhances the binding of an isoform ($VEGF_{165}$) to VEGFR-2 and chemotaxis by this mechanism. NRP-1 was also shown to be essential for vascular development in the zebrafish as a mediator of VEGF-dependent angiogenesis.

VEGF also stimulates the growth of collateral vessels with promising therapeutic use in inadequate tissue perfusion. This has been shown by several studies in which VEGF enhanced collateral vessel formation, increased capillary density, and promoted resting and vasodilator induced blood flow.[131] Stone and co-workers sought to improve vascular grafts by coating prosthetic surfaces with VEGF bound to a "basecoat" albumin.[132] They hypothesized that endothelial cell proliferation and migration would be supported by VEGF-coated grafts. Thus, VEGF may be of value in graft patency.[132] A soft-tissue model, which consisted of human adipose tissue, stromal cells, and umbilical vein endothelial cells in a fibrin-microcarrier scaffold has also been developed.[133] The application of VEGF resulted in a high initial angiogenic response.[133] Capillary-like structures could not be stabilized in the long term; however, supplementation with IGF-I resulted in optimal maintenance and a decrease in the length of capillary-like structures.[133] Additionally, several studies have also reported synergistic effects of VEGF and FGF-2 in the production of new blood vessels.[134] The expression of VEGF splicing isoforms and its receptors may play a role in the healing process after rat femoral drill-hole injury.[135] Adenoviral transfer of the vegf-A gene may induce bone formation *in vivo* by increasing osteoblast activity.[136] Hence, VEGF appears to be a promising candidate as a signaling molecule in neovascularization and bone tissue engineering.

Vascularity and angiogenesis are significant concerns that need to be addressed when designing tissue-engineered therapies. In this regard, angiogenic factors such as VEGF, FGF, and, possibly, IGF-I are indispensable in designing tissue-engineered therapies. It is likely that more than one molecule is required to achieve optimal therapeutic effects *in vivo*. To determine the proper combination of proteins, the distinct biological mechanisms, and how they intertwine to create, maintain, and regenerate the complete tissue must be elucidated.

INTRACELLULAR SIGNALING

The tissue microenvironment contains a unique composition of growth factors, cells, and ECM proteins that collaboratively function in a manner that is specific to each tissue type. Hormonal signals are relayed to the cell via cell surface receptors and integrins. This is known as "outside-in" signaling. Binding of a growth factor to its receptor or an ECM protein to an integrin can initiate intracellular signaling pathways, resulting in tissue-specific gene expression. IGF-I, FGF-2, TGF-β1, and BMP-2 are present within the extracellular matrix of bone and influence bone development, remodeling, and repair. For tissue engineering applications, it is desirable to include signaling molecules that initiate the expression of tissue-specific genes that enhance cell differentiation and tissue remodeling. To design a safe and effective therapy, we need to understand the underlying biological mechanisms transduced by signaling molecules from outside to inside the cell. The downstream effects of cell interactions with signaling molecules of the bone microenvironment are described below.

The most abundant growth factor in bone is IGF-II. However, the function of IGF-I in bone is more characterized. Alterations in IGF-I receptor abundance, affinity, or ratio of type I to type II receptors may affect IGF activity in bone cells. Distinct signaling pathways activated by each receptor contribute to preferential signaling cascades in early-and late-stage differentiated bone cells.[137] IGF-I requires additional intracellular signaling molecules to regulate bone metabolism. For example, insulin receptor substrate 1 (IRS-1) was shown to be a necessary co-factor of IGF-I signal transduction, mediating tyrosine receptor phosphorylation.[138] IRS-1 mutant mice showed reduced osteoblast proliferation and differentiation. Further, IRS-1 mediates signaling by vitamin D_3. IGF-I as well as FGF-2 are also known to activate osteocalcin expression via PKA and PKC pathways.[139]

FGF-2 is an important regulator of bone growth and development having dual roles in both osteoblast proliferation and differentiation. FGF-2 treatment of immature osteoblasts increases proliferation. However, treatment of later-stage osteoblasts leads to increased cell death.[37] Therefore, FGF-2 may be important in tissue remodeling and maintaining the delicate balance between bone-forming cells (osteoblasts) and bone-degrading cells (osteoclasts). Osteoblasts synthesize FGF-2 and store it in a bioactive form in the ECM. The control of gene expression by FGF-2 is biphasic and dependent on the stage of maturation. Recently, it was demonstrated that osteocalcin was upregulated by FGF-2 via the MAP kinase pathway.[88]

TGF-β and growth factors of its superfamily are multifunctional hormones. The nature of their effects is based on the composition of the cellular microenvironment. Although TGF-β1 elicits a variety of effects, relatively simple signaling pathways can account for the diverse cell responses to TGF-β1. Binding to its receptor initiates complex formation of Smad proteins (Smad2 and Smad3) in the cytoplasm. Smad partners and regulators that vary among cell types contribute to cell-specific responses to TGF-β1.

TGF-β1 is one of the major growth factors in the bone microenvironment and functions as a potential coupler between bone formation and resorption. TGF-β1 signals via Smads to regulate bone gene expression. Specifically, TGF-β1 activates runx2 expression.[140] Runx2 is also activated via BMP-2 and is thought to be a mediator between BMP-2 and TGF-β1 activities in the bone microenvironment.

BMP-2, a member of the TGF-β superfamily, is a popular molecule of choice for bone regeneration. BMP-2 has been shown to regulate osteoblast differentiation. It has been shown to stimulate bone formation *in vitro* and *in vivo* and has shown promise for clinical applications.[141] However, one serious concern is that the dose of BMP-2 required for effective healing is exceedingly large (1.7–3.4 mg).[142–144] One potential solution is to identify novel additional signaling molecules that are important for bone formation either alone or in conjunction with BMP-2. To identify such proteins, we must understand BMP-2's mechanism of action in osteoblast differentiation, bone development, and remodeling.

To initiate intracellular signaling, BMP binds as a dimer to a complex of two transmembrane receptors (type I and II) and starts a phosphorylation cascade upregulating bone-specific differentiation genes (e.g., runx2, osteocalcin, bone sialoprotein, type I collagen, and alkaline phosphatase). These genes regulate progression toward an osteoblastic phenotype and matrix mineralization. According to a recent report, BMP-2 receptor oligomerization controls two independent signaling pathways to regulated bone-specific genes. (Figure 5.1).[145]

- The first pathway involves BMP-2 dimerization and binding to a preformed receptor complex, initiating a phosphorylation cascade. The type I receptor phosphorylates Smad1, Smad5, and Smad8. A complex of these three is formed with Smad4 in the cytoplasm. The complex translocates to the nucleus and functions as a transcriptional regulator of bone-specific genes.

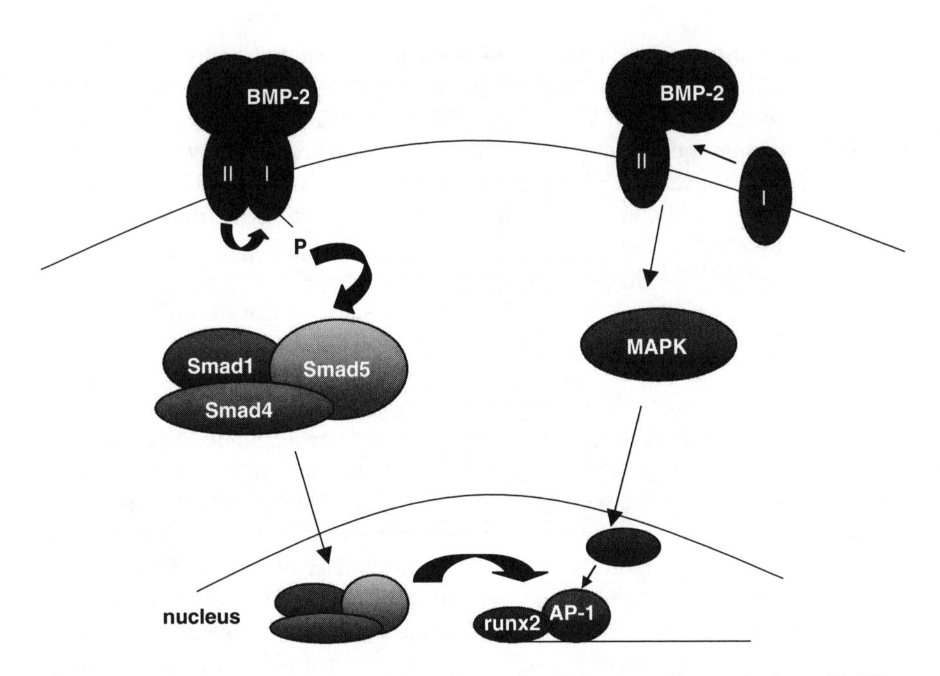

Figure 5.1 BMP-2 signaling is based on the mode of receptor oligomerization. (1) When BMP-2 dimer binds to a preformed complex of type I and type II receptors, a phosphorylation cascade involving the Smad proteins is transduced, resulting in the activation of bone-specific genes. (2) When BMP-2 binds to type II receptor, type I receptor initiates signal transduction via the MAP kinase pathway.

Smad6 and Smad7 inhibit the formation of this complex by binding to the type I receptor, thus blocking phosphorylation of Smad1, Smad5, and Smad8.[146]

- The second signaling pathway is activated when the BMP-2 dimer binds to receptor type II only. This binding event recruits receptor type I and initiates a phosphorylation cascade via the MAPK pathway involving p38 and Erk1/2. This pathway culminates with activation of AP-1 in the nucleus.

There is potential for crosstalk between the two pathways. Smad proteins have been shown to modulate AP-1 activities. Within the promoters of bone-specific genes, AP-1 responsive elements initiate gene expression for osteoblast differentiation. Signaling molecules of the MAP kinase pathway such as Erk have also been shown to inhibit actions of the Smad proteins. The MAP kinase pathway can be activated via other receptor tyrosine kinases (RTK) such as epidermal growth factor or hepatocyte growth factor receptor kinase. Erk-mediated phosphorylation of Smad1 inhibits nuclear translocation.[147] The preferential activation of Smad1 suggests a role for Smad1 in cell fate.[146]

Runx2, a transcriptional regulator of osteoblast differentiation genes, can be activated by Smad translocation into the nucleus.[148] This activation of Runx2 occurs via phosphorylation.[148] Smad proteins can activate transcription via direct binding to DNA or through interactions with other DNA-binding proteins. BMP-2-signaled Smad proteins directly interact with Runx2, suggesting a role for Smads in Runx2 activation.[140]

Osteoblast differentiation via Smad-dependent signaling is the more widely accepted mechanism of action for BMP-2-induced osteogenesis. However, BMP-2 is also known to activate the MAP kinase pathway. MAP kinases have been shown to upregulate osteoblast differentiation factors such as alkaline phosphatase and osteocalcin.[149] Runx2 activation by phosphorylation can also occur via the MAP kinase pathway.[150] Interestingly, an increase in transcriptional activity of runx2 via phosphorylation by MAP kinase was not accompanied by a significant change in runx2 message or protein.[150] Perhaps runx2 protein is regulated either by post-translational modification or an accessory factor. Recently, it was shown that the nuclear protein junB could be important for induction of runx2 by BMP-2 and TGF-β.[151] This and other studies indicate that junB is an upstream regulator of runx2 and the p38 MAP kinase component is an additional requirement for the induction of runx2.[148,151]

The MAP kinase pathway responds to a number of signaling molecules, including growth factors, integrins, and mechanical stresses. Different components of the MAP kinase pathway are utilized, based on the mode of activation. Erk, one of the MAP kinases, was also shown to be essential for osteoblast growth and differentiation, adhesion, migration, and integrin expression.[152] Interestingly, one contradictory report claimed that inhibition of the MAP kinase pathway via the p38 component increased early osteoblast differentiation by upregulating the expression of alkaline phosphatase and promoting mineralization.[153] The authors suggest that osteoblast differentiation requires a "fine-tuning" of the MAP kinase pathway in gene expression.[153]

The induction of MAP kinase via integrins is prolonged, whereas growth factor-stimulated MAP is more transient and may require integrin-mediated cell adhesion.[152]

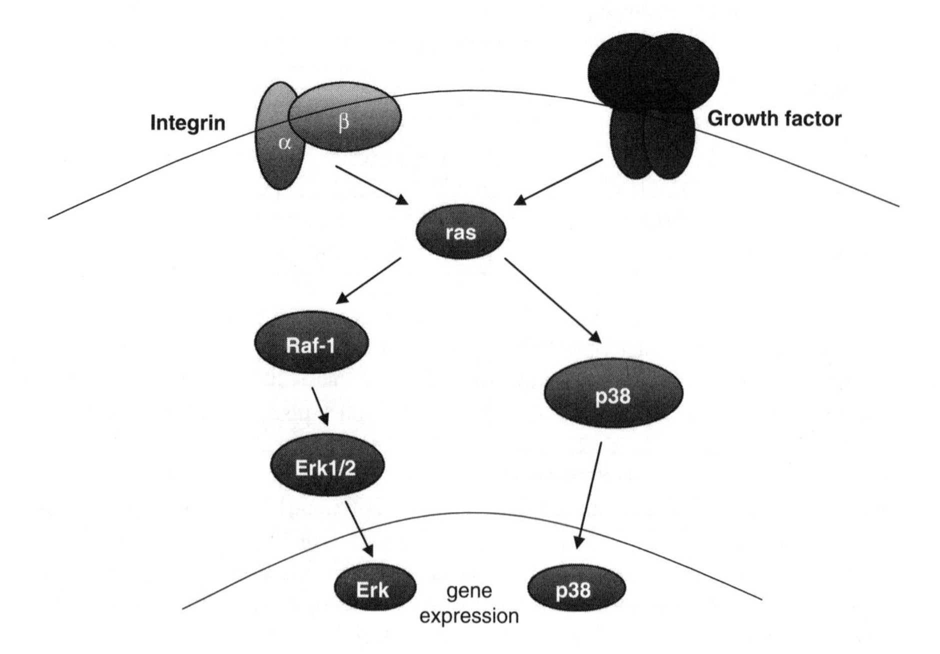

Figure 5.2 Integrin and growth factors synergistically activate tissue-specific gene expression for differentiation via the MAP kinase pathway.

It has been documented that a type I collagen interaction with cells via $\alpha_1\beta_1$ or $\alpha_2\beta_1$ integrins activates the MAP kinase pathway.[150,154] BMP-7 activation of osteocalcin and bone sialoprotein mRNA was synergistically enhanced by collagen–integrin interactions.[155] Although well established in systems other than bone, growth factor–ECM synergy is not well documented in the bone microenvironment (Figure 5.2). However, it is possible that ECM proteins may synergize with BMP-2 via convergence of the MAP kinase pathway to enhance osteoblast differentiation. Inclusion of a combination of signaling molecules within a tissue-engineered scaffold would be advantageous for wound healing.

It is becoming increasingly clear that one factor alone will not successfully heal a wound. The natural tissue microenvironment is a complex mixture of growth factors and ECM proteins that are unique to each tissue. The proper concentrations, temporal and spatial expression are important factors in determining function. The ideal tissue-engineered therapy should closely mimic the natural make–up of the bone healing environment. This will likely require both growth factors and ECM proteins that work together to enhance osteoblastic gene expression that is responsible for tissue remodeling.

CONCLUSION

The above studies are beginning to demonstrate the potential of targeted growth factor delivery to contribute to the regeneration of diseased tissues, and enhance the healing potential of grafted tissues. In the future, a better understanding of the molecular mechanisms of disease states, the molecules involved in the regeneration process, the microenvironment of healing tissues, and the signaling pathways will enable researchers to design better devices and therapies. It is also critical to understand the relationships among growth factors and between growth factors and extracellular matrices. This knowledge will provide the foundation of fabricating devices that could deliver several growth factors in a temporal manner and at optimal concentrations to achieve more effective therapies. The biological principles will serve as the blueprint for device designs to play a physiologic modulator of the tissue to be regenerated.

References

1. Anderson, J.M. (1998). Biocompatability of Tissue Engineered Implants, in *Frontiers in Tissue Engineering*, L.V. McIntire, Ed., Pergamon: Oxford, U.K.; New York, NY. p. 152, 165.
2. Nimni, M.E. (1997). Polypeptide growth factors: targeted delivery systems. *Biomaterials*, 18 (18): 1201–1225.
3. McKay, I. and Brown, K.D. (1998). *Growth Factors and Receptor: A Practical Approach*. Practical Approach Series; 194. Oxford, England ; New York: Oxford University Press. xxii, 266.
4. Tabata, T. (2001). Genetics of morphogen gradients. *Nat. Rev. Genet.* 2(8): 620–630.
5. Monteiro, A. et al. (2001). Butterfly eyespot patterns: evidence for specification by a morphogen diffusion gradient. *Acta Biotheor.* 49(2): 77–88.
6. Singer, A.J. and Clark, R.A. (1999). Cutaneous wound healing. *N. Engl. J. Med.* 341: 738–746.

7. Ferrara, N. and Gerber, H.P. (2001). The role of vascular endothelial growth factor in angiogenesis. *Acta Haematol.*106(4): 148-156.
8. Rifkin, D.B. et al. (1999). Proteolytic control of growth factor availability. *Apmis.* 107(1): 80–85.
9. Mohan, S. and Baylink, D.J. (2002). IGF-binding proteins are multifunctional and act via IGF-dependent and -independent mechanisms. *J. Endocrinol.* 2002. 175(1): 19–31.
10. Nunes, I. et al. (1997). Latent transforming growth factor-beta binding protein domains involved in activation and transglutaminase-dependent cross-linking of latent transforming growth factor-beta. *J, Cell Biol.* 136(5): 1151–1163.
11. Balemans, W. and Van Hul, W. (2002). Extracellular regulation of BMP signaling in vertebrates: a cocktail of modulators. *Dev. Biol.* 250(2): 231–250.
12. Shimomura, T. et al. (1995). Activation of hepatocyte growth factor by two homologous proteases, blood-coagulation factor XIIa and hepatocyte growth factor activator. *Eur J Biochem.* 229(1): 257–261.
13. LaRochelle, W.J. et al. (1991). A novel mechanism regulating growth factor association with the cell surface: identification of a PDGF retention domain. *Genes Dev.* 1991. 5(7): 1191–1199.
14. Sahni, A., Odrljin, T. and Francis, C.W. (1998). Binding of basic fibroblast growth factor to fibrinogen and fibrin. *J Biol Chem.* 273(13): 7554–7559.
15. Sahni, A. and Francis, C.W. (2000). Vascular endothelial growth factor binds to fibrinogen and fibrin and stimulates endothelial cell proliferation. *Blood* 96(12): 3772–3778.
16. Mroczkowski, B. et al. (1989). Recombinant human epidermal growth factor precursor is a glycosylated membrane protein with biological activity. *Mol. Cell. Biol.* 9(7): 2771–2778.
17. Anderson, D.M. et al. (1990). Molecular cloning of mast cell growth factor, a hematopoietin that is active in both membrane bound and soluble forms. *Cell* 63(1): 235–243.
18. Derynck, R. et al. (1984). Human transforming growth factor-alpha: precursor structure and expression in E. coli. *Cell.* 1984. 38 (1): 287–297.
19. Baxter, R.C. (2000). Insulin-like growth factor (IGF)-binding proteins: interactions with IGFs and intrinsic bioactivities. *Am. J, Physiol. Endocrinol. Metab.* 278 (6): E967– E976.
20. Chan, K. and Spencer, E.M. (1998). Megakaryocytes endocytose insulin-like growth factor (IGF) I and IGF-binding protein-3: a novel mechanism directing them into alpha granules of platelets. *Endocrinology.* 139(2): 559–565.
21. Najean, Y., Ardaillou, N. and Dresch, C. (1969). Platelet lifespan. *Annu. Rev. Med.*20: 47–62.
22. Campbell, P.G. et al. (1992). Involvement of the plasmin system in dissociation of the insulin-like growth factor-binding protein complex. *Endocrinology* 130(3): 1401–1412.
23. Saksela, O. and Rifkin, D.B. (1990). Release of basic fibroblast growth factor-heparan sulfate complexes from endothelial cells by plasminogen activator-mediated proteolytic activity. *J. Cell Biol.* 110(3): 767–775.
24. Houck, K.A. et al. (1992). Dual regulation of vascular endothelial growth factor bioavailability by genetic and proteolytic mechanisms. *J. Biol. Chem.* 267(36): 26031–26037.
25. Naldini, L. et al. 1992. Extracellular proteolytic cleavage by urokinase is required for activation of hepatocyte growth factor/scatter factor. *EMBO J.* 11(13): 4825–4833.
26. Saltzman, W.M. and Olbreicht, W.L. (2002). Building drug delivery into tissue engineering. *Nature Reviews: Drug Discovery*,. 1 March pp. (177–186).

27. Poynton, A.R. and Lane, J.M. (2002). Safety profile for the clinical use of bone morphogenetic proteins in the spine. *Spine*. 27(16 Suppl 1): S40–S48.
28. Richardson, T.P. et al. (2001). Polymeric system for dual growth factor delivery. *Nat. Biotechnol.* 19(11): 1029–1034.
29. Lai, C.F. and Cheng, S.L. (2002). Signal transductions induced by bone morphogenetic protein-2 and transforming growth factor-beta in normal human osteoblastic cells. *J. Biol. Chem.* 277(18): 15514–15522.
30. Li, R.W. and Wozney, J.M. (2001). Delivering on the promise of bone morphogenetic proteins. *Trends Biotechnology*. 19(7): 255–265.
31. Gittens, S.A. and Uludag, H. (2001). Growth factor delivery for bone tissue engineering. *J. Drug Target* 9(6): 407–429.
32. Chen, Y. (2001). Orthopedic applications of gene therapy. *J. Orthop. Sci.* 6(2): 199–207.
33. Massague, J. and Wotton, D. (2000). Transcriptional control by the TGF-beta/Smad signaling system. *EMBO J* 19(8): 1745–1754.
34. Zhang, M. et al. (2002). Osteoblast-specific knockout of the IGF receptor gene reveals an essential role of IGF signaling in bone matrix mineralization. *J. Biol. Chem.* 277(46): 44005–44012
35. Nixon, A.J. et al. (2000). Insulinlike growth factor-I gene therapy applications for cartilage repair. *Clin. Orthop*. 379: S201–S213.
36. Kurtz, C.A. et al. (1999). Insulin-like growth factor I accelerates functional recovery from Achilles tendon injury in a rat model. *Am. J. Sports Med.* 27(3): 363–369.
37. Mansukhani, A. et al. (2000). Signaling by fibroblast growth factors (FGF) and fibroblast growth factor receptor 2 (FGFR2)-activating mutations blocks mineralization and induces apoptosis in osteoblasts. *J. Cell Biol.* 149(6): 1297–1308.
38. Olofsson, B. et al. (1999). Current biology of VEGF-B and VEGF-C. *Curr. Opin. Biotechnol.* 10(6): 528–535.
39. Giannobile, W.V. (1996). Periodontal tissue engineering by growth factors. *Bone* 19(1 Suppl): 23S–37S.
40. Soker, S., Machado, M. and Atala, A. (2000). Systems for therapeutic angiogenesis in tissue engineering. *World. J. Urol.* 18(1): 10–18.
41. Reddi, A.H. (1994). Symbiosis of biotechnology and biomaterials: applications in tissue engineering of bone and cartilage. *J. Cell. Biochem.* 56(2): 192–195.
42. Abe, N., et al. (2002). Enhancement of bone repair with a helper-dependent adenoviral transfer of bone morphogenetic protein-2. *Biochem. Biophys. Res. Commun.* 297(3): 523–527.
43. Joyce, M.E., Jingushi, S. and Bolander, M.E. (1990). Transforming growth factor-beta in the regulation of fracture repair. *Orthop. Clin. North Am.* 21(1): 199–209.
44. Bourque, W.T., Gross, M. and Hall, B.K. (1993). Expression of four growth factors during fracture repair. *Int. J. Dev. Biol.* 37(4): 573–579.
45. Robey, P.G. et al. (1987). Osteoblasts synthesize and respond to transforming growth factor-type beta (TGF-beta) *in vitro. J. Cell Biol.* 105(1): 457–463.
46. Kale, S. et al. (2000). Three-dimensional cellular development is essential for *ex vivo* formation of human bone. *Nat. Biotechnol.* 18(9): 954–958.
47. Critchlow, M.A., Bland, Y.S. and Ashhurst, D.E. (1995). The effect of exogenous transforming growth factor-beta 2 on healing fractures in the rabbit. *Bone* 16(5): 521–527.
48. Centrella, M., McCarthy, T.L. and Canalis, E. (1991). Transforming growth factor-beta and remodeling of bone. *J. Bone Joint Surg. Am.* 73(9): 1418–1428.
49. Zhang, H., Ahmad, M. and Gronowicz, G. (2003). Effects of transforming growth factor-beta 1 (TGF-beta1) on *in vitro* mineralization of human osteoblasts on implant materials. *Biomaterials* 24(12): 2013–2020.

50. Park, S.H., O'Connor, K.M. and McKellop, H. (2003). Interaction between active motion and exogenous transforming growth factor Beta during tibial fracture repair. *J. Orthop. Trauma*. 17(1): 2–10.
51. Aspenberg, P. et al. (1996). Transforming growth factor beta and bone morphogenetic protein 2 for bone ingrowth: a comparison using bone chambers in rats. *Bone* 19(5): 499–503.
52. Ripamonti, U. et al. (1997). Recombinant transforming growth factor-beta1 induces endochondral bone in the baboon and synergizes with recombinant osteogenic protein-1 (bone morphogenetic protein-7) to initiate rapid bone formation. *J Bone Miner. Res*. 12(10): 1584–1595.
53. Lind, M. et al. (1993). Transforming growth factor-beta enhances fracture healing in rabbit tibiae. *Acta Orthop. Scand*. 64(5): 553–556.
54. Boden, S.D. (1999). Bioactive factors for bone tissue engineering. *Clin. Orthop*. (367 Suppl): S84–S94.
55. Shi, Y. and Massague, J. (2003). Mechanisms of TGF-beta signaling from cell membrane to the nucleus. *Cell*.. 113(6): 685–700.
56. Schmitt, J.M. et al. (1999). Bone morphogenetic proteins: an update on basic biology and clinical relevance. *J. Orthop. Res*.1999. 17(2): 269–278.
57. Urist, M.R. (1965). Bone: formation by autoinduction. *Science* 150(698): 893–899.
58. Wozney, J.M. et al. (1988). Novel regulators of bone formation: molecular clones and activities. *Science* 242(4885): 1528–1534.
59. Pittenger, M.F. et al. (1999). Multilineage potential of adult human mesenchymal stem cells. *Science* 284(5411): 143–147.
60. Prockop, D.J. (1997). Marrow stromal cells as stem cells for nonhematopoietic tissues. *Science*276(5309): 71–74.
61. Hay, E. et al. (2001). Bone morphogenetic protein-2 promotes osteoblast apoptosis through a Smad-independent, protein kinase C-dependent signaling pathway. *J. Biol. Chem*. 276(31): 29028–29036.
62. Shi, W. et al. (2001). Gremlin negatively modulates BMP-4 induction of embryonic mouse lung branching morphogenesis. *Am. J. Physiol. Lung Cell Mol. Physiol*. 280(5): L1030–L1039.
63. Trousse, F., Esteve, P. and Bovolenta, P. (2001). Bmp4 mediates apoptotic cell death in the developing chick eye. *J. Neurosci*. 21(4): 1292–1301.
64. Lieberman, J.R., Daluiski, A. and Einhorn, T.A. (2002). The role of growth factors in the repair of bone. Biology and clinical applications. *J. Bone Joint Surg. Am*. 84-A(6): 1032–1044.
65. Bahamonde, M.E. and Lyons, K.M. (2001). BMP3: to be or not to be a BMP. *J. Bone Joint Surg. Am*. 83-A(Suppl 1(Pt 1)): p. S56–S62.
66. Han, B. et al. (2002). Collagen-targeted BMP3 fusion proteins arrayed on collagen matrices or porous ceramics impregnated with Type I collagen enhance osteogenesis in a rat cranial defect model. *J. Orthop. Res*. 20(4): 747–755.
67. Campisi, P. et al. (2003). Expression of bone morphogenetic proteins during mandibular distraction osteogenesis. *Plast. Reconstr. Surg*. 111(1): 201-8; discussion 209–210.
68. Saito, N. et al. (2003). Local bone formation by injection of recombinant human bone morphogenetic protein-2 contained in polymer carriers. *Bone* 32(4): 381–386.
69. Breitbart, A.S. et al. (1999). Gene-enhanced tissue engineering: applications for bone healing using cultured periosteal cells transduced retrovirally with the BMP-7 gene. *Ann. Plast. Surg*. 42(5): 488–495.
70. Turgeman, G. et al. (2001). Engineered human mesenchymal stem cells: a novel platform for skeletal cell mediated gene therapy. *J. Gene Med*. 3(3): 240–251.

71. Yasko, A.W. et al. (1992). The healing of segmental bone defects, induced by recombinant human bone morphogenetic protein (rhBMP-2). A radiographic, histological, and biomechanical study in rats. *J. Bone Joint Surg. Am.* 74(5): 659–670.
72. Bostrom, M. et al. (1996). Use of bone morphogenetic protein-2 in the rabbit ulnar nonunion model. *Clin. Orthop.* (327): 272–282.
73. Gerhart, T.N. et al. (1993). Healing segmental femoral defects in sheep using recombinant human bone morphogenetic protein. *Clin. Orthop.* (293): 317–326.
74. Sciadini, M.F. and Johnson, K.D. (2000). Evaluation of recombinant human bone morphogenetic protein-2 as a bone-graft substitute in a canine segmental defect model. *J. Orthop. Res.* 18(2): 289–302.
75. Boden, S.D. et al. (1998). Laparoscopic anterior spinal arthrodesis with rhBMP-2 in a titanium interbody threaded cage. *J. Spinal Disord.* 11(2): 95–101.
76. Hollinger, J.O., Winn, S. and Bonadio, J. (2000). Options for tissue engineering to address challenges of the aging skeleton. *Tissue Eng.* 6(4): 341–350.
77. Baskin, D.S. et al. (2003). A prospective, randomized, controlled cervical fusion study using recombinant human bone morphogenetic protein-2 with the CORNERSTONE-SR trade mark allograft ring and the ATLANTIS trade mark anterior cervical plate. *Spine* 28(12): 1219–1225.
78. Burkus, J.K. et al. (2003). Is INFUSE bone graft superior to autograft bone? An integrated analysis of clinical trials using the LT-CAGE lumbar tapered fusion device. *J. Spinal Disord. Tech.* 16(2): 113–122.
79. Baffour, R. et al. (1992). Enhanced angiogenesis and growth of collaterals by in vivo administration of recombinant basic fibroblast growth factor in a rabbit model of acute lower limb ischemia: dose-response effect of basic fibroblast growth factor. *J. Vasc. Surg.* 16(2): 181–191.
80. Flores-Sarnat, L. (2002). New insights into craniosynostosis. *Semin. Pediatr. Neurol.* 9(4): 274–291.
81. Marie, P.J., Debiais, F. and Hay, E. (2002). Regulation of human cranial osteoblast phenotype by FGF-2, FGFR-2 and BMP-2 signaling. *Histol. Histopathol.* 17(3): 877–885.
82. Canalis, E., McCarthy, T.L. and Centrella, M. (1989). Effects of platelet-derived growth factor on bone formation *in vitro*. *J. Cell. Physiol.* 140(3): 530–537.
83. Jingushi, S. et al. (1990). Acidic fibroblast growth factor (aFGF) injection stimulates cartilage enlargement and inhibits cartilage gene expression in rat fracture healing. *J. Orthop. Res.* 8(3): 364–371.
84. Zhang, X., Sobue, T. and Hurley, M.M. (2002). FGF-2 increases colony formation, PTH receptor, and IGF-1 mRNA in mouse marrow stromal cells. *Biochem. Biophys. Res. Commun.* 290(1): 526–531.
85. Burgess, W.H. and Maciag, T. (1989). The heparin-binding (fibroblast) growth factor family of proteins. *Annu. Rev.Biochem.* 58: 575–606.
86. Burgess, W.H. et al. (1990). Characterization and cDNA cloning of phospholipase C-gamma, a major substrate for heparin-binding growth factor 1 (acidic fibroblast growth factor)-activated tyrosine kinase. *Mol. Cell. Biol.* 10(9): 4770–4777.
87. Orban, J.M., Marra, K.G. and Hollinger, J.O. (2002). Composition options for tissue-engineered bone. *Tissue Eng.* 8(4): 529–539.
88. Xiao, G. et al. (2002). Fibroblast growth factor 2 induction of the osteocalcin gene requires MAPK activity and phosphorylation of the osteoblast transcription factor, Cbfa1/Runx2. *J. Biol. Chem.* 277(39): 36181–36187.
89. Murakami, S. et al. (2003). Recombinant human basic fibroblast growth factor (bFGF) stimulates periodontal regeneration in class II furcation defects created in beagle dogs. *J. Periodontal Res.* 38(1): 97–103.

90. Goodman, S.B. et al. (2003). Local infusion of FGF-2 enhances bone ingrowth in rabbit chambers in the presence of polyethylene particles. *J. Biomed. Mater. Res.* 65A(4): 454–461.
91. Franke Stenport, V. et al. (2003). FGF-4 and titanium implants: a pilot study in rabbit bone. *Clin. Oral Implants Res.* 14(3): 363–368.
92. Dunstan, C.R. et al. (1999). Systemic administration of acidic fibroblast growth factor (FGF-1) prevents bone loss and increases new bone formation in ovariectomized rats. *J. Bone Miner. Res.* 14(6): 953–959.
93. Power, R.A., Iwaniec, U.T. and Wronski, T.J. (2002). Changes in gene expression associated with the bone anabolic effects of basic fibroblast growth factor in aged ovariectomized rats. *Bone* 31(1): 143–148.
94. Pandit, A.S. et al. (1998). Stimulation of angiogenesis by FGF-1 delivered through a modified fibrin scaffold. *Growth Factors* 15(2): 113–123.
95. Montero, A. et al. (2000). Disruption of the fibroblast growth factor-2 gene results in decreased bone mass and bone formation. *J. Clin. Invest.* 105(8): 1085–1093.
96. Wergedal, J.E. et al. (1990). Skeletal growth factor and other growth factors known to be present in bone matrix stimulate proliferation and protein synthesis in human bone cells. *J. Bone Miner. Res.* 5(2): 179–186.
97. Canalis, E. (1980). Effect of insulinlike growth factor I on DNA and protein synthesis in cultured rat calvaria. *J. Clin. Invest.* 66(4): 709–719.
98. McCarthy, T.L., Centrella, M. and Canalis, E. (1989). Regulatory effects of insulin-like growth factors I and II on bone collagen synthesis in rat calvarial cultures. *Endocrinology* 124(1): 301–309.
99. Middleton, J. et al. (1995). Osteoblasts and osteoclasts in adult human osteophyte tissue express the mRNAs for insulin-like growth factors I and II and the type 1 IGF receptor. *Bone* 16(3): 287–293.
100. Andrew, J.G. et al. (1993). Insulinlike growth factor gene expression in human fracture callus. *Calcif. Tissue Int.* 53(2): 97–102.
101. Raile, K. et al. (1994). Human osteosarcoma (U-2 OS) cells express both insulin-like growth factor-I (IGF-I) receptors and insulin-like growth factor-II/mannose-6- phosphate (IGF-II/M6P) receptors and synthesize IGF-II: autocrine growth stimulation by IGF-II via the IGF-I receptor. *J. Cell. Physiol.* 159(3): 531–541.
102. Lazowski, D.A. et al. (1994). Regional variation of insulin-like growth factor-I gene expression in mature rat bone and cartilage. *Bone* 15(5): 563–576.
103. Rubin, J. et al. (2002). IGF-I regulates osteoprotegerin (OPG) and receptor activator of nuclear factor-kappaB ligand *in vitro* and OPG *in vivo*. *J. Clin. Endocrinol. Metab.* 87(9): 4273–4279.
104. Mochizuki, H. et al. (1992). Insulin-like growth factor-I supports formation and activation of osteoclasts. *Endocrinology* 131(3): 1075–1080.
105. Guicheux, J. et al. (1998). Growth hormone stimulatory effects on osteoclastic resorption are partly mediated by insulin-like growth factor I: an *in vitro* study. *Bone* 22(1): 25–31.
106. Zhang, M. et al. (2003). Paracrine overexpression of IGFBP-4 in osteoblasts of transgenic mice decreases bone turnover and causes global growth retardation. *J. Bone Miner. Res.* 18(5): 836–843.
107. Okazaki, K. et al. (20003). Expression of parathyroid hormone-related peptide and insulin-like growth factor I during rat fracture healing. *J. Orthop. Res.* 21(3): 511–520.
108. Banu, J., Wang, L. and Kalu, D.N. (2003). Effects of Increased Muscle Mass on Bone in Male Mice Overexpressing IGF-I in Skeletal Muscles. *Calcif. Tissue Int.* 73(2) 196–201.

109. McCarthy, T.L. et al. (1991). Prostaglandin E2 stimulates insulin-like growth factor I synthesis in osteoblast-enriched cultures from fetal rat bone. *Endocrinology* 128(6): 2895–900.
110. Bikle, D.D. et al. (2002). Insulin-like growth factor I is required for the anabolic actions of parathyroid hormone on mouse bone. *J. Bone Miner. Res.* 17(9): 1570–1578.
111. Pfeilschifter, J. et al. (1990). Stimulation of bone matrix apposition in vitro by local growth factors: a comparison between insulin-like growth factor I, platelet-derived growth factor, and transforming growth factor beta. *Endocrinology* 127(1): 69–75.
112. Thaller, S.R., Dart, A. and Tesluk, H. (1993). The effects of insulin-like growth factor-1 on critical-size calvarial defects in Sprague-Dawley rats. *Ann. Plast. Surg.* 31(5): 429–433.
113. Tanaka, H. et al. (1994). *In vivo* and *in vitro* effects of insulin-like growth factor-I (IGF-I) on femoral mRNA expression in old rats. *Bone* 15(6): 647–653.
114. Shen, F.H. et al. (2002). Systemically administered mesenchymal stromal cells transduced with insulin-like growth factor-I localize to a fracture site and potentiate healing. *J. Orthop. Trauma* 16(9): 651–659.
115. Lee, C.W. et al. (2002). Muscle-based gene therapy and tissue engineering for treatment of growth plate injuries. *J. Pediatr. Orthop.* 22(5): 565–572.
116. Madry, H. et al. (2002). Gene transfer of a human insulin-like growth factor I cDNA enhances tissue engineering of cartilage. *Hum. Gene Ther.* 13(13): 1621–1630.
117. Tanaka, H. et al. (2002). Effect of IGF-I and PDGF administered *in vivo* on the expression of osteoblast-related genes in old rats. *J. Endocrinol.* 174(1): 63–70.
118. Thomas, T. et al. (1999). Response of bipotential human marrow stromal cells to insulin-like growth factors: effect on binding protein production, proliferation, and commitment to osteoblasts and adipocytes. *Endocrinology* 140(11): 5036–5044.
119. Trippel, S.B. (1997). Growth factors as therapeutic agents. *Instr. Course Lect.*46: 473–476.
120. Andrew, J.G. et al. (1995). Platelet-derived growth factor expression in normally healing human fractures. *Bone* 16(4): 455–460.
121. Giannobile, W.V. et al. (2001). Platelet-derived growth factor (PDGF) gene delivery for application in periodontal tissue engineering. *J. Periodontol.* 72(6): 815–823.
122. Fiedler, J. et al. (2002). BMP-2, BMP-4, and PDGF-bb stimulate chemotactic migration of primary human mesenchymal progenitor cells. *J. Cell. Biochem.* 87(3): 305–312.
123. Kubota, K. et al. (2002). Platelet-derived growth factor BB secreted from osteoclasts acts as an osteoblastogenesis inhibitory factor. *J. Bone Miner. Res.* 17(2): 257–265.
124. Bouletreau, P.J. et al. (2002). Factors in the fracture microenvironment induce primary osteoblast angiogenic cytokine production. *Plast. Reconstr. Surg.* 110(1): 139–148.
125. Tyndall, W.A. et al. (2003). Decreased platelet derived growth factor expression during fracture healing in diabetic animals. *Clin Orthop.* 48: 319–330.
126. Im, S.Y. et al. (2003). Growth factor releasing porous poly (epsilon-caprolactone)-chitosan matrices for enhanced bone regenerative therapy. *Arch. Pharm. Res.* 26(1): 76–82.
127. Carmeliet, P. (2000). Mechanisms of angiogenesis and arteriogenesis. *Nat. Med.* 6(4): 389–395.
128. Bates, D.O., Lodwick, D. and Williams, B. (1999). Vascular endothelial growth factor and microvascular permeability. *Microcirculation* 6(2): 83–96.
129. Yancopoulos, G.D. et al. (2000). Vascular-specific growth factors and blood vessel formation. *Nature*. 407(6801): 242–248.
130. Ferrara, N., Gerber, H.P. and LeCouter, J. (2003). The biology of VEGF and its receptors. *Nat. Med.* 9(6): 669–676.

131. Takeshita, S. et al. (1997). Microangiographic assessment of collateral vessel formation following direct gene transfer of vascular endothelial growth factor in rats. *Cardiovasc. Res*. 35(3): 547–552.
132. Stone, D. et al. (2002). A biologically active VEGF construct *in vitro*: implications for bioengineering-improved prosthetic vascular grafts. *J. Biomed. Mater. Res*. 59(1): 160–165.
133. Frerich, B. et al. (2001). *In vitro* model of a vascular stroma for the engineering of vascularized tissues. *Int. J. Oral Maxillofac. Surg* 30(5): 414–420.
134. Nehls, V., Herrmann, R. and M. Huhnken, M. (1998). Guided migration as a novel mechanism of capillary network remodeling is regulated by basic fibroblast growth factor. *Histochem. Cell Biol*. 109(4): 319–329.
135. Uchida, S. et al. (2003). Vascular endothelial growth factor is expressed along with its receptors during the healing process of bone and bone marrow after drill-hole injury in rats. *Bone* 32(5): 491–501.
136. Hiltunen, M.O. et al. (2003). Adenovirus-mediated VEGF-A gene transfer induces bone formation *in vivo*. *FASEB J*. 17(9): 1147–1149.
137. McCarthy, T.L. et al. (1998). Alternate signaling pathways selectively regulate binding of insulin-like growth factor I and II on fetal rat bone cells. *J. Cell. Biochem*. 68(4): 446–456.
138. Ogata, N. et al. (2000). Insulin receptor substrate-1 in osteoblast is indispensable for maintaining bone turnover. *J. Clin. Invest*. 105(7): 935–943.
139. Boguslawski, G. et al. (2000). Activation of osteocalcin transcription involves interaction of protein kinase A- and protein kinase C-dependent pathways. *J. Biol. Chem*. 275(2): 999–1006.
140. Bae, S.C. et al. (2001). Intimate relationship between TGF-beta/BMP signaling and runt domain transcription factor, PEBP2/CBF. *J Bone Joint Surg Am*. 83-A(Suppl 1)(Part 1): S48–S55.
141. Barboza, E., Caula, A. and Machado, F. (1999). Potential of recombinant human bone morphogenetic protein-2 in bone regeneration. *Implant Dent* .8(4): 360–367.
142. Boden, S.D. et al. (2000). The use of rhBMP-2 in interbody fusion cages. Definitive evidence of osteoinduction in humans: a preliminary report. *Spine* 25(3): 376–381.
143. Cochran, D.L. et al. (2000). Evaluation of recombinant human bone morphogenetic protein-2 in oral applications including the use of endosseous implants: 3-year results of a pilot study in humans. *J. Periodontol*. 71(8): 1241–1257.
144. Boyne, P.J. et al. (1997). A feasibility study evaluating rhBMP-2/absorbable collagen sponge for maxillary sinus floor augmentation. *Int. J. Periodontics Restorative Dent*. 17(1): 11–25.
145. Nohe, A. et al. (2002). The mode of bone morphogenetic protein (BMP) receptor oligomerization determines different BMP-2 signaling pathways. *J. Biol. Chem*. 277(7): 5330–5338.
146. Ebara, S. and Nakayama, K. (2002). Mechanism for the action of bone morphogenetic proteins and regulation of their activity. *Spine*. 27(16 Suppl 1): S10–S15.
147. Kretzschmar, M., Doody, J. and Massague, J. (1997). Opposing BMP and EGF signalling pathways converge on the TGF-beta family mediator Smad1. *Nature* 389(6651): 618–622.
148. Lee, K.S. et al. (2000). Runx2 is a common target of transforming growth factor beta1 and bone morphogenetic protein 2, and cooperation between Runx2 and Smad5 induces osteoblast-specific gene expression in the pluripotent mesenchymal precursor cell line C2C12. *Mol. Cell. Biol*. 20(23): 8783–8792.

149. Gallea, S. et al. (2001). Activation of mitogen-activated protein kinase cascades is involved in regulation of bone morphogenetic protein-2-induced osteoblast differentiation in pluripotent C2C12 cells. *Bone* 28(5): 491–498.
150. Xiao, G. et al. (2000). MAPK pathways activate and phosphorylate the osteoblast-specific transcription factor, Cbfa1. *J. Biol. Chem.* 275(6): 4453–4459.
151. Lee, K.S., Hong, S.H. and Bae, S.C. (2002). Both the Smad and p38 MAPK pathways play a crucial role in Runx2 expression following induction by transforming growth factor-beta and bone morphogenetic protein. *Oncogene* 21(47): 7156–7163.
152. Lai, C.F. et al. (2001). Erk is essential for growth, differentiation, integrin expression, and cell function in human osteoblastic cells. *J. Biol. Chem.* 276(17): p. 14443–14450.
153. Higuchi, C. et al. (2002). Continuous inhibition of MAPK signaling promotes the early osteoblastic differentiation and mineralization of the extracellular matrix. *J. Bone Miner. Res.* 17(10): 1785–1794.
154. Mizuno, M., Fujisawa, R. and Kuboki, Y. (2000). Type I collagen-induced osteoblastic differentiation of bone-marrow cells mediated by collagen-alpha2beta1 integrin interaction. *J. Cell. Physiol.* 184(2): 207–213.
155. Xiao, G. et al. (2002). Bone morphogenetic proteins, extracellular matrix, and mitogen-activated protein kinase signaling pathways are required for osteoblast-specific gene expression and differentiation in MC3T3-E1 cells. *J. Bone Miner. Res.* 17(1): 101–110.

6 Biodegradable Polymers and Microspheres in Tissue Engineering

Kacey G. Marra

Contents

0-8493-1621-9/05/$0.00+$1.50
© 2005 by CRC Press LLC

INTRODUCTION

In the United States, there are over 500,000 bone grafts per year to replace or repair diseased or damaged bone. Autologous bone graft has long been considered the clinical "gold standard." Harvest of autograft, however, can lead to complications including chronic harvest site pain, infection, nerve damage, cosmetic deformity, and hemorrhage. In addition, autograft harvest increases operative time and cost. Allograft (e.g., cadaver bone), has been proposed as an effective alternative; yet, this material is also plagued by problems including immunogenicity, viral transmission, compromised physiologic and biomechanical properties, and potentially limited supply. Metal implants are frequently used for these purposes, but they cannot perform as efficiently as a healthy bone, and metallic structures cannot remodel with time. To help address the need for better bone substitutes, bone tissue engineers seek to create synthetic, three-dimensional bone scaffolds made from polymeric materials incorporating cells or growth factors to induce the growth of normal bone tissue. The following chapter describes the role of polymers in tissue regeneration, with an emphasis on bone, nerve, and skin regeneration. This chapter also describes the fabrication of tissue-engineered scaffolds, and the use of polymer microspheres to deliver growth factors in bone tissue engineering applications.

POLYMERIC SCAFFOLDS IN BONE TISSUE ENGINEERING

Polymeric scaffold materials are typically biodegradable (e.g., absorbable) and should support cell attachment and growth. The family of poly(α-hydroxy acids), for example, poly(lactic acid) (PLA) and poly(glycolic acid) (PGA), have been widely useful toward this end. Furthermore, copolymers of PLA and PGA have been modified with hydroxyapatite or tricalcium phosphate to impart mechanical strength and osteoconductivity.

The matrix, or scaffold, of a tissue-engineered construct is of significant consequence to bone regeneration. Variables such as pore size, mechanical properties, degradation processes, and surface chemistry, all contribute to overall scaffold viability. Biodegradable polyesters have had a vast impact in medicine during the last century. Particularly, PLA and PGA have been utilized in a countless number of clinical applications. Their structures are depicted in Figure 6.1. These polymers are degraded hydrolytically, resulting in a breakdown of the polymer chain, and they also undergo enzymatic breakdown *in vivo*.

POLY(GLYCOLIC ACID)

PGA (Figure 6.1a) is synthesized by the ring opening of glycolide; PGA is also referred to as poly(glycolide). PGA is a highly insoluble polymer; hexafluoroiso-

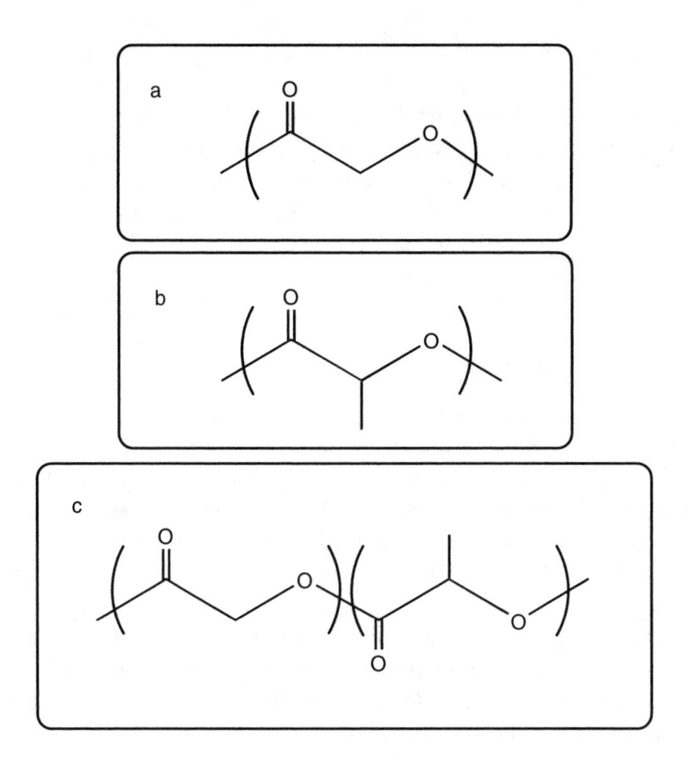

Figure 6.1 Structures of (a) PGA, (b) PLA, and (c) PLGA.

propanol is one of PGA's only solvents, and this solvent is highly toxic. Therefore, PGA is more commonly utilized as a copolymer with PLA. PGA can be extruded into fibers, which can then be braided into woven structures. PGA degrades to glycolic acid, which is then metabolized in the body.

Poly(Lactic Acid)

PLA (Figure 6.1b) is synthesized by the ring opening of lactide. As two enantiomeric isomers of lactide exist (e.g., D and L), both poly(L-lactic acid) (PLLA) and poly(D-lactic acid) (PDLA), have been examined as biomaterials. PLLA is more crystalline, and degrades more slowly than PDLA. PLA's degradation product is lactic acid, which is also metabolized, yielding CO_2 and water. A copolymerization of PDLA with PLLA yields PDLLA, which is being used clinically in craniofacial fixation applications. PDLLA is comprised of 70% PLLA and 30% PDLA, and is manufactured by Bionx Implants (as Biosorb), by Synthes Maxillofacial (as the Synthes Resorbable Fixation System), and by Macropore (Macrosorb FX). These implants degrade *in vivo* from 1 to 6 years.[1] PDLLA is manufactured in sheets, plates, screws, and pins.

Poly(Lactic-*co*-Glycolic Acid)

The copolymer of PLA and PGA, poly(lactic-*co*-glycolic acid) (PLGA), (Figure 6.1c) has been extensively studied in a variety of tissue engineering applications.

The PLGA copolymer that consists of 82% PLLA and 18% PGA is referred to as LactoSorb. LactoSorb was introduced in 1996 and is being sold by Lorenz Surgical Inc. LactoSorb was shown to completely resorb in 12 months,[2] and has been used clinically in pediatric craniofacial reconstruction for six years.

Another clinically useful PLGA polymer is DeltaSystem, manufactured by Stryker-Leibinger. This tripolymer consists of 85% PLLA, 5% PDLLA, and 10% PGA, and maintains 81% of its mechanical strength 8 weeks after implantation; complete resorption will occur between 1.5 and 3 years.[1]

PLGA/Ceramic Composites

Marra and colleagues have examined blending biodegradable polymers with ceramics in an attempt to improve the osteoconductivity of the scaffolds.[3,4] Combining calcium-based ceramics with biodegradable polymers is an approach that yields composites with desirable properties such as controlled degradation, improved osteoblast adhesion, and tailored mechanical strength. Figure 6.2a is a scanning electron micrograph of a physically blended composite of PLGA and hydroxyapatite, the natural mineral component of bone. In a higher magnification, Figure 6.2b, the mineral particles are evident within the polymer. We have examined a physical blend of PLGA, PCL, and hydroxyapatite in both rabbit heterotopic models and rat calvarial defects. Our preliminary results demonstrate that PLGA/PCL/HA composites are promising scaffolds for bone regeneration.

Polypropylene Fumarate

Poly(propylene fumarate) (Figure 6.3a) is used as an injectable, expandable porous implant for *in situ* regeneration of critical-sized defects.[5] This polymer was developed by Mikos' group in 1996[6] and has shown promising results in several *in vivo* studies.[7,8] Poly(propylene fumarate) can be photo-crosslinked to form hydrogels or injectable implants.

Other Biodegradable Polymers

Other biodegradable polymers that are being examined as bone tissue engineering scaffolds include poly(caprolactone) (PCL) (Figure 6.3b). Poly(caprolactone) is a polyester that has been examined in a porcine orbital defect model.[9] However, the most significant bone growth was observed when the polycaprolactone scaffold was seeded with bone marrow cells. The role of cells in bone biology is discussed further in chapter 3.

Kohn's group has extensively studied polycarbonates modified with amino acids for potential applications in bone tissue engineering.[10,11] Figure 6.3c shows the structure of poly(dioxanone) (manufactured by Ethicon), which has been examined clinically in cranial vault procedures with promising results.[12]

Recently under evaluation are novel polymers, termed dendrimers, for potential use in bone tissue engineering. Dendrimers differ from traditional straight-chain polymers in that their synthesis is initiated with a central core molecule. Polymerization of more branched molecules ultimately leads to a globular structure with multiple

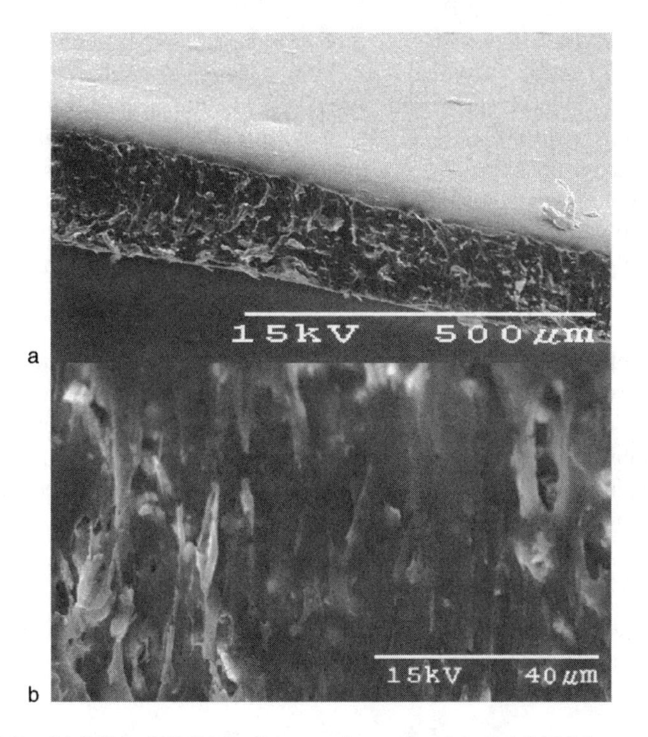

Figure 6.2 (a) SEM of PLGA/hydroxyapatite composite, and (b) higher magnification.

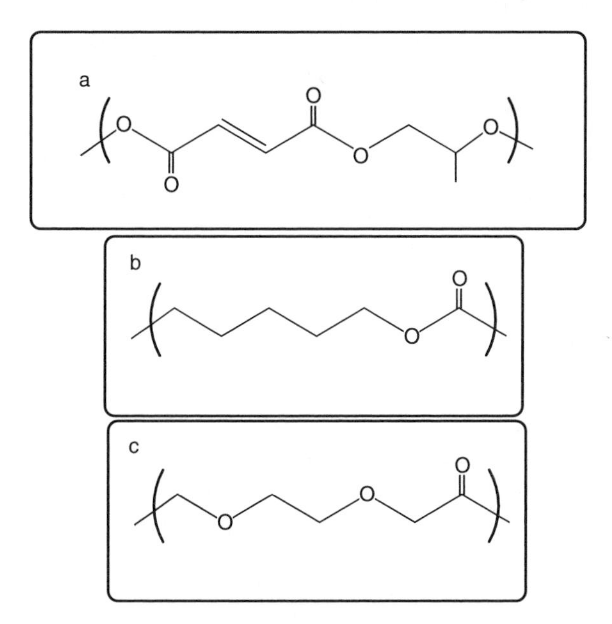

Figure 6.3 Structures of (a) poly(propylene fumarate), (b) poly(caprolactone), and (c) poly(dioxanone).

terminal functionalities. Dendrimers can be tailored to include ceramics, drugs, or growth factors via covalent bonding to the termini. Figure 6.4 shows the structure of a poly(ethylene glycol)/lysine-based dendrimer. Blending of this polymer with ceramics, such as hydroxyapatite, results in a scaffold with film-forming properties and osteoconductive potential. Figure 6.5 shows an SEM depicting the striated nature of the dendrimer and the ceramic particles within the composite. These novel polymers are of particular interest for their degradation properties (e.g., bulk vs. surface) and are currently under investigation in the laboratory of Toby Chapman.[13]

Electrically conductive polymers are being investigated as scaffold materials for electroresponsive cell types, including the bone and nerve. Hyaluronic acid/poly(pyrrole) composites have been found to stimulate angiogenesis in a subcutaneous rat model, although the wound healing benefits of this material have not been conclusively demonstrated.[14] Hyaluronic acid and collagen have been shown to exhibit promising osteoconductive properties.[15,16] Electrically conductive polymers are also being studied in nerve regeneration, and are described in Section Polymeric scaffolds in neuronal tissue engineering below.

Future of Polymers in Bone Regeneration

The use of polymers in bone tissue engineering has been, and remains to be widely studied. The design of novel biomaterials has a promising future in bone tissue engi-

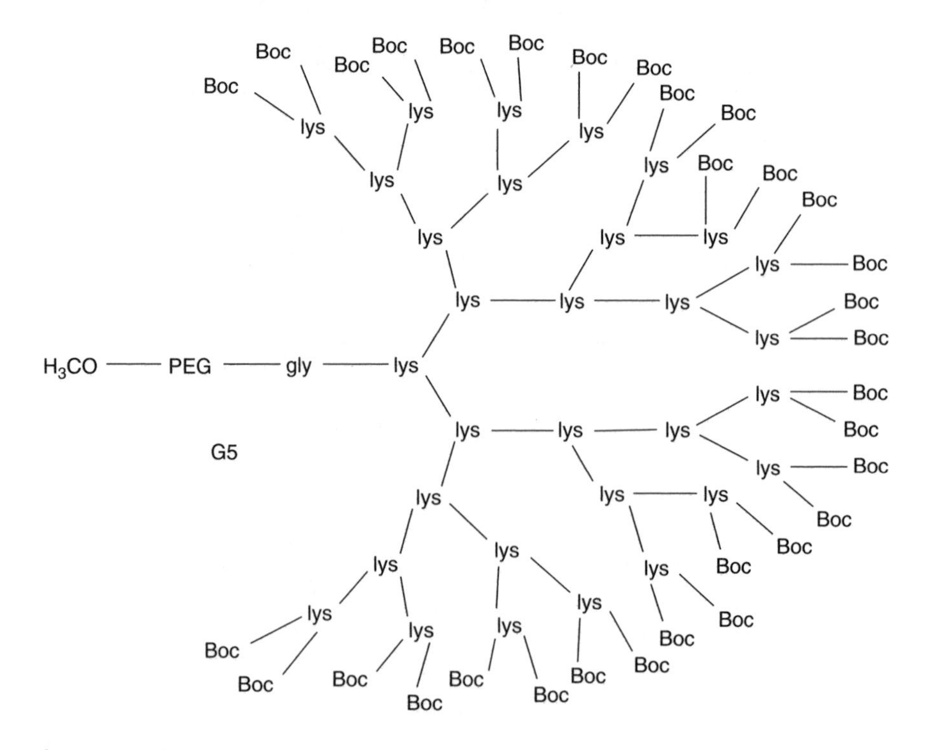

Figure 6.4 Structure of generation 5 polylysine-based dendrimer. PEG=poly(ethylene glycol); lys=lysine; BOC=protecting group.

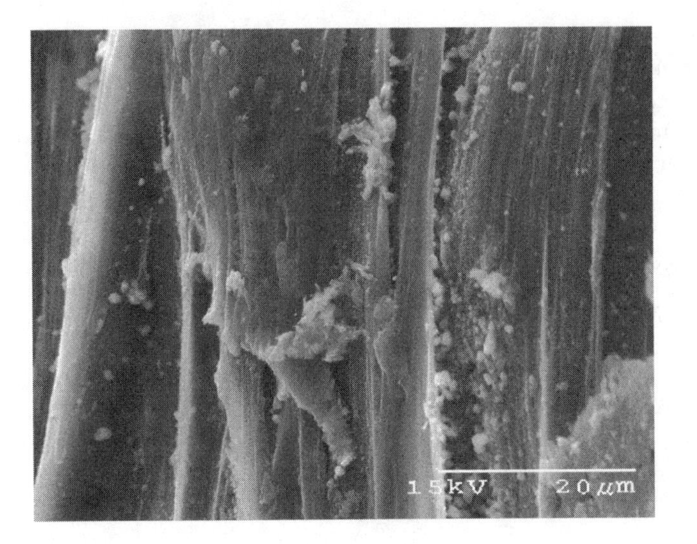

Figure 6.5 SEM of generation 6 polylysine-based dendrimer blended with 20% hydroxyapatite.

neering. The inclusion of growth factors and stem cells within various polymeric structures is also promising. Perhaps most promising are polymer/ceramic composites. Through the combination of ceramics with a polymeric material, the brittleness and nonresorbability of the ceramic would be offset, while at the same time the osteoconductivity and mechanical properties of the polymer would be improved. The interdisciplinary nature of tissue engineering requires that the most promising future for bone regeneration could be a strong collaboration between polymer chemists, clinicians, biologists, and engineers.

POLYMERIC SCAFFOLDS IN NEURONAL TISSUE ENGINEERING

Replacement of bone often includes nerve repair, especially in cases of trauma or tumor resection. Tissue-engineered nerve guides have consisted of biodegradable, FDA-approved polymers, such as PLA, PGA, PCL, and copolymers thereof,[17–19] as is the tradition of many tissue-engineered scaffolds. Those polymers as well as other polymeric matrices are described below.

POLY(α-HYDROXY ACIDS)

Rutkowski and Heath have published extensively in the area of polymeric, biodegradable nerve guides.[20–22] They have fabricated synthetic conduits of PLA following den Dunnen's method of fabrication.[23] Briefly, water-soluble poly(vinyl alcohol) (PVA) rods are utilized as a sacrificial mandrel, and the PLA is coated on the rods. The rods are dissolved in water, resulting in a PLA conduit. Axon regeneration based on varying parameters of the conduits, such as porosity, wall thickness, and Schwann cell seeding density, has been examined.[20] Maquet et al. have prepared porous PLA conduits, subsequently coated with PVA, and implanted into a rat sciatic nerve defect model, with fair results.[17] PLA has been modified with PCL, by several groups, including Rodriguez et al., in an attempt to improve the maintenance of mechanical properties during degradation and decrease the inflammatory response (e.g., by decreasing the PLA content).[18] Also, PLA–PCL copolymers are highly permeable, which permits the necessary exchange of nutrients and molecules in the wound site.[24] Caprolactone has also been copolymerized with trimethylene carbonate to produce nerve guides with promising properties.[25]

Polyphosphoesters

Polyphosphoesters are being examined as potential nerve guides due to their numerous attractive properties, such as biodegradability, biocompatibility, and flexibility, in coupling biomolecules under physiological conditions.[26,27] These polymers were examined in a rat sciatic nerve model, resulting in fairly successful regeneration through the 10-mm gap.[26,27] Biodegradable polyphosphazenes have also been examined in a 10-mm rat sciatic nerve model, with satisfactory results.[28]

Nondegradable Polymers

Silicone is the current clinical alternative to nerve autografts, and has been extensively studied.[29–31] Other nondegradable polymers, such as polyethylene[32] and polyurethanes,[33] have also been examined in peripheral nerve repair. Madison et al. compared polyethylene vs. plasticized PLA tubes in a 4-mm mouse sciatic nerve defect model.[32] The tubes were further modified by the inclusion of the extracellular matrix protein laminin within the conduits, which enhanced axonal outgrowth. Both teflon and polysulfone have also been examined as nerve guides, with satisfactory results.[34] Conducting polymers such as poly(pyrrole) and poly(pyrrole)/hyaluronic acid composites have shown promise.[14,35] In an attempt to attain mechanical properties comparable to those of native nerves, hydrogel tubes of poly(2-hydroxyethyl methacrylate-*co*-methyl methacrylate) [p(HEMA-*co*-MMA)] have been prepared by Dalton et al.[36] The conduits are flexible, containing an interconnected macroporous, inner layer. The p(HEMA-*co*-MMA) tubes had similar mechanical properties (e.g., modulus up to 400 kPa) to those of the spinal cord, which has a reported elastic modulus range between 200 and 600 kPa. To improve the flexibility of the nerve guides, plasticizers have been incorporated into PLA tubes.[37] However, implantation typically results in a rapid loss of plasticizer, which in turn quickly alters the mechanical properties.

Polymer/Collagen Composites

Marra's laboratory has been examining collagen microcarriers as vehicles to deliver neural cells within polymeric biomaterials for nerve repair. CultiSphers are unique collagen-based macroporous microcarriers (100–300 μm in diameter), and are manufactured by Percell Biolytica from pharmaceutical-grade porcine gelatin by a process that yields a highly cross-linked gelatin matrix with a high mechanical and thermal stability (Figure 6.6). The augmented surface area of the CultiSphers as well as the protection from stress provided to the cells in the interior of the CultiSphers

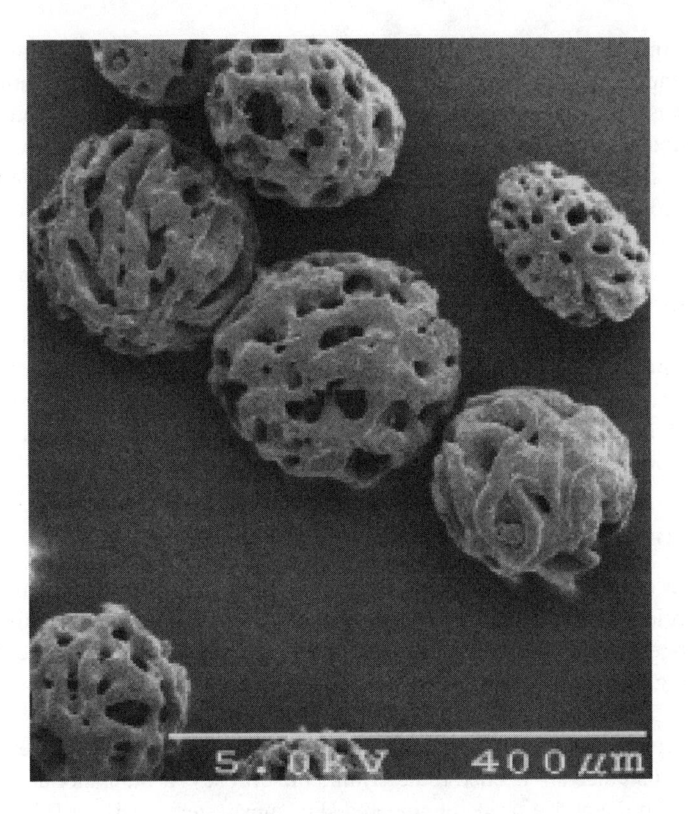

Figure 6.6 SEM of CultiSphers.

results in enhanced cell culture capabilities. CultiSphers are used in the culture of a variety of cell types, including the *in vitro* culture and differentiation of human adult mesenchymal stem cells, as well as several bone cell lines.[38,39] The effect of incorporating collagen microcarriers into PCL on the mechanical properties as well as the cytocompatibility is currently being examined.[39]

Future of Polymers in Nerve Regeneration

Both biodegradable and nondegradable polymers have been examined for nerve regeneration therapies. The future of clinically viable, commercially practical guides is likely a combination of native and synthetic materials, as well as the necessary inclusion of neurotrophins. Of utmost promise is the utilization of stem cells in nerve regeneration, particularly for spinal cord injury repair. It is a challenge to create a tissue-engineered nerve guide that will be conducive to cell seeding as well as to attain the desired mechanical properties.

POLYMERS IN SKIN REGENERATION

Tissue-engineered skin is perhaps the most commercially advanced tissue-engineered product to date. There are several tissue-engineered products for skin regeneration on the market. The first product, Apligraf®, was produced by Organogenesis (http://www.organogenesis.com/), and consisted of human skin cells, that is, epidermal keratinocytes and dermal fibroblasts, seeded into a bovine collagen matrix. Although expensive, this product was promising for burn patients. Another tissue-engineered product, Dermagraft®, now marketed by Smith and Nephew, consists of human fibroblasts seeded in PLGA. Smith and Nephew also markets TransCyte®, which is similar to Dermagraft®; however, TransCyte® is composed of human newborn fibroblasts, which are then cultured on the nylon mesh of Biobrane, and a thin silicone membrane is bonded to the mesh, which provides a moisture vapor barrier for the wound. OrCel® is a new tissue-engineered skin substitute consisting of a scaffold of bovine collagen, seeded with fibroblasts and keratinocytes, and is currently being used in clinical trials for both venous leg ulcers and diabetic foot ulcers. Although tissue-engineered skin options are expensive, they are promising.

Native Polymers for Skin Regeneration

Also under examination is small intestinal submucosa (SIS). Oasis™ is marketed by Cook Surgical and consists of the collagen middle layer of the pig small intestine. SIS has been widely used in clinical applications, including cardiovascular tissue engineering applications, with promising results.[40–43] Another porcine collagen derivative is E-Z-Derm™, which consists of cross-linked collagen pig skin. Both of these porcine materials are in the early stages of research and development and further clinical studies are needed to determine the efficacy of the materials.

Although the above tissue-engineered skin substitutes consist of FDA-approved polymeric materials, there is considerable research being conducted in the synthesis of other polymers for skin regeneration.

Future of Polymers in Skin Regeneration

Although skin is further ahead commercially than bone or nerve substitutes, the ideal skin graft remains elusive. Among the most promising areas in skin tissue engineering is the combination of cells with biodegradable polymers. Of particular promise are native polymer-based materials, such as SIS. The future of these substitutes must address their high costs, as compared to allograft.

SCAFFOLD FABRICATION

As discussed above, polymers are a very important aspect of designing a tissue-engineered therapy, thus the need to describe scaffold fabrication and especially the important factor of synthesizing a porous scaffold for bone and nerve regenerative therapy. Porosity can be introduced into a polymeric scaffold using a variety of techniques. One of the most common techniques is particulate-leaching, also known as solvent-leaching. Other techniques include emulsion freeze-drying, phase separation, 3D printing, and gas foaming. Each technique has both advantages and disadvantages, which are described in the following sections.

PARTICULATE-LEACHING

The porogen leaching was first patented by Mikos et al. in 1996.[44] This technique involves dispersing water-soluble particles, such as salt, sugars, or polymer spheres, in a matrix consisting of the scaffold material dissolved in an organic solvent. After solvent evaporation, a composite of the polymer and porogen remains. The composite is then immersed in water until complete dissolution of the porogen occurs, resulting in a porous scaffold. This technique may yield the following:

(1) "skin" of nonporous polymer at the surface,
(2) nonhomogeneous dispersion of pores,
(3) lack of inner connectivity of the pores, and
(4) remaining porogen within the scaffold after porogen leaching.

Many of these shortcomings have been addressed by Agrawal et al.,[45,46] which include sonication of the porogen/polymer matrix during solvent evaporation,[45] as well as the development of the following techniques to invoke porosity.

EMULSION FREEZE-DRYING

The emulsion freeze-drying technique involves creating an emulsion by homogenizing a polymer solvent solution and water. The mixture is then rapidly quenched in liquid nitrogen, and the solvent and water are removed by freeze-drying. Control of processing parameters, such as volume fraction of the dispersed phase, results in control of the porosity. Advantages of this technique include the ability to control pore sizes from 15 to 200 microns, the ability to obtain porosity >90%, and the possibility of incorporating growth factors within the scaffolds.

PHASE SEPARATION

The phase separation technique involves thermodynamic demixing of a homogeneous polymer/solvent solution into both a polymer-rich phase and a polymer-poor phase. This occurs either by cooling the solution below a bimodal solubility curve, or by exposing the solution to another immiscible solvent. This procedure is similar to the emulsion freeze-drying technique; however, phase separation does not involve the homogenization of emulsion freeze-drying. Initially, a pore size range of 1–20 μm was obtained, limiting the use of the resulting scaffolds. However, a modification of the procedure, termed a "coarsening effect" (e.g., heating above the cloud point), was used to create macropores (>100 μm). A disadvantage of this technique is the use of dioxane as a solvent, which is a suspected carcinogen.

3D PRINTING

3D printing is a solid free-form fabrication technique that involves selectively directing a solvent onto polymer powder packed with salt particles to build complex 3D structures as a series of very thin two-dimensional slices. The polymer/salt composites are then immersed in water to dissolve the salt particles, resulting in porous

(60%) devices with micropores (45–150 μm in diameter). This is a relatively new technique, and applications in tissue engineering are being explored. This technique is further described in a later chapter entitled: "Design and Fabrication of Bone Tissue Engineering Scaffolds" by Dr. Hollister.

Gas Foaming

The use of carbon dioxide to create porosity in polymers has been studied by Mooney et al.[47] This technique avoids the use of organic solvents and high temperatures, which permits incorporation of growth factors during fabrication. Polymers are subjected to high-pressure carbon dioxide (800 psi) for 48 h to saturate the polymer with the gas. When the pressure is slowly reduced to atmospheric pressure, carbon dioxide nucleates and grows within the polymer, forming pores. However, one disadvantage of this technique is that many of the resulting pores are closed (e.g., there is a lack of pore inner connectivity). Modifications to this technique include combining gas foaming with particulate-leaching, and this has resulted in more pore inner connectivity.[47]

Future of Polymer Fabrication Techniques

New techniques to develop improved scaffolds are consistently being developed. A procedure that results in a reproducible scaffold is most desirable, in addition to a protocol that is both simple and inexpensive. Resultant scaffold properties such as pore size and porosity must be controlled for specific applications. Furthermore, fabrication techniques that will permit the inclusion of cells or growth factors are highly desired.

POLYMER MICROSPHERES AND GROWTH FACTOR DELIVERY

As previously discussed, polymers play an important structural role allowing cells to migrate and proliferate to facilitate tissue regeneration at the implanted site. In addition to the structural role, they play an equally important role in the delivery of growth factors that could be regulated by the degradation properties of the polymer. The delivery of cells or growth factors to the wound site requires a suitable scaffold material, and both natural and synthetic biodegradable polymers have been examined, thus making the development of tissue-engineered matrices an area of intense interest. The use of cells in tissue-engineered scaffolds is further detailed in chapter 3. We will initially discuss the importance of growth factor delivery by using biodegradable scaffolds.

Growth Factors

Growth factors are natural hormones, usually proteins, and often they exert their effect via cell surface receptors. Growth factors could stimulate and increase the production of connective tissue, promote remodeling, and create a new supply of blood vessels to nourish the site to be regenerated. The choice of growth factor to be delivered will provide the appropriate cues for the cells to differentiate to a specific lineage such as bone and nerve (the role of growth factors is further discussed in the chapter entitled: "signaling molecules"). One method of growth factor delivery is the adsorption of growth factors to a polymer surface prior to implantation.[48] This method has many disadvantages such as a lack of a precise controlled delivery over time as well as a lack of specific placement of varying growth factors within the scaffold. Thus, researchers have developed alternative methods to address these shortcomings. One delivery technique is the encapsulation of drugs into polymeric microspheres and susbsequent incorporation of the microspheres within scaffolds. Microsphere encapsulation is a proven method of controlled delivery, and growth factors can be encapsulated within microspheres. This technique is further described below.

Polymer Microspheres

Polymer microspheres (i.e., hollow capsules) are characteristically fabricated from synthetic polymers such as PLGA, or natural polymers (e.g., gelatin or alginate). Fabrication of polymer microspheres includes a number of techniques, such as water/oil/water emulsion, oil/water emulsion, and spray-drying. The growth factors are encapsulated within these degradable spheres. As the capsule degrades, the growth factor is released, resulting in a controlled, extended release. Drugs and other growth factors have successfully been incorporated into polymer microspheres, such as human growth hormone, Japanese encephalitis virus vaccine, vascular endothelial growth factor, transforming growth factor β_1 (TGF-β_1) fibroblast growth factor, and cisplatin. Specifically, Lu et al. have examined the release of TGF-β_1 from PLGA microspheres,[49] and the subsequent effect of TGF-β_1 on bone marrow stromal cells.[50] The controlled release of TGF-β_1 was determined to enhance the proliferation and osteoblastic differentiation of marrow stromal cells cultured on poly(propylene fumarate) substrates. Babensee et al. have written a comprehensive review on growth factor delivery.[51]

Polymer Microspheres Within Polymer Scaffolds

In an attempt to control the release of growth factors from tissue-engineered scaffolds, Hu et al. have reported the controlled release of proteins from coated polymer microspheres embedded in tissue-engineered scaffolds.[52] A model protein, bovine serum albumin (BSA), was encapsulated within PLGA microspheres. The microspheres were coated with poly(vinyl alcohol), and incorporated into PLGA tissue-engineered scaffolds during fabrication. The release of BSA from PLGA microspheres, coated PLGA microspheres, and microspheres embedded in a porous PLGA scaffold was measured; this study demonstrated the feasibility of this approach. An additional study demonstrated the effect of modifying the coating on BSA release.[53]

Sintering PLGA microspheres to form a porous scaffold is a novel technique recently developed by Laurencin's group.[54] The resulting scaffolds were ~75% porous, and the average pore size was 100 μm. However, the use of heat in this procedure limits the inclusion of growth factors during fabrication.

Microspheres containing growth factors have been injected into a porous scaffold post-fabrication. Recently, Richardson et al. described the fabrication and evaluation of a polymer scaffold that delivers two growth factors for bone regeneration.[55] The use of growth factors in nerve regeneration includes examination of the incorporation of glial growth factor within a polymer conduit, and more extensively, the neurotrophin NGF has been demonstrated to promote the differentiation of several classes of neurons. NGF is a survival factor in both neuronal cell culture and *in vivo*, and has been encapsulated and delivered from polymer microspheres.

Future of Growth Factor Delivery

The role of growth factors in bone, nerve, and skin tissue engineering must be fully understood and characterized prior to any inclusion of growth factors in biomaterials. Many researchers are attempting to control the delivery of the growth factors within the wound site, which is of utmost importance. Gene therapy has a strong role in the future of growth factor delivery, but many further studies are needed to determine the effect of viral and nonviral delivery in humans.

CONCLUSIONS

The role of biodegradable polymers in tissue engineering has been described in this chapter, and the importance of polymers has been emphasized. From the synthesis of novel biomaterials to the use of FDA-approved polyesters, polymers are widely studied in bone, nerve, and skin regeneration. Composites of native and synthetic polymers as well as composites of ceramics and polymers have been outlined. Although FDA-approved poly(α-hydroxy acids) remain the scaffold material of choice for many tissue engineers, the limitations of these polymers have resulted in the synthesis of numerous novel polymers, including highly branched dendrimers.

The properties of the polymers have many requirements to fulfill prior to clinical utility. Parameters such as pore size, mechanical strength, degradation kinetics, and surface chemistry contribute to the overall scaffold viability.

The use of biodegradable microspheres to deliver growth factors is a well-studied area. Although the use of polymer microspheres in drug delivery is a well-established field, the use of microspheres in tissue engineering is nascent, and quite promising.

The future of tissue-engineered scaffolds includes materials that are conducive to stem cell seeding as well as scaffolds that can deliver growth factors in a controlled fashion.

References

1. Imola, M.J. and Schramm, V.L. 2002. Resorbable internal fixation in pediatric cranial base surgery. *Laryngoscope*, 112(10): 1897–1901.
2. Eppley, B.L. and Reilly, M. (1997). Degradation characteristics of PLLA-PGA bone fixation devices. *J. Craniofac. Surg*. 8(2): 116–120.
3. Dunn, A.S., Campbell, P.G. and Marra, K.G. (2001). The influence of polymer blend composition on the degradation of polymer/hydroxyapatite biomaterials. *J. Mater. Sci. Mater. Med.* 12(8): 673–677.
4. Marra, K.G., Campbell, P.G., DiMilla, P.A., Kumta, P.N., Mooney, M.P., Szem, J.W., and Weiss, L.E. (1999). Novel three dimensional biodegradable scaffolds for bone tissue engineering. in Materials Research Society Symposium Proceedings, Biomedical Materials: Drug Delivery, Implants and Tissue Engineering, Boston, MA.

5. Lewandrowski, K.U., et al. (1999). Effect of a poly(propylene fumarate) foaming cement on the healing of bone defects. *Tissue Eng.* 5(4): 305–316.
6. Yaszemski, M.J. et al. (1996). *In vitro* degradation of a poly(propylene fumarate)-based composite material. *Biomaterials* 17(22): 2127–2130.
7. Vehof, J.W. et al. (2002). Bone formation in transforming growth factor beta-1-coated porous poly(propylene fumarate) scaffolds. *J. Biomed. Mater. Res.* 60(2): 241–251.
8. Fisher, J.P. et al. (2002) Soft and hard tissue response to photocrosslinked poly(propylene fumarate) scaffolds in a rabbit model. *J. Biomed. Mater. Res.* 59(3): 547–556.
9. Rohner, D. et al. (2002). [Individually CAD–CAM technique designed, bioresorbable 3-dimensional polycaprolactone framework for experimental reconstruction of craniofacial defects in the pig]. *Mund. Kiefer Gesichtschir* 6(3): 162–167.
10. Fiordeliso, J., Bron, S. and Kohn, J. (1994). Design, synthesis, and preliminary characterization of tyrosine-containing polyarylates: new biomaterials for medical applications. *J. Biomater. Sci. Polym. Ed.* 5(6): 497–510.
11. Choueka, J. et al. (1996). Canine bone response to tyrosine-derived polycarbonates and poly(L-lactic acid). *J. Biomed. Mater. Res.* 31(1): 35–41.
12. Fearon, J.A. (2003). Rigid fixation of the calvaria in craniosynostosis without using "rigid" fixation. *Plast. Reconstr. Surg.* 111(1): 27–38.
13. Petricca, S.E., Marra, K.G., Kumta, P.N. (2004). Chemical synthesis of poly(lactic-*co*-glycolic acid)/hydroxyapatite composites for orthopaedic applications, *J. Biomed. Mater. Res.* Submitted.
14. Collier, J.H., Camp, J.P., Hudson, T.W. and Schmidt, C.E. (2000). Synthesis and characterization of polypyrrole-hyaluronic acid composite biomaterials for tissue engineering applications. *J. Biomed. Mater. Res.* 50(4): 574–584.
15. Liu, L.S. et al. (1999). An osteoconductive collagen/lhyaluronate matrix for bone regeneration. *Biomaterials* 20(12): 1097–1108.
16. Bakos, D., Soldan, M. and Hernandez-Fuentes, I. (1999). Hydroxyapatite–collagen–hyaluronic acid composite. *Biomaterials* 20(2): 191–195.
17. Maquet, V., Martin, D., Malgrange, B., Franzen, R., Schoenen, J., Moonen, G., and Jerome, R. (2000) Peripheral nerve regeneration using bioresorbable macroporous scaffolds. *J. Biomed. Mater. Res.* 52: 639–651.
18. Rodriguez, F.J., Gomez, N., Perego, G. and Navarro, X. (1999). Highly permeable polylactide-caprolactone nerve guides enhance peripheral nerve regeneration through long gaps. *Biomaterials* 20: 1489–1500.
19. Hadlock, T., Sundback, C., Koka, R., Hunter, D., Cheney, M. and Vacanti, J.P. (1999). A novel, biodegradable polymer conduit delivers neurotrophins and promotes nerve regeneration. *Laryngoscope* 109(9): 1412–1416.
20. Rutkowski, G.E. and Heath, C.A. (2002). Development of a bioartificial nerve graft. I. Design based on a reaction–diffusion model. *Biotechnol. Progr.* 18(2): 362–372.
21. Rutkowski, G.E. and Heath, C.A. (2002). Development of a bioartificial nerve graft. II. Nerve regeneration *in vitro*. *Biotechnol. Progr.* 18(2): 373–379.
22. Heath, C.A. and Rutkowski, G.E. (1998). The development of bioartificial nerve grafts for peripheral-nerve regeneration. *Trends Biotechnol.* 16(4): 163–168.
23. den Dunnen, W.F. et al. (1996). Light-microscopic and electron-microscopic evaluation of short-term nerve regeneration using a biodegradable poly(DL-lactide-epsilon-caprolacton) nerve guide. *J. Biomed. Mater. Res.* 31(1): 105–115.
24. den Dunnen, W.F. and Meek, M.F. (2001). Sensory nerve function and auto-mutilation after reconstruction of various gap lengths with nerve guides and autologous nerve grafts. *Biomaterials*, 22(10): 1171–1176.

25. Pego, A.P. et al. (2001). Copolymers of trimethylene carbonate and epsilon-caprolactone for porous nerve guides: synthesis and properties. *J. Biomater. Sci. Polym. Ed.* 12(1): 35–53.
26. Wang, S. et al. (2001). A new nerve guide conduit material composed of a biodegradable poly(phosphoester). *Biomaterials*, 22(10): 1157–1169.
27. Xu, X. et al. (2002). Polyphosphoester microspheres for sustained release of biologically active nerve growth factor. *Biomaterials*, 23: 3765–3772.
28. Langone, F. et al. (1995). Peripheral nerve repair using a poly(organo)phosphazene tubular prosthesis. *Biomaterials*, 16(5): 347–353.
29. Kakinoki, R. et al. (1995). Relationship between axonal regeneration and vascularity in tubulation—an experimental study in rats. *Neurosci. Res.* 23(1): 35–45.
30. Lundborg, G. et al. (1982). Nerve regeneration in silicone chambers: influence of gap length and of distal stump components. *Exp. Neurol.*, 76(2): 361–375.
31. Madison, R.D., Da Silva, C.F., Dikkes, P. (1988). Entubulation repair with protein additives increases the maximum nerve gap distance successfully bridged with tubular prostheses. *Brain Res.* 447(2): 325–334.
32. Madison, R.D. et al. (1987). Peripheral nerve regeneration with entubulation repair: comparison of biodegradeable nerve guides versus polyethylene tubes and the effects of a laminin-containing gel. *Exp. Neurol.* 95(2): 378–390.
33. Robinson, P.H. et al. (1991). Nerve regeneration through a two-ply biodegradable nerve guide in the rat and the influence of ACTH4-9 nerve growth factor. *Microsurgery* 12(6): 412–419.
34. Navarro, X. et al. (1996). Peripheral nerve regeneration through bioresorbable and durable nerve guides. *J. Peripher. Nerv. Syst.* (1): 1.
35. Schmidt, C.E., Shastri, V.R., Vacanti, J.P., Langer, R. (1997). Stimulation of neurite outgrowth using an electrically conducting polymer. *Proc. Natl. Acad. Sci. USA*, 94: 8948–8953.
36. Dalton, P.D., Flynn, L. and Shoichet, M.S. (2002). Manufacture of poly(2-hydroxyethyl methacrylate-co-methyl methacrylate) hydrogel tubes for use as nerve guidance channels. *Biomaterials*, 23(18): 3843–3851.
37. Luciano, R.M., de Carvalho Zavaglia, C.A. and de Rezende Duek, E.A. (2000). Preparation of bioabsorbable nerve guide tubes. *Artif. Organs* 24(3): 206–208.
38. Doctor, J.S. et al. (2001). Using collagen microcarriers to deliver cells for bone tissue engineering. *Dev. Biol.* 235: 245.
39. Waddell, R.L. et al. (2003). Using PC12 cells to evaluate poly(caprolactone) and collagenous microcarriers for applications in nerve guide fabrication. *Biotechnol. Progr.*, 19(6):1767–1774.
40. Nerem, R.M. and Seliktar, D. (2001). Vascular tissue engineering. *Annu. Rev. Biomed. Eng.* 3: 225–243.
41. Roeder, R. et al. (1999). Compliance, elastic modulus, and burst pressure of small-intestine submucosa (SIS), small-diameter vascular grafts. *J. Biomed. Mater. Res.* 47(1): 65–70.
42. Badylak, S.F. et al. (1998). Small intestinal submucosa: a substrate for *in vitro* cell growth. *J. Biomater. Sci. Polym. Ed.* 9(8): 863–878.
43. Lindberg, K. and Badylak, S.F. (2001). Porcine small intestinal submucosa (SIS): a bioscaffold supporting *in vitro* primary human epidermal cell differentiation and synthesis of basement membrane proteins. *Burns* 27(3): 254–266.
44. Mikos, A.G., Sarakinos, G., Vacanti, J.P., Langer, R.S., and Cima, L.G., (1996). Biocompatible Polymer Membranes and Methods of Preparation of Three Dimensional Membrane Structures. U.S. Patent, 5514378.

45. Agrawal, C.M., McKinney, J.S., Huang, D., and Athanasiou, K.A. (2000), The use of the vibrating particle technique to fabricate highly permeable biodegradable scaffolds, in *Synthetic Bioabsorbable Polymers for Implants*, C.M. Agrawal, Parr, J., and Lin, S., Eds., American Society for Testing and Materials: Philadelphia, PA. p. STP 1396.
46. Lin, H.R. et al. (2002). Preparation of macroporous biodegradable PLGA scaffolds for cell attachment with the use of mixed salts as porogen additives. *J. Biomed. Mater. Res.* 63(3): 271–279.
47. Harris, L.D., Kim, B.S., and Mooney, D.J. (1998). Open pore biodegradable matrices formed with gas foaming. *J. Biomed. Mater. Res.* 42: 396–402.
48. Winn, S.R. et al. (1999). Tissue-engineered bone biomimetic to regenerate calvarial critical-sized defects in athymic rats. *J. Biomed. Mater. Res.* 45(4): 414–421.
49. Lu, L., Stamatas, G.N. and Mikos, A.G. (2000). Controlled release of transforming growth factor beta-1 from biodegradable polymer microparticles. *J. Biomed. Mater. Res.* 50: 440–451.
50. Peter, S.J. et al. (2000). Effects of transforming growth factor beta-1 released from biodegradable polymer microparticles on marrow stromal osteoblasts cultured on poly(propylene fumarate) substrates. *J. Biomed. Mater. Res.* 50: 452–462.
51. Babensee, J.E., McIntire, L.V. and Mikos, A.G. (2000). Growth factor delivery for tissue engineering. *Pharm. Res.* 17(5): 497–504.
52. Hu, Y., Hollinger, J.O., and Marra, K.G., (2001). Controlled release from coated polymer microparticles embedded in tissue-engineered scaffolds. *J. Drug Targeting*, 9(6): 431–438.
53. Meese, T.M. et al. (2002). Surface studies of coated polymer microspheres and protein release from tissue-engineered scaffolds. *J. Biomater. Sci. Polym. Ed.* 13(2): 141–151.
54. Ambrosio, A.M.A. et al. (2001). A novel amorphous calcium phosphate polymer ceramic for bone repair: I. Synthesis and characterization. *J. Biomed. Mater. Res.* 58(3): 295–301.
55. Richardson, T.P. et al. (2001). Polymeric system for dual growth factor delivery. *Nat. Biotech.*, 19(11): 1029–1034.

7 Design and Fabrication of Bone Tissue Engineering Scaffolds

Scott J. Hollister, Juan M. Taboas, Rachel M. Schek, Cheng-Yu Lin and Tien Min Chu

Contents

0-8493-1621-9/05/$0.00+$1.50
© 2005 by CRC Press LLC

INTRODUCTION

Bone tissue engineering utilizes scaffolds to deliver biofactors including cells, genes, and proteins to regenerate bone. The scaffold itself must fulfill three primary functions to ensure successful treatment of bone defects. First, the scaffold must provide the correct anatomic geometry to define and maintain the space for tissue regeneration. Second, the scaffold must provide temporary mechanical load bearing within the tissue defect. This second function is especially critical if the scaffold/biofactor construct is directly implanted without prior incubation in a bioreactor. Third, the scaffold should enhance the regenerative capability of the chosen biofactor. The difficulty in designing scaffolds is that the three primary functions often pose conflicting design requirements.

A primary motivation for computational design is to achieve a balance between the load bearing and tissue regeneration requirements of scaffolds. For load-bearing purposes, achieving stiffness and strength equivalent to bone tissue for load-bearing purposes requires minimally porous scaffolds. Conversely, enhanced delivery of biofactors requires more highly connected porous scaffolds[1,2] to allow cell migration, vascularization, and connected tissue formation within scaffolds. Computational design and fabrication can be employed to optimize scaffolds to best fulfill the conflicting functions.

A second motivation for computational scaffold design/fabrication is based on studies demonstrating that a scaffold's internal architecture influences tissue regeneration. Kühne et al.[3] found that corraline hydroxyapatite (HA) with an average 500 μm pore diameter showed good bone infiltration in a rabbit model, while HA with an average 200 μm pore diameter showed poor bone infiltration. Grenga et al.[4] noted that HA architecture affected vascularization concurrent with bone formation. Kuboki[5] found different bone development pathways associated with different average HA pore diameters. Pores averaging 90–100 μm demonstrated endochondral ossification with significant cartilage, while pores averaging 300–400 μm demonstrated intramembranous ossification with no cartilage. Hui et al.[6] noted that increased bone formation in trabecular bone graft in rabbits increased with higher fluid conductance, a parameter dependent on pore architecture and connectivity. These studies, however, all have the limitation that scaffold pore size represents an average, with a wide variation. Also, pore connectivity is not fixed. Therefore, the ability to design and fabricate scaffolds with fixed architectures is needed to better understand how bone regeneration is influenced by scaffold architecture.

A third motivation for computational scaffold design/fabrication is the need to match anatomic shape. The potential sites for anatomic reconstruction range from complex craniofacial and orthopedic trauma to joint reconstruction and spinal fusion. These complex geometries are represented digitally by voxels in either Computed Tomography (CT) or Magnetic Resonance Imaging (MRI) scans. The digital information must be interpreted and converted into information that can be used to create an exterior scaffold shape.

In summary, there are both basic and applied research issues in bone tissue engineering that require controlled scaffold design and fabrication. For clinical applications, it is obvious that scaffolds matching complex anatomic defects are necessary. In addition, the scaffolds will need to have specific load-bearing capabilities as well as biofactor delivery capabilities. For basic research, scaffolds with precisely controlled architecture are needed to test hypotheses concerning scaffold effects on basic bone regeneration biology. Only by integrating basic bone biology with precisely engineered scaffolds can we optimize bone tissue engineering treatments. In this chapter, we present integrated design and fabrication methods that can be used to engineer scaffolds with anatomic shape and controlled architecture.

HIERARCHICAL IMAGE-BASED SCAFFOLD DESIGN

THE NEED FOR HIERARCHICAL DESIGN

Bone tissue is a hierarchical structure, with features ranging from nanometers to centimeters in size, a 10^8 magnitude change. This structural hierarchy enables a tremendous variation in both function and physiology. Moving up the hierarchical scale, packing of mineral into collagen, organization of this mineral/collagen composite into lamellar structures, and finally organization of lamellar structures into trabeculae or osteons provides a wide range of mechanical stiffness and strength. Moving down the hierarchical scale, it has been hypothesized that the same organization provides specific nutrients as well as environmental signals to osteocytes embedded in bone tissue.[7–9] Although unproven, a widely held design paradigm is that tissue engineering scaffolds should mimic natural tissue structure and function as closely as possible. Under this paradigm, tissue engineering scaffolds should have a hierarchical structure.

The functional and regenerative scaffold requirements also suggest the need for a hierarchical design. At a global anatomic level, where bone features range from mm to cm, the scaffold must replicate anatomic shape and provide sufficient stiffness and strength for physiologic loading. Pores on a scale of 0.1 mm (100 μm) are necessary to either seed sufficient densities of cells or enable a sufficient density of invading host cells to regenerate bone tissue. Additionally, structures in the 0.1–1 mm will determine the global-level mechanical properties. Finally, feature sizes ranging from .001 to .01 mm (1–10 μm) will influence both individual cell attachment and activity. Controlling hierarchical features in scaffolds will be necessary to test fundamental scaffold design hypotheses and produce scaffolds optimized for both function and tissue regeneration.

IMAGE-BASED METHODS FOR DESIGNING HIERARCHICAL FEATURES

Designing scaffolds requires computational methods to represent hierarchical features. The most commonly used computational design methods are those grouped under the title Computer Aided Design (CAD). CAD methods allow the user to create objects using wireframe, surface, and solid models. These models are based on specific mathematical equations for geometric shapes, and can thus represent shapes exactly. Using such models, complicated features can be made by performing Boolean operations of intersection and union. The drawback of CAD techniques is that complicated anatomic and microstructural features are arduous and time consuming to represent due to the fact that CAD objects are built from base representations of geometric objects like cylinders, spheres, and cubes. Although it is possible to create complex CAD designs using facets or wireframe data, such files can in themselves become extremely memory intensive and complicated to use. This makes hierarchical features especially difficult to represent.

An alternative to CAD techniques for feature representation is density distribution within a grid of pixels. Since this is how all imaging modalities represent features, we term this approach an image-based design. The two major advantages of image-based design are (1) the ability to represent features with any degree of complexity and (2) the commonality of data structures with any imaging modality. The first advantage allows complex features at any scale to be represented without difficulty, in many cases with considerably less work than with CAD techniques. The second advantage allows scaffold designs to easily incorporate the shape of anatomic defects from CT or MRI images. The disadvantage of image-based methods compared to CAD is the necessity of utilizing a large dataset to represent features, depending on the desired spatial resolution. However, CAD files can also become quite large when representing complex objects, especially complex anatomic shapes using faceted surfaces. We have chosen image-based methods for scaffold design, based on the ease in representing and performing Boolean operations with complex data, coupled with the ease of directly using anatomic image data.

To represent hierarchical features using image-based design, each distinct design feature on each level of hierarchy is embodied using a separate image database. For example, consider a scaffold replacing a femoral diaphyseal defect with three hierarchical scales: (1) the femoral defect shape, at a scale of centimeters, (2) microstructural pores at a scale of hundreds of microns, and (3) cell attachment features at a scale of tens of microns. In this case, there would be one femoral defect image database, one image database for each different microstructure pore design, and one image database for each cell attachment feature design.

The key component of hierarchical image-based design is specification of a density ρ for each pixel in each hierarchical image database. A global anatomic-level pixel density is denoted as ρ_i^0, where the superscript denotes the hierarchical level and the subscript i denotes the pixel location within the database. The first-level microstructure density is denoted as ρ_i^1, the second-level microstructure density is denoted as ρ_i^2, and so forth. The spatial relationship among the hierarchical density values is shown in Figure 7.1.

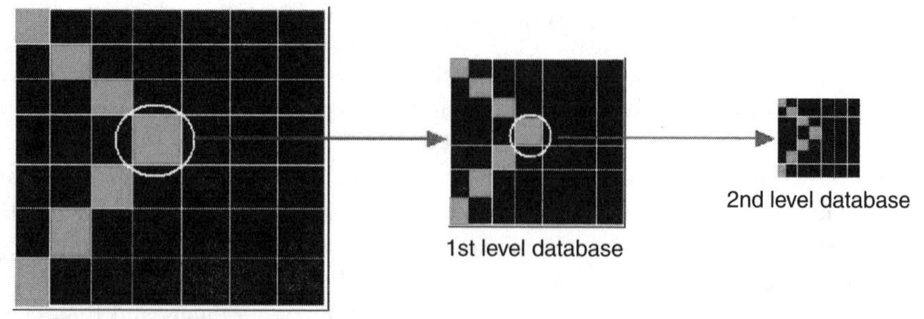

Figure 7.1 Spatial relationship between hierarchical densities. A global anatomic pixel contains all microstructural pixels at levels 1, 2, etc. Likewise, the level 1 microstructural pixel contains all microstructural pixels at levels 2 and so forth.

Mathematically, each hierarchical pixel density is a function of factors, which could include anatomy, stiffness, stress, strain, nutrient transport, cell adhesion, diffusion, permeability, etc. at its own level and the density at more macroscopic levels. Material density cannot exist at a microscopic level in a scaffold if it does not exist at a more macroscopic level. For example, it is obvious that if a pixel at the 0th level is outside the reconstruction area, then subsequent hierarchical pixels cannot contain material, and thus must have a density of zero. To account for the interaction of hierarchical densities, we define an indicator function χ_i^k for each pixel location i of each hierarchical level k with a value of one for pixel densities greater than zero, otherwise zero,

$$\begin{aligned} \chi_i^k &= 1 \quad \text{if} \quad \rho_i^k > 0, \\ \chi_i^k &= 0 \quad \text{if} \quad \rho_i^k > 0. \end{aligned} \tag{7.1}$$

This allows us to define the density at any given level as

$$\begin{aligned} \rho_i^0 &= \rho_i^0 \quad (\text{anatomy}, \varepsilon, \sigma, \cdots), \\ \rho_i^1 &= \chi_i^0\, \rho_i^1 \quad (C^0, K^0, \text{tissue microstructure}, \cdots), \\ \rho_i^2 &= \chi_i^0\, \chi_i^1\, \rho_i^2 \quad (C^1, K^1, \text{cell attachment, tissue ultrastructure}, \cdots), \end{aligned} \tag{7.2}$$

where anatomy denotes pixel density defined by anatomic image scans, ε is the global-level strain, σ is the global-level stress, C^0 is the global-level mechanical stiffness, and K^0 is the global-level permeability; tissue microstructure indicates pixel densities defined by microstructure tissue scans (i.e., micro-computed tomography [μCT] or micro-MRI [μMRI], C^1 is the 1st level microstructure stiffness, K^1 is the 1st level microstructure permeability; cell attachment denotes the features designed to enhance cell attachment, and tissue ultrastructure indicates pixel densities defined by ultrastructural imaging techniques like confocal microscopy. Equation (7.2) can be written compactly as

$$\rho_i^k = \chi_i^0\, \chi_i^1 \cdots \chi_i^{k-2}\, \chi_i^{k-1}\, \rho_i^k = \rho_i^k \prod_{n=1}^{k} \chi_i^{k-n}. \tag{7.3}$$

To fabricate scaffolds from image-based data, we need ways to define design pixel densities at each level and convert the resulting image design database into fabrication-specific data.

0th Level Global Anatomic Design Pixel Definition

One of the three primary scaffold requirements is the necessity to match anatomic deficit shape. Creating scaffolds to match anatomic defects requires the ability to incorporate CT or MRI image data into the hierarchical image-based design methodology. CT or MRI images must first be processed and segmented. The most robust way is to have an experienced user outline the tissue defect using region of interest (ROI) techniques. These methods allow the user to outline a selected region on a 2D image slice. This procedure is repeated until the entire 3D scaffold shape is created.

Our research group performs this procedure using the ROI tools in the Interactive Data Language software (IDL, Research Systems Inc., Boulder, CO).

First, each slice image is read into IDL. The image may be defined in a wide variety of formats including JPEG, TIFF, PNG, BMP, as well as raw image data. Next, the user selects either the DEFROI (DEFine Region Of Interest) or the CW_DEFROI (a graphic user interface, GUI) tool in IDL to select the region of interest. All pixels within the ROI are selected and the tools return the one-dimensional (1D) pixel indices identifying the selected pixels. 1D indices are related to the traditional two-dimensional (2D) image indices by $1D_{ind} = X_{tot}Y_{ind} + X_{ind}$, where $1D_{ind}$ is the 1D index, X_{tot} is the total number of pixels in the x image direction, Y_{ind} is the y image pixel index, X_{ind} is the x image pixel index, and all indices start from 0.

The ROI technique may be used to define heterogeneous density regions for each given hierarchical scale. The purpose of heterogeneous density regions is to map the location of the first-level microstructure. The density values serve as a flag that denotes into which global pixel locations the first-level microstructure pixels should be mapped. For example, one region defined with a pixel density of 1 would indicate one type of microstructure, while a second region defined with a pixel density of 2 would indicate a second type of microstructure.

1st Level Microstructure Design Pixel Definition

The next hierarchical level in scaffold design contains features ranging in size from 0.1 to 1 mm (100–1000 μm). We denote this as 1st level scaffold microstructure, also referred to as scaffold architecture. Features at this scale will affect the behavior of cell multitudes, rather than individual cells. There are three basic methods by which 1st level scaffold microstructure design pixels may be defined. These three methods are denoted as (1) material process design, (2) periodic cell design, and (3) biomimetic design.

Material Process Design

Material process design is so named because the architecture design is determined by the method used to process the material. A prime example is the use of porogens like salt embedded in dissolved polymer solutions [see, e.g., [10–13]]. The architecture design is determined completely by the density and packing of salt (Figure 7.2).

The user has limited control over architecture design when using material process methods. The bulk porosity may be changed by altering the amount of porogen, but design variables cannot easily be parameterized and utilized in a mathematical design model.

Periodic Cell Design

Periodic cell design utilizes a basic unit cell (here, we refer to "unit cell" as a design unit, not a biological cell) that is repeated to create an architecture volume. As such, only the basic unit cell pixels need be designed. Neighboring design pixels are automatically defined through the repetition of the unit cell at a fixed period. The periodic unit cell design pixels may be defined in numerous ways. A straightforward design that ensures adequate connected pore channels for cell migration and permeability for nutrient flow is that of interconnected orthogonal cylinders (Figure 7.3a, b).

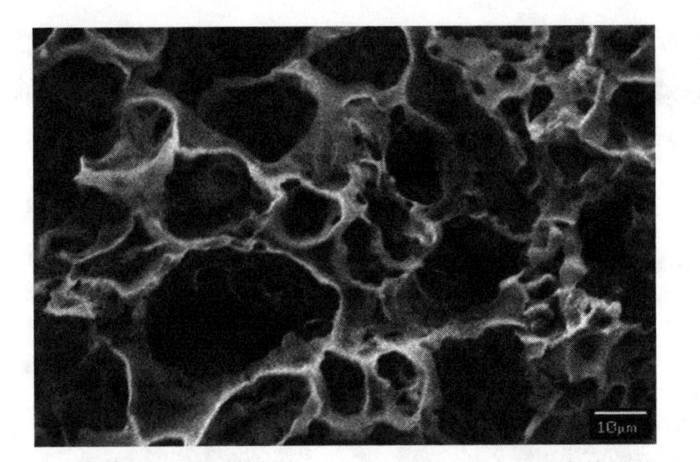

Figure 7.2 Example of polymer scaffold architecture created by porogen leaching. The architecture is completely determined by the porogen packing density. (Reprinted from Biomaterials, 24, Taboas, JM, Maddox, RD, Krebsbach, PH, Hollister, SJ, "Indirect solid free form fabrication of local and global porous, biomimetic, and composite 3D polymer-ceramics scaffolds," 181–194, Copyright 2002, with permission from Elsevier.)

The unit cell design in Figure 7.3a, b is created easily using image-based techniques. Three 3D design image databases are created, one for each channel. Each 3D database is created on a slice-by-slice basis where pixels are assigned density based on the general equation for an ellipse

$$\frac{x_i^n}{a^n} + \frac{y_i^n}{b^n} \leq 1 \Rightarrow \rho_i^1 = 0,$$
$$\frac{x_i^n}{a^n} + \frac{y_i^n}{b^n} > 1 \Rightarrow \rho_i^1 = 1, \quad (7.4)$$

where x_i denotes the x coordinate at the pixel centroid location i, y_i denotes the y coordinate at the pixel centroid location i, a is the ellipse radius along the x-axis and b is the ellipse radius along the y-axis, ρ_i^1 is the pixel density at location i, and n is an even exponent. Note that if $n = 2$, an ellipse is generated. As $n \rightarrow \infty$, the geometric figure approaches a rectangle. Equation (7.4) is verified for each pixel in the 3D design image dataset for each direction. The ellipse diameters and exponent n may be different for each direction to create channels with different diameters and shapes. Once each database is created, they are combined using a Boolean union operation to create a final interconnected porous design unit cell (Figure 7.3a, b).

A significant advantage of the periodic cell design is the ability to calculate effective properties like stiffness and permeability using homogenization theory.[14,15] Using the image database directly as a finite element model, the following local equilibrium equation is solved:

$$\int_V \hat{\varepsilon}^{\mathrm{T}} C \varepsilon^k \,\mathrm{d}V = \int_V \hat{\varepsilon}^{\mathrm{T}} C^k \,\mathrm{d}V; \; k = 1\text{–}6 \quad (7.5)$$

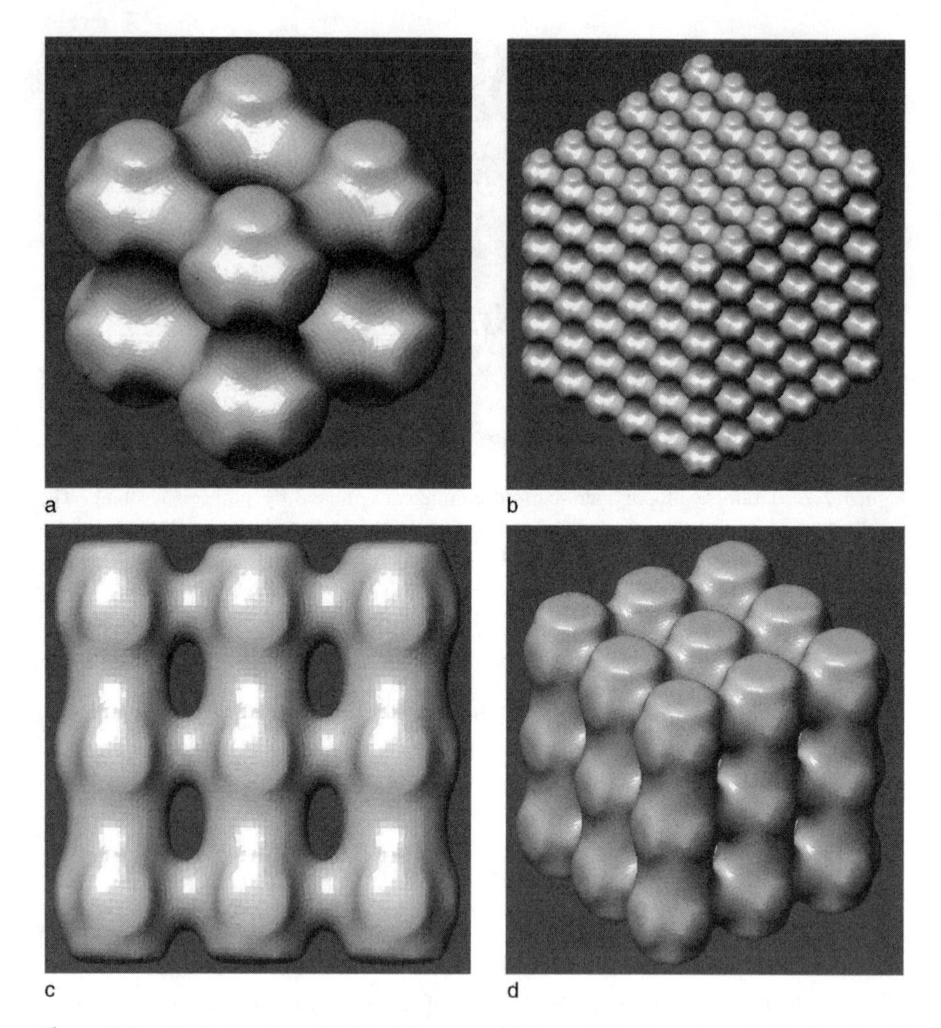

Figure 7.3 3D designed scaffold architecture with interconnecting pores. (a) Basic unit cell design. (b) Repeated unit cells. (c) Periodic architecture optimized to match bone properties, side view. (d) Optimized periodic architecture, trimetric view.

where C is the base scaffold microstructure stiffness matrix at a given hierarchical level, C^k is the kth column of the microstructure stiffness matrix, ε^k are characteristic strains in the microstructure, and $\hat{\varepsilon}$ is a virtual strain, with superscript T denoting the transpose. Equation (7.5) is solved six times, once for each of C. The effective scaffold stiffness can be calculated directly from the base scaffold material stiffness and the scaffold architecture as

$$[C]^{\text{eff}} = \sum_{n=1}^{N} [C]^n \left([I] - \left[\{\varepsilon^1\}\,\{\varepsilon^2\}\,\{\varepsilon^3\}\,\{\varepsilon^4\}\,\{\varepsilon^5\}\,\{\varepsilon^6\}\right]\right)^n = \sum_{n=1}^{N} [C]^n\,[M]^n,$$
$$[M]^n = \left([I] - \left[\{\varepsilon^1\}\,\{\varepsilon^2\}\,\{\varepsilon^3\}\,\{\varepsilon^4\}\,\{\varepsilon^5\}\,\{\varepsilon^6\}\right]\right)^n, \tag{7.6}$$

where $[C]^{\text{eff}}$ is the effective scaffold stiffness, $[C]^n$ is the stiffness of the nth scaffold phase material at a given hierarchical level, $\{\varepsilon^i\}$ are the six strain vectors obtained by solving Equation (7.5) and $[M]^n$ is the local structure matrix[14] that characterizes the architecture of the nth scaffold phase at a given hierarchical level. Equation (7.6) essentially shows how the overall scaffold mechanical stiffness depends on the base scaffold mechanical properties and the scaffold architecture. Thus, once a periodic cell design is created, its mechanical stiffness can be directly calculated using Equations (7.5) and (7.6).

A more powerful approach to periodic cell design is to embed Equation (7.5) in an optimization scheme. Then, instead of creating a periodic design and a posteriori evaluating its effective properties, we can iteratively design the scaffold architecture to match the desired effective properties. We have developed two approaches to optimize scaffold architecture such that the effective mechanical stiffness matches bone stiffness while a desired porosity for biofactor delivery is retained.

The first approach is termed restricted topology optimization [16] since we assume a class of scaffold architectures and optimize a *restricted* set of design variables. We previously chose the interconnected orthogonal cylindrical pore design (Figure 7.3) as the base topology for scaffold architecture. The three pore diameters were chosen as design variables for the restricted topology optimization. The local structure matrix M is computed (Equation 7.6) for permutations of all pore channels between 0.2 and 1.0 mm. The dependence of M on pore diameters is then fit using a cubic polynomial:

$$\begin{aligned}\left[M_{ij}(d_1,d_2,d_3)\right] = {} & a_{ij}^0 + a_{ij}^1 d_1 + a_{ij}^2 d_2 + a_{ij}^3 d_3 \\ & + a_{ij}^4 d_1^2 + a_{ij}^5 d_1 d_2 + a_{ij}^6 d_1 d_3 + a_{ij}^7 d_2^2 + a_{ij}^8 d_2 d_3 + a_{ij}^9 d_3^2 \\ & + a_{ij}^{10} d_1^3 + a_{ij}^{11} d_1^2 d_2 + a_{ij}^{12} d_1^2 d_3 + a_{ij}^{13} d_1 d_2^2 + a_{ij}^{14} d_1 d_2 d_3 + a_{ij}^{15} d_1 d_3^2 \\ & + a_{ij}^{16} d_2^3 + a_{ij}^{17} d_2^2 d_3 + a_{ij}^{18} d_2 d_3^2 + a_{ij}^{19} d_3^3; \quad i = 1\text{–}6, \quad j = 1\text{–}6,\end{aligned} \tag{7.7}$$

where d_i are the three pore diameters, M_{ij} is the i,j element of the local structure matrix, and a_{ij}^n are the coefficients of the polynomial fit. Note that M is a 6×6 matrix, and the coefficients a are fit for each element of M. Once M is parameterized in Equation (7.7), we can then pose a least-squares optimization problem to determine the optimal pore diameter to ensure that the scaffold-effective properties will match native bone tissue; the regenerate bone tissue properties will match native bone tissue properties (assuming complete regeneration filling available pores), and a desired scaffold porosity will be maintained:

Objective function:

$$\underset{d_1,d_2,d_3}{\text{Min}} \left\{ \begin{array}{l} \displaystyle\sum_{i=1}^{n} \left(\frac{C_i^{\text{bone eff}} - C_i^{\text{tissue eff}}}{C_i^{\text{bone eff}}} \right)^2 + \\ \displaystyle\sum_{i=1}^{n} \left(\frac{C_i^{\text{bone eff}} - C_i^{\text{scaffold eff}}}{C_i^{\text{bone eff}}} \right)^2 \end{array} \right\} \quad \text{where } n = 1\text{–}9, \tag{7.8}$$

Constraints:

$$d_1, d_2, d_3 \geq \text{minimum buildable feature},$$

$$\frac{V_{\text{pore}}}{V_{\text{total}}} \geq \%\ \text{Porosity},$$

where d_i are the pore diameters, V_{pore} is the scaffold architecture pore volume, V_{total} is the total scaffold architecture volume, $C_i^{\text{tissue eff}}$ are the nine orthotropic native bone elastic constants, $C_i^{\text{bone eff}}$ are the nine orthotropic regenerate bone elastic constants, $C_i^{\text{scaffold eff}}$ and are the nine orthotropic scaffold elastic constants. The constraints ensure that a minimum porosity is maintained and that the smallest feature size can be built using the fabrication methods discussed in Section Hierarchical scaffold fabrication. The optimization problem (Equation (7.8)) is solved using the MATLAB™ fmincon nonlinear programming module, assuming that the regenerate tissue has a fixed stiffness and completely fills the available pore space. Using this formulation, we have been able to design scaffold architectures such that both the scaffold and the regenerate bone tissue matched native minipig mandibular bone stiffness.[16] An example of optimized architecture is shown in Figure 7.3c, d.

The advantage of restricted topology optimization is the ability to determine the best architecture while at the same time maintaining a constraint on feature size. Maintaining a minimum feature size is important for fabrication purposes. The most significant disadvantage of restricted topology optimization is its inability to match all effective property constants. For example, bone is generally regarded as orthotropic and has nine elastic constants. Having only three design parameters makes it difficult to match all nine elastic constants.

An alternative to restricted topology optimization is full microstructure topology optimization.[17,18] The full microstructure topology optimization does not utilize a fixed topology, but rather creates a topology *de novo* by designing each pixel density in 3D space. Each pixel density is itself a design variable, meaning that the total number of design variables is equal to the total number of pixels in the 3D dataset. Note that this method also designs topology within a unit design cell, which is then repeated within a volume to create scaffold architecture. The increased number of design variables makes matching of all effective property parameters more likely.

The mathematical formulation for full microstructure topology optimization may be written as

$$\min_{\rho} \quad w_1 \cdot \left\| C_{11}^{\text{scaffold}} - C_{11}^{\text{bone}} \right\|_{L_2} + w_2 \cdot \left\| C_{22}^{\text{scaffold}} - C_{22}^{\text{bone}} \right\|_{L_2} + \cdots + w_9 \cdot \left\| C_{66}^{\text{scaffold}} - C_{66}^{\text{bone}} \right\|_{L_2}$$

subject to:

$$0 < \rho \leq 1$$

$$\int_V \rho \, dV \leq \text{Vol} \Rightarrow \text{Minimum Percent Porosity Constraint},$$

$$\text{where: } \left\| C_{ij}^{\text{scaffold}} - C_{ij}^{\text{bone}} \right\|_{L_2} = \left(C_{ij}^{\text{scaffold}} - C_{ij}^{\text{bone}} \right)^2,$$

where C_{ij}^{scaffold} are bone elastic constants, C_{ij}^{scaffold} are target bone elastic constants, ρ is the density, and Vol is the maximum scaffold volume. Since the microstructure changes cannot be mapped *a priori* as in restricted topology optimization, the local homogenization Equation (7.5) must be solved in each optimization iteration. Lin et al.[18] used the Method of Moving Asymptotes (MMA[19]) to update the pixel densities at each iteration. The final result is a free-form structure that can differ from architectures created by material process design or restricted topology optimization. Figure 7.4 shows an example the designed scaffold unit cell architecture that matches minipig mandibular condyle trabecular bone effective elastic properties.

Biomimetic Design

Biomimetic is a term denoting mimicry of biologic structure and processes. In terms of scaffold design, biomimetic refers to scaffolds that copy bone tissue structure and function, although it has been most widely applied to describe surface protein mimicry.[20] Biomimetics is an intuitively attractive design principal for tissue engineering scaffolds. Yet, there is no proof that biomimetic designs provide superior tissue regeneration results compared to other engineering designs. Indeed, it is not currently possible to replicate all aspects of bone tissue structure and function on all hierarchical levels. This begs the question as to which aspects of bone structure/function and what hierarchical level are most critical to replicate in a scaffold for enhanced bone regeneration. Given such ambiguity in deciding the most critical biomimetic characteristics, it is difficult to develop specific requirements for scaffold engineering. For example, the optimization approach in Section Periodic Cell Design provides a

Figure 7.4 Example of unit cell designed using full microstructure topology optimization methods. This cell is repeated in 3D space to create the complete scaffold architecture.

method to reproduce trabecular bone-effective elastic constants for most commonly used tissue engineering materials. However, as demonstrated in Figure 7.4, the resulting designed microstructure in most cases does not resemble trabecular architecture. Thus, while the mechanical function is replicated, the structure is not replicated.

For porous bone tissue engineering scaffolds, mimicking trabecular bone architecture is the default biomimetic scaffold structure at the 1st level microstructure. It is also the most straightforward biomimetic structure design, requiring only a 3D micro-CT volume dataset for replication. The image database for any micro-CT scanned trabecular architecture can be used as an image-based design dataset. Furthermore, using image-based design techniques, the trabecular architecture may be manipulated in 3D space to create a scaffold architecture design.

Despite the intuitive attractiveness of making direct trabecular architecture scaffolds, there are at least four significant disadvantages to this approach. First, the base biomaterial stiffness needs to match individual trabeculae stiffness if one desires the scaffold to match effective bone properties. Second, the scaffold porosity would not be flexible unless the native trabecular architecture changes. Third, the regenerate tissue would be the opposite of the original trabecular architecture, thereby matching neither the native tissue effective stiffness nor structure. Fourth, it is very difficult to build such architectures at the desired resolution (see Section Solid Free-form Fabrication). Therefore, most trabeculae in the biomimetic design would have to be scaled up in size.

2nd Level Microstructure Design Pixel Definition

The second-level hierarchical microstructure covers features ranging from 0.01 to 0.1 mm (10–100 μm). Features at the low end of this range may affect the behavior of individual cells or a small group of cells. Theoretically, 2nd level microstructure design pixels may be defined using the same three methods for 1st level microstructure design pixel definition: (1) material process design, (2) periodic cell design, and (3) biomimetic design. In reality, the limitations of fabrication methods necessary for periodic cell and biomimetic designs make it unfeasible to implement these approaches at the current time. Thus, features at this level are currently created by the method in which the material is processed. These include surface etching, abrasion, and coating.

HIERARCHICAL SCAFFOLD FABRICATION

Scaffold Materials

Bone tissue engineering requires scaffolds fabricated from materials that are osteoconductive while providing adequate stiffness and strength. Osteoconductivity is difficult to define and can only be determined by *in vitro* and *in vivo* experiments. Candidate bone scaffold materials are hydroxyapatite (HP), tri-calcium phosphate (TCP), polylactic acid (PLA), polyglycolic acid (PGA), polylactic–polyglycolic acid copolymers (PLGA), and polypropylene fumarate (PPF). Successful bone tissue engineering will require synthesis of these materials alone or in composites based on the designs created using the hierarchical computational techniques. The challenge is to fabricate these candidate scaffold materials in the desired configuration.

SOLID FREE-FORM FABRICATION

Material process design creates a *de facto* design over which there is a limited control. Periodic cell Design and Biomimetic designs produce complex 3D structures that must be fabricated with controlled precision. There are two methods to fabricate such complex 3D structures. The most common is *subtractive fabrication* in which one starts from a block of material and machines away excess material until the final structure is obtained. The second is *additive fabrication*, in which material is added where no material previously existed to create the final 3D structure. Of the two, additive fabrication is more versatile in the range of 3D structures that can be created. Additive fabrication is more commonly known as Solid Free-Form Fabrication (SFF), or as Rapid Prototyping (RP). Rapid Prototyping reflects the early application of SFF where design prototypes were created, not necessarily final products. However, since the goal of scaffold fabrication is to create final products, the term SFF will be used to denote the capability of creating complex 3D free-form biomaterial scaffolds directly or the molds for such scaffolds. There are a variety of SFF systems currently on the market. These systems are described in brief below; however, the reader may find more detailed information and comparisons between SFF systems at the Castle Island's Worldwide Guide to Rapid Prototyping (http:// home.att.net/~castleisland/), or in the reference text by Cooper.[21]

All current commercially available SFF techniques build structures on a layer-by-layer basis. These systems use either surface triangular facet representation, called a .STL file, or a layer-by-layer contour file, whose format varies between systems, to define the object to be fabricated. Thus, any design, whether CAD or image-based, must be transformed into a surface or contour representation.

There are a number of SFF technologies (Figure 7.5); but most fall into two general classes. The first class utilizes material, either powdered or liquid, on a platform. The material is then processed by a device that is moved over the material. The second class actually feeds material directly through a nozzle onto a platform. The first class of techniques is the basis for three commercial systems, although each is different in practice. The stereolithography apparatus (SLA) developed in 1987 by 3DSystems™ (www.3dsystems.com) utilizes a laser beam to photopolymerize liquid monomer held in a vat (Figure 7.5a). Once one layer is polymerized, the platform on which it sits is lowered and the next layer of monomer is polymerized on top. If no structure exists beneath the current layer, a support must be built to prevent the overhanging layer from collapsing. The advantage of SLA is the speed and surface finish of the parts; also, photopolymerizable biomaterial polymers can be built directly using SLA. Disadvantages include a very limited ability to directly build ceramic parts, the laborious procedure of removing supports, and the rough surface produced at support sites.

Selective Laser Sintering (SLS) uses a laser to sinter together powdered materials in a bed (Figure 7.5b). SLS was originally developed by DTM Corporation in Austin, Texas. The rights to the SLS system were recently purchased by 3D Systems (www.3dsystems.com). The SLS has two reservoir areas. One holds the powder supply and brings powder up to be spread over the second build reservoir. The laser sinters, or melts powder particles together. Following completion of one layer, the

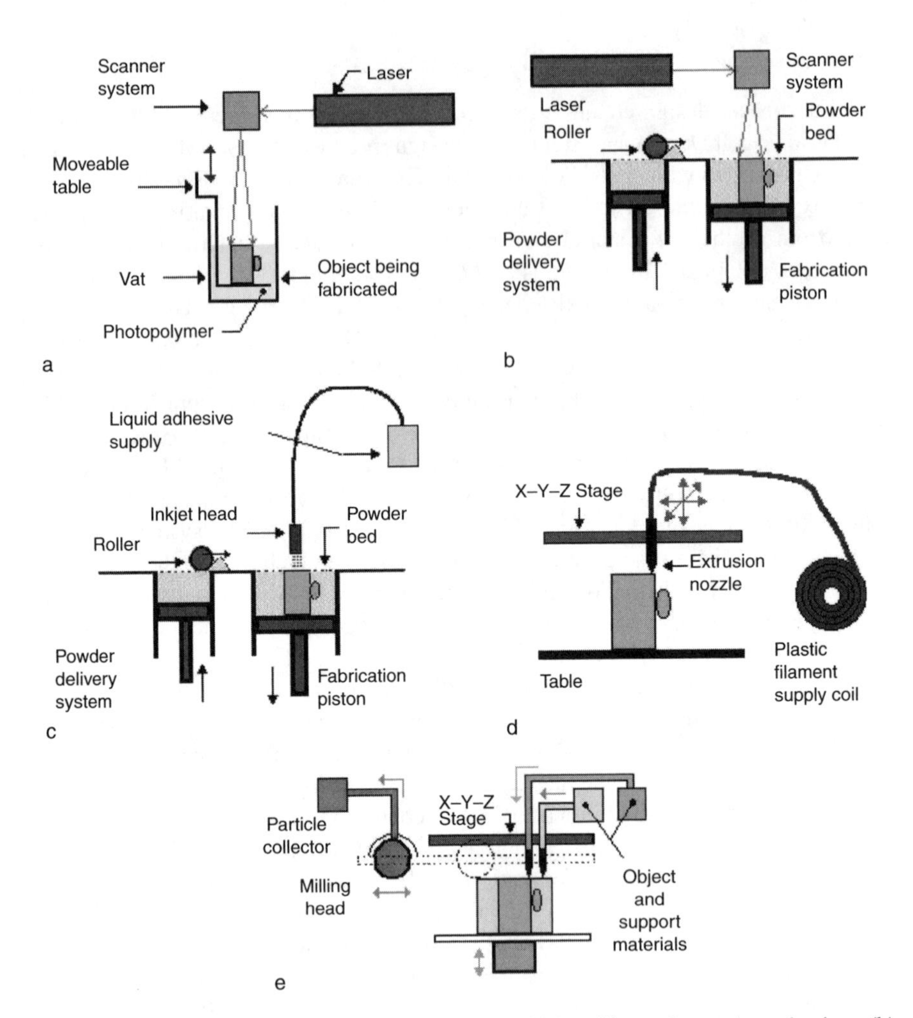

Figure 7.5 (a) Schematic of the SLA process, which utilizes photopolymerization. (b) Schematic of the SLS process that uses a laser to sinter-powdered material. (c) Schematic of the 3D printing technology, which prints a binder onto powdered material. (d) Schematic of the FDM system that deposits material through a heated nozzle. (e) Schematic of the MM2 system that prints wax. Copyright Castle Island's Worldwide Guide to Rapid Prototyping (http://home.att.net/~castleisland/), used with permission.

build reservoir piston is lowered and powder is spread for the next layer. The advantage of SLS is the ability to build materials in powder form without support. This includes thermoplastic polymers, ceramics, and metals. The disadvantage of SLS is the coarser surface finish of the final part and the cost of the system. The resolution of both SLA and SLS depends primarily on the spot size of the laser beam, although in SLS resolution will also depend on the material powder size.

The third SFF machine that processes material on a platform is the 3D Printing (3DP) technology developed at the Massachusetts Institute of Technology (MIT).

Unlike SLA or SLS, 3DP does not use a laser to process material, but rather prints a chemical binder or solvent onto powdered material (Figure 7.5c). As in the SLS system, there are two reservoirs: one for the powdered material and one for the build part. Once a layer is finished, the platform in the build reservoir is lowered and another layer of powder is deposited in the build reservoir. The advantage of 3DP is the capability to build parts very rapidly and without supports. The disadvantage is its surface finish and part fragility. Z Corporation (www.zcorp.com) sells commercial versions of 3DP technology that prints using either a starch/cellulose or plaster powder. Therics Inc. (www.therics.com) has adapted the 3DP technology to build directly with both polyesters and hydroxyapatite/polymer composites for tissue engineering and drug delivery applications, but does not sell these modified 3DP systems.

The second general class of SFF technology processes material using heat as it flows through a nozzle. The nozzle is guided by the same surface or contour structure description that is used by material platform-based SFF systems. The most popular of the second class of systems is the Fused Deposition Modeling™ (FDM) system made by Stratasys, Inc. (www.stratsys.com). FDM uses a thin filament of material unwound from a coil and fed through a nozzle. The filament is heated as it passes through the nozzle and deposited following a pattern (Figure 7.5d). The flow of material may be turned off or on through the nozzle. The major advantage of FDM is the material strength of the base polycarbonate or acrylonitrile butadiene stryrene (ABS) material with which the system builds. The disadvantage compared to overhead laser-based or binder printing systems is speed. For biomaterial fabrication, materials that can be produced in thin filaments and that have melting points within the nozzle heating range can be used to build the system.

The other widely used approach that utilizes a nozzle to deposit material is the thermo wax inkjet technology from Solidscape™ (www.solid-scape.com). There are three machines within this line, the ModelMaker2™ (MM2), PatternMaster™ (PM), and the T66™. The Solidscape systems use two nozzles, one that prints a polysulfonamide build and a second that prints a red wax support (Figure 7.5e). Finally, once a layer is printed, that layer is milled to achieve a fine uniform thickness. The significant advantage of the Solidscape inkjet systems is the detailed resolution that can be achieved, finer than other available SFF systems. In addition, either the wax or polysulfonamide material is relatively easy to remove and can be used to cast bioceramics. The disadvantage of the inkjet technology is its very slow build speed compared to other SFF systems.

The overwhelming advantage of using the SFF technique for scaffold fabrication is the precise control over 3D geometry. This control makes it possible to realize the complex image-based designs that match anatomic shape incorporating 1st level microstructures that are either biomimetic or created by topology optimization. The obstacles to SFF scaffold fabrication include (1) the expense of SFF machines, ranging from $30,000 to over $200,000, (2) the necessity of converting complex image-based designs into data that can be utilized by SFF systems, (3) the difficulty in adapting biomaterials to SFF fabrication, and (4) the feature size limits that can be built using SFF, typically ranging between 300 and 600 μm. Our research group has addressed the second obstacle by developing customized software that can convert image-based designs into SFF data (see the next Section). A few research groups have addressed the difficulty in utilizing biomaterials with SFF by adapting a limited number of biomaterials for SFF fabrication (section Direct sacffold fabrication via SFF). Our own and other research groups have also overcome SFF material limitations by creating molds via SFF into which biomaterials are cast (section Scaffold fabrication via indirect SFF and casting). This approach retains the fabrication advantages of SFF while expanding the number of biomaterials from which complex 3D scaffolds may be made.

Creating SFF Data Input from Image-based Designs

As noted in the previous Section all SFF machines require a data file that describes the structure surface to drive the fabrication mechanism. This input data file may describe the structure surface using either triangular facets in true 3D or as polyline contours for each fabricated layer. The most commonly used approach is to represent the structure surface using triangular facets in 3D. This format, denoted with the file extension .stl and originally developed for the SLA™ system, has become a universal standard that is accepted by all SFF machines. The basic format represents each triangular facet in the surface mesh by the three vertex coordinates and a normal that points outward from the surface.

The key to fabricating designed scaffolds is transitioning from the 3D image data set to the triangular facet surface representation. This can be done automatically using marching cubes methods that generate a triangular mesh isosurface directly from 3D image data. The image density level must be specified for surface extraction. We have utilized the shade_volume command in IDL software within a specially written program to automatically generate .stl triangular surface facet data from 3D image design data.

A disadvantage of creating.stl surface data from image-based designs is that very complex hierarchical designs will generate extremely large datasets, often over 1 Gigabyte. For such large files, it is much more advantageous to split the design down into layers and process each layer separately into contours. Contour files, however, are not as universally accepted as .stl files, although Slice Contour format, .slc, and AutoCAD™.DXF format are accepted by many machines. We have developed software that runs within IDL to directly extract contours from image-slice designs and directly write the resulting data to .SLF, a proprietary contour format for the Solidscape inkjet nozzle systems. In this software, each image design slice is processed using the Image_Contour command in IDL to write polyline vertices and connectivity.

Direct Scaffold Fabrication via SFF

It is, of course, most desirable to build scaffolds directly on an SFF machine. Theoretically, this would provide automated scaffold engineering from design through fabrication. In reality, it is very difficult to adapt SFF machines to directly build a single biomaterial, let alone the multitude of composite biomaterials that may be needed for tissue engineering. Still, some scaffold biomaterials have been adapted for direct fabrication using SLA, SLS, FDM, and 3DP.

Zien et al.[22] used poly (ε-caprolactone) (PCL) for mesh scaffold fabrication using the FDM method. PCL filaments were extruded to a 1.7 mm diameter using a one-shot extruder (Alex James & Associates, Greensville, SC). The PCL filament nozzle moved at a rate of 6.35 mm/sec and the FDM liquefier temperature was set at 125°C. Scaffold designs were created using traditional CAD design in Unigraphics software and output in .stl format for slicing in Stratasys Quickslice software. They created alternating ply designs (Figure 7.6) that had pore dimensions ranging between 160 and 700 μm, and porosity ranging from 48 to 77%. The designed scaffold stiffness ranged from 40 to 70 MPa, and the yield strength ranged from 2 to 3.5 MPa. The base PCL stiffness and strength were reported to be 510 and 16 MPa, respectively. These results demonstrated that FDM could be used to produce scaffolds having mechanical properties equal the lower range of reported trabecular bone properties (trabecular bone stiffness ranges from 10 to 1000 MPa, and

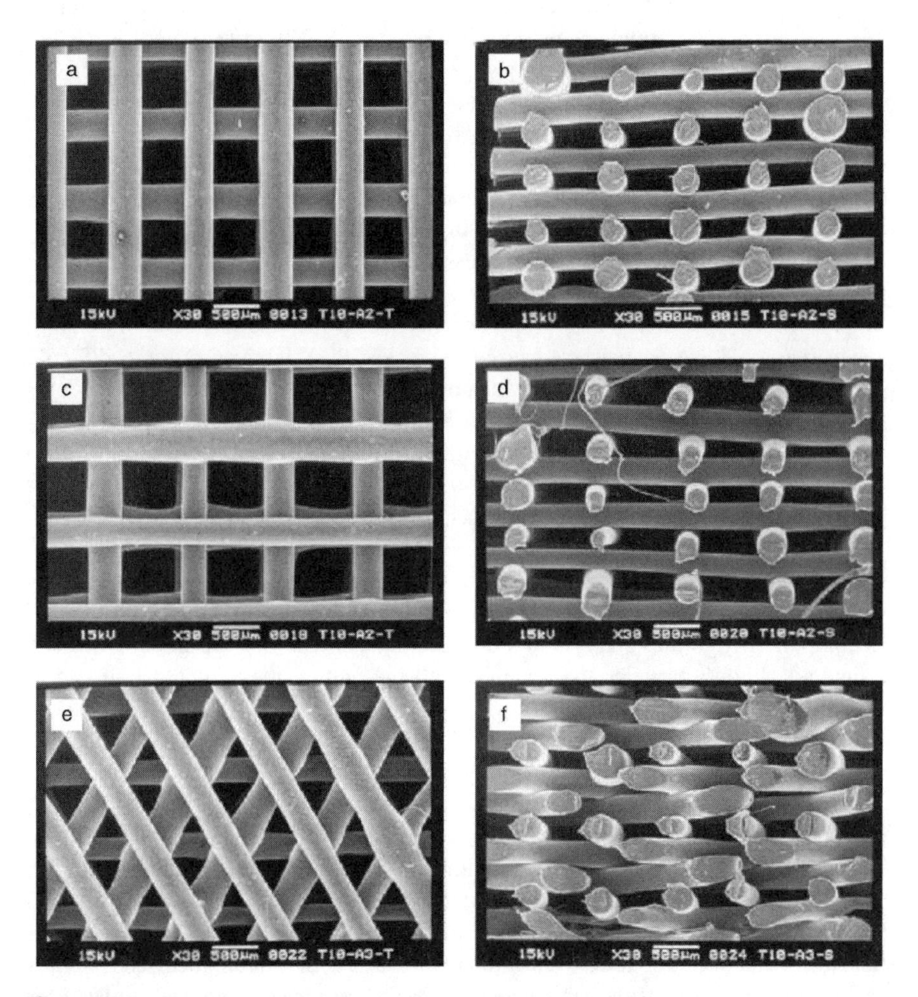

Figure 7.6 Example of alternating ply design PCL scaffolds created by Zien et al. (Reprinted from Biomaterials, 23, Zien, I, Hutmacher, DW, Tan, KC, Teoh, SH, "Fused deposition modeling of novel scaffold architectures for tissue engineering applications," 1169–1185, Copyright 2002, with permission from Elsevier.)

compressive and shear strength ranges from 1 to 20 MPa[23]) and porosity for bone tissue engineering. It also demonstrates that directly producing biomaterial scaffolds using FDM requires the ability to create thin filaments from the desired biomaterial.

Sherwood et al.[24] used the Therics proprietary version of the 3DP SFF process to build an osteochondral composite consisting of 90% porous D,L-PLGA/L-PLA for the cartilage portion and 55% porous L-PLGA/TCP composite for the bone portion. NaCl was used as a porogen for both portions of the scaffold. Powders for each material portion were spread across the build reservoir, and chloroform was printed to bind the powders. Elastic moduli ranged from 50 to 450 MPa, with the compressive yield strength ranging from 2.5 to 14 MPa, again within the range of human trabecular bone.

SLA has been used to directly fabricate PPF in a collaborative effort between Case Western Reserve University and Rice University.[25–27] A photopolymerizable version of PPF was created by adding a biascylphosphine oxide (Irgacure, CIBA Speciality Chemicals Additives Division, Tarrytown, NY) as a photoinitiator. The material was processed on an SLA 250/40 system with a 325 nm ultraviolet laser. Fisher et al. [25,26] constructed 6.3 mm scaffolds with porosity ranging from 57 to 75% and pore diameters ranging from 300 to 800 μm. The scaffolds were implanted in rabbit cranial defects and followed for 2 and 8 weeks. While direct bone contact was noted with all formulations of the PPF scaffolds, the percent bone fill in the PPF scaffolds was less than 6% in all cases. SLA-fabricated PPF had base stiffness ranging from 20 to 200 MPa and fracture strength ranging from 20 to 70 MPa, again within the range of trabecular bone mechanical properties.[28] The mechanical properties increase as the viscosity of the PPF solution increases. However, increased viscosity makes SLA fabrication more difficult, presenting a challenging trade-off between manufacturability and function.

Das et al.[28] investigated the use of SLS-fabricated polyamide 6 (Nylon6) scaffolds for bone tissue engineering. They utilized Atofina Orgasol (www.atofina.com) 1002 ES4 powdered Nylon6 with 38–42 μm particle size. The scaffolds were built on a Sinterstation 2000 using 200°C preheat of the powder, with 7 W laser power, a 1257.3 mm/sec scan speed and a 100 μm build layer thickness. Using this technique, Das et al. were able to build scaffolds with 800 μm pore diameters. In addition, they built biomimetic scaffolds that directly replicated trabecular bone architecture. Biocompatibility studies on Nylon6 indicated that cell viability was good. However, unsintered Nylon6 particles that leached into the culture media could be toxic to cells. Pilot *in vivo* results in minipigs demonstrated that the Nylon6 scaffolds could support bone regeneration, but not as well as traditional osteoconductive materials like HA. This again could reflect unsintered particles coming off the scaffold.

These results demonstrate that direct scaffold fabrication is possible on SLA, SLS, 3DP, and FDM systems. In most cases, however, significant modification of the base biomaterial is necessary for utilization with SFF machines. These modifications, for example, reducing PPF viscosity for better SLA fabrication, may lead to reduced mechanical properties. In addition, results of Fisher et al.[25,26] and Das et al.[28] demonstrate that materials introduced to allow direct biomaterial SFF or unprocessed materials may have negative effects on bone regeneration. Furthermore, all materials utilized for SFF fabrication, except for the work by Sherwood et al.,[24] have been polymers. Even in the case of Sherwood et al.,[24] the scaffold was a polymer/ceramic blend. It is often necessary to sinter ceramics to achieve desirable mechanical properties, a task that to date has not been demonstrated with current direct SFF techniques. Finally, it is difficult to build discrete polymer–ceramic composites using SFF techniques.

SCAFFOLD FABRICATION VIA INDIRECT SFF AND CASTING

Indirect SFF is an alternate approach to scaffold fabrication that preserves 3D complexity while increasing material selection. This approach uses SFF to fabricate a mold into which biomaterials are cast rather than building the scaffold directly. This technique was first discussed by Chu et al.,[29,30] who used SLA fabricated epoxy resin molds into which hydroxyapatite slurry was cast. Later, the authors used wax molds built on the MM2 thermowax inkjet printing machine to cast ceramic slurries. Charriere and colleagues used this approach to create calcium phosphate ceramic scaffolds.[29,30] Taboas et al. [31] later extended the indirect SFF/Casting technique of Chu and colleagues to make polymer scaffolds, polymer–ceramic composite scaffolds, and biomimetic scaffolds. Casting provides for manufacturing from a wider material selection than current direct SFF techniques, and the base material does not need to be modified.

Hydroxyapatite Scaffolds Fabricated by Indirect SFF and Casting

Chu et al.[29] first utilized molds for HA casting built from Ciba-Geigy epoxy 5170 on a 3D systems SLA 250/40 system. The slurry consisted of a suspension vehicle, hydroxyapatite (HA) powder from plasma-biotal (www.plasma-biotal.com), a dispersant, and a thermal initiator. The suspension vehicle was a 50/50% by weight mixture of propoxylated neopentolglycol diacrylate (PNPGDA) and isobornyl acrylate (IBA). The dispersant is a 50/50% mixture of quartenary ammonium acetate and aromatic phosphate ester. The dispersant helps to decrease the mixture viscosity for casting. To this dispersant/suspension mixture, up to 40% by weight HA powder may be added. The dispersant dose should be about 4% weight for minimal viscosity. Benzoyl peroxide (BPO) is added at 0.15% weight to the suspension/dispersant/HA mixture to initiate thermal curing. This complete mixture may then be cast into a mold, and then allowed to be thermally cured. Once the mixture cures, the mold must be removed. This is most readily done by heating the mold to burn it out. Once the mold and acrylic are burnt out, the ceramic mixture, now known as a "green body" must be sintered at high temperatures to create a dense structure with good mechanical properties. Chu et al. found that sintering at temperatures between 1300 and 1350°C were sufficient to produce over 90% dense structures while avoiding HA decomposition.

Chu and colleagues[30] later used the indirect SFF HA technique to produce scaffolds for both mechanical and *in vivo* testing. Scaffold molds were designed using the image-based design technique, following the section Image-based methods for designing hierarchical features. The same HA slurry mixture described by Chu et al.[29] was then cast into the epoxy molds. Pore diameters ranging from 334 to 450 μm were achieved in the scaffolds. Chu et al. measured an average mechanical stiffness and strength of 1.4 GPa and 30 MPa, respectively, at the high end of trabecular bone properties. They also found significant bone regeneration with between 30 and 60% of pore volume filled by bone within the designed scaffolds in an *in vivo* minipig mandibular defect model. A difficulty with the epoxy resin is that it may undergo significant thermal expansion during sintering, often cracking the HA scaffolds. To address resin mold fabrication difficulties, our research group has switched to creating wax molds using inkjet wax printing machines. The advantage is that the wax material has a low melting point, thereby avoiding significant thermal expansion during burnout. In addition, support removal is much easier. The disadvantage is that the wax printing machine is significantly slower than the SLA machine.

Calcium Phosphate Cement Scaffolds Fabricated by Indirect SFF and Casting

The process for creating sintered hydroxyapatite scaffolds can readily be extended to creating scaffolds from calcium phosphate cements. Charriere and colleagues [32,33] developed a technique to create designed calcium phosphate scaffolds using indirect SFF casting techniques. They used the periodic cell design technique to create an orthogonal pore design with 22% porosity. The scaffold design was then built using an inkjet wax printing technique. The hydroxyapatite cement was created by reacting monetite (DCP, Merck) with calcite (CC, Merck) in water to precipitate HA. For casting purposes, a dispersing agent polyacrylic acid was added to the mixture. The mixture was then cast into the wax mold (Figure 7.7). In another example of hierarchical scaffold fabrication, these hydroxyapatite cement scaffolds also contain microporosity. This microporosity raises the total porosity to 56%. The microporosity in cements is typically less than 10 μm, which is too small for cell invasion, but still potentially useful for biofactor delivery.

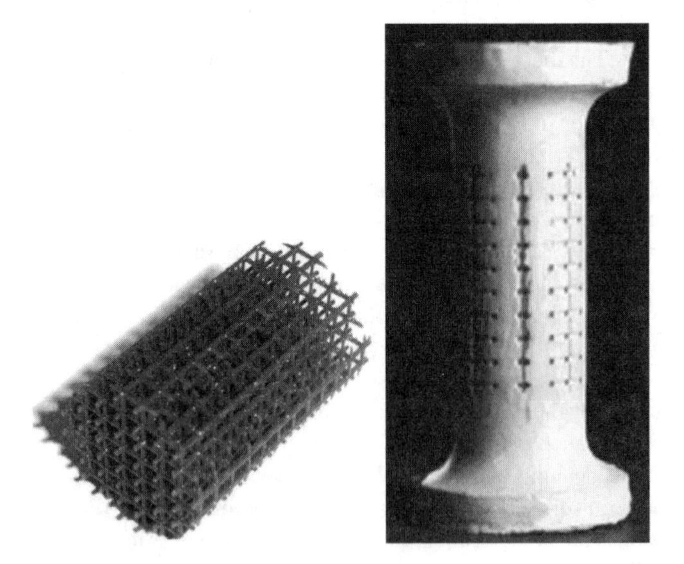

Figure 7.7 Calcium phosphate ceramic cement scaffolds made by casting in a inkjet fabricated wax mold (a) wax mold. (b) Resulting mechanical testing specimen. (Reprinted from Biomaterials, 24, Charriere, E., Lemaitre, J., Zysset, Ph, "Hydroxyapatite cement scaffolds with controlled macroporosity: fabrication protocol and mechanical properties," 809–817, Copyright 2003, with permission from Elsevier.)

Charriere performed extensive mechanical testing of the scaffolds, measuring Young's and Shear Moduli, as well as compressive, tensile, and shear strength. In addition, Charriere et al. used homogenization theory [14] to compute the anisotropic elastic properties of the periodic cell design. The measured Young's moduli and shear moduli were 7.4 and 2.4 GPa, respectively. Measured strength was 12.6 MPa in compression, 1.3 MPa in shear, and 0.7 MPa in tension. Stiffness computed using homogenization theory compared very well with experimental measures, confirming its use as a valuable aid in designing scaffold microstructure. Scaffold stiffness was between cancellous and cortical bone. Although compressive strength measures were greater than cancellous bone, tensile and shear strength were considerably less than both cancellous and cortical bone, reflecting the brittle behavior of ceramic cements.

Polymer and Composite Ceramic/Polymer Scaffold Fabrication via Indirect SFF/Casting

Although ceramic scaffolds can be cast into SFF fabricated molds, casting tissue engineering polymers, like PLA and PGA, is difficult because most solvents for these polymers will dissolve or degrade mold materials. In addition, polymer casting via solvents can create porous scaffolds whose mechanical properties do not match bone properties. Taboas et al. [31] addressed these issues by developing HA ceramic-based molds for polymer casting. Taboas et al. used the process developed by Chu et al.[29,30] to create HA molds. The HA slurry was cast into molds built on a Solidscape MM2™ machine from wax and polysulfonamide. The polysulfonamide is removed and ceramic slurry is cast into the wax. After the ceramic is sintered, biopolymers like PLA or PGA may either be solvent cast or melt cast in the ceramic mold. The commonly used polymer solvents do not affect the ceramic. In addition, the ceramic mold can withstand temperatures beyond the melting point of commonly used polymers like PLA and PGA. This indirect SFF/casting fabrication method creates numerous possibilities for creating not only complex 3D polymer scaffold architectures but also polymer scaffolds with a true hierarchy in feature size as well as both discrete ceramic/polymer composites and blend ceramic/polymer composites. The indirect SFF method provides the most versatile method with which to create hierarchical scaffolds with designed features on the 1st level microstructure in the 100–1000 μm range and a 2nd level microstructure, albeit with limited control, in the 5–20 μm feature range. In fact, layering artifacts created by SFF will create features in the 10–40 μm range.

To create polymer scaffolds, Taboas et al. used both solvent casting and melt casting. For melt casting, polymers are heated to 10–20°C above the polymer melting point (120°C for PLA; 150°C for PGA). PLLA and PGA were purchased from Birmingham Polymers, Inc. (Birmingham, AL; (205) 917-2231; www.birminghampolymers.com). The molten polymer is placed in a reservoir, after which the ceramic mold is pushed into the molten polymer, which infiltrates through the mold pores. Using this technique, Taboas et al. were able to create discrete composites of PLA and PGA. For the composite, PGA was allowed to infiltrate one half of the mold. PGA was cooled to 130°C, still above the melting point of PLA. Molten PLA was then pressurized into the remaining half of the ceramic mold. The mixture was cooled to 100°C for 30 min to allow crystal formation. Following polymer formation, molds are removed using an acid solvent RDO (APEX Engineering Products Corp., Plainfield, IL; (815) 436-2200). The scaffold is agitated in RDO for 1–6 h. In addition to melt casting, solvent casting can be easily utilized with ceramic molds. For solvent casting, Taboas et al. used chloroform to dissolve PLA in 25% w/v. The polymer was cast and the solvent was evaporated. This process was repeated until full infiltration of the mold was achieved.

Discrete ceramic/polymer scaffolds can be readily created using the ceramic mold casting technique. There are two approaches for creating ceramic/polymer discrete scaffolds. In the first method, the HA ceramic mold is dipped into molten polymer heated as previously described and allowed to penetrate to the desired depth and fill the HA pores. The composite is then cooled using the same protocol as for the polymer only scaffolds. Only the polymer/HA portion of the scaffold is then submerged in RDO following the previously described protocol to yield a bonded ceramic/polymer scaffold. The second approach to composite scaffold fabrication is to fabricate the ceramic and polymer scaffold portions separately. The ceramic portion is then heated and the polymer portion is mated to the ceramic portion. Figure 7.8 shows an example of a composite HA/PLA scaffold and a micro-CT scan showing the bonded interface between the two materials.

Unique two-scale hierarchical scaffolds can be created by combining designed ceramic molds with solvent casting techniques (Figure 7.9). Taboas et al.[31] used both porogen leaching and emulsion diffusion with ceramic-based casting techniques to create scaffolds that had 1st level microstructure pores between 400 and 800 μm in diameter and 2nd level microstructure pores ranging from 5 to 120 μm in size. For porogen leaching, sieved NaCl grains ranging from 104 to 125 μm are packed into the ceramic mold pores. PLA dissolved in chloroform in a 7.5% w/v ratio is then cast into the ceramic mold and evaporated under 15″ Hg vacuum overnight.

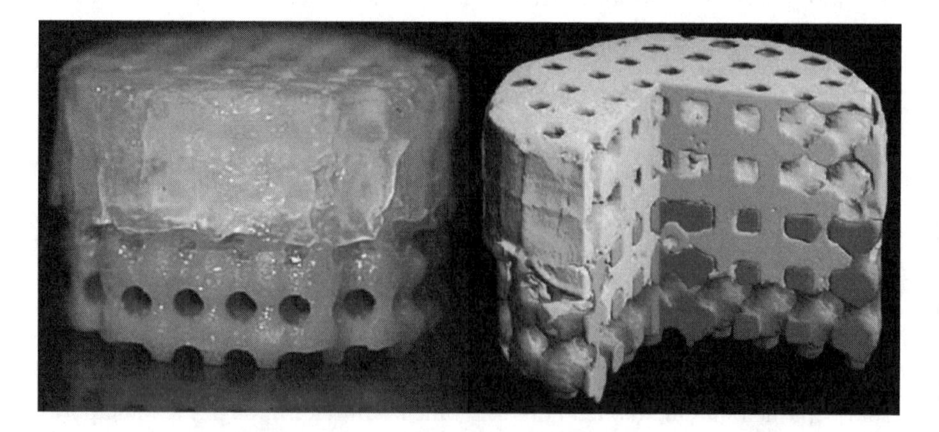

Figure 7.8 Discrete composite ceramic/polymer scaffolds. (a) HA/PLA scaffold. (b) colorized micro-CT image showing a bonded interface between HA and PLA. (Reprinted from Biomaterials, 24, Taboas, JM, Maddox, RD, Krebsbach, PH, Hollister, SJ, "Indirect solid free form fabrication of local and global porous, biomimetic and composite 3D polymer-ceramics scaffolds," 181–194, Copyright 2002, with permission from Elsevier.)

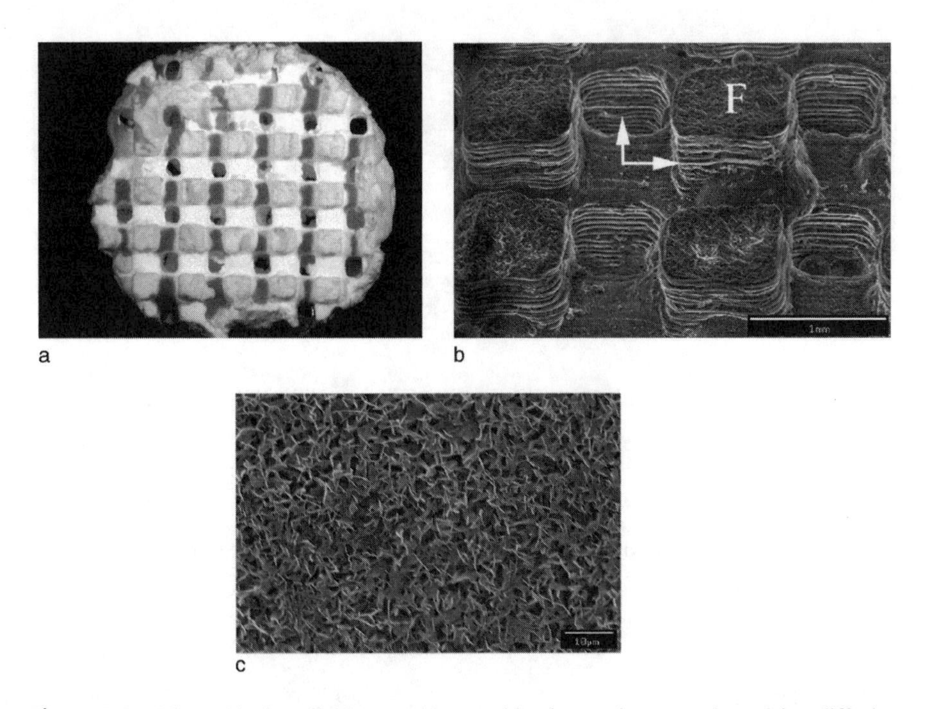

Figure 7.9 Hierarchical scaffold created by combined ceramic cast and emulsion diffusion technique. (a) Complete PLA scaffold. (b) 36.8× SEM view showing global rectangular pores, including a layering artifact from the mold fabrication process. (c) 2400× SEM view of 5 – 11 μm 2nd level microstructure pores. (Reprinted from Biomaterials, 24, Taboas, JM, Maddox, RD, Krebsbach, PH, Hollister, SJ, "Indirect solid free form fabrication of local and global porous, biomimetic and composite 3D polymer-ceramics scaffolds," 181–194, Copyright 2002, with permission from Elsevier.)

Even finer 2nd level microstructure features can be created using emulsion solvent diffusion or emulsion freeze drying. For emulsion solvent diffusion, PLA is dissolved in tetrahydrofuran and cast into the ceramic mold at 60ºC. The mold is then cooled to room temperature (22.5ºC), soaked for two days in ethanol, and then air-dried overnight. Figure 7.9 shows scaffolds created using this technique with 1st level microstructure pores of 600 μm and 2nd level microstructure pores on the order of 5–11 μm. For emulsion freeze drying, the cast mold is snap frozen in liquid nitrogen, then freeze dried under a dry ice and ethanol slurry at 30″ Hg vacuum.

In addition to creating polymer scaffolds that have 1st level microstructure pores and 2nd level microstructure pores outside of the larger pores, it is possible to create 2nd level microstructure pores within the 1st level microstructure pores. This type of scaffold allows the use of 2nd level microstructure in the form of a polymer sponge for cell/gene seeding and retention, while allowing 1st level microstructural features for structural reinforcement. Furthermore, the 1st level microstructure may be made from either ceramic or polymer. The HA scaffold is created as described in the section Calcium phosphate cement scaffolds fabricated by indirect SFF and casting. Following sintering of the HA, NaCl particles are packed into the HA pores. The

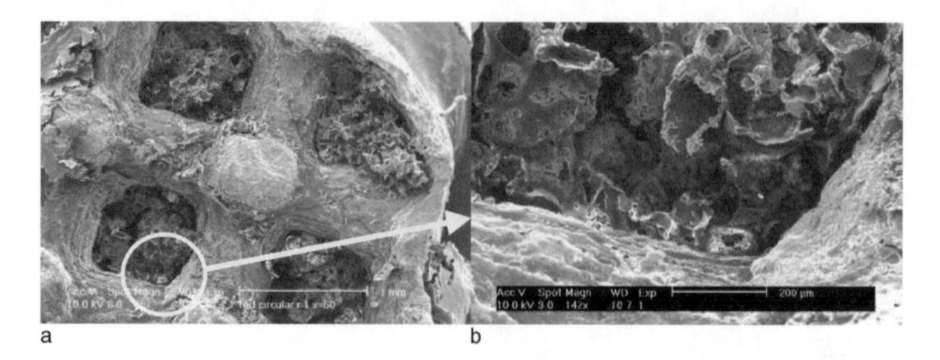

Figure 7.10 Example of a hierarchical HA scaffold with a PLA sponge inside the HA pores. (a) global view showing PLLA sponge within HA pores. (b) Localized view showing sponge features.

dissolved PLA is then cast into the pores, and the particles are removed with water to complete the sponge. Figure 7.10 shows an example of an HA 1st level microstructure scaffold with a polymer sponge inside the HA pores.

CONCLUSIONS

Bone tissue engineering scaffolds must meet a number of often conflicting requirements for mechanical function and enhance bone regeneration. In addition, the scaffold must define and maintain the potential space for bone regeneration within complex 3D defects. Creating scaffolds to fulfill this multitude of requirements requires precise control over the scaffold external shape and internal architecture. Controlled scaffold architecture is also fundamental for performing *in vitro* and *in vivo* experiments testing numerous scaffold design variables about which there is limited quantitative data. Creating these controlled architectures will require continued integration of computational design techniques with biomaterial free-form fabrication. This chapter has described one such approach coupling image-based design with both direct and indirect SFF scaffold fabrication. Continued research in both computational design, to include other scaffold characteristics like mass transport, coupled with continued fabrication research, expanding the range of biomaterial fabrication, is needed to advance bone tissue engineering.

ACKNOWLEDGMENTS

The authors gratefully acknowledge the support and input of their colleagues from the Skeletal Engineering Group at the University of Michigan. In addition, we would like to recognize the important contributions of Drs. Paul Krebsbach, Stephen Feinberg, Noboru Kikuchi, John Halloran, and Suman Das to this ongoing research effort. Finally, we are grateful to the NIH NIDCR for financial support of this research through grants DE R01 13608 (a Bioengineering Research Partnership) and DE R01 13416.

References

1. Bruder, S. P., Kraus, K. H. et al. (1998). Critical-sized canine segmental femoral defects are healed by autologous mesenchymal stem cell therapy. *Trans. Annu Meeting Orthopaedic Res. Soc.* 44, 147.
2. Mikos, A. G., Sarakinos, G. et al. (1993). Prevascularization of porous biodegradable polymer. *Biotechnol. Bioeng.* 42, 716–723.
3. Kühne, J.H., Bartl, R., Frisch, B., Hammer, C., Jansson, V. and Zimmer, M. (1994). Bone formation in coralline hydroxyapatite: effects of pore size studied in rabbits. *Acta Orthop. Scand.* 65, 2246–2252.
4. Grenga, T.E., Zins, J.E. and Bauer, T.W. (1989). The rate of vascularization of coralline hydroxyapatite. *Plast. Reconstr. Surg.* 84, 245–249.
5. Kuboki, Y., Jin, Q., Kikuchi, M., Mamood, J. and Takita, H. (2002). Geometry of artificial ECM: sizes of pores controlling phenotype expression in BMP-induced osteogenesis and chondrogenesis. *Connect. Tissue Res.* 43, 520–523.
6. Hui, P.W., Leung, P.C. and Sher, A. (1996). Fluid conductance of cancellous bone graft as a predictor for graft–host healing response. *J. Biomech.* 29,123–132.
7. Brand, R.A., Stanford, C.M. and Nicolella, D.P. (2001). Primary adult human bone cells do not respond to tissue (continuum) level strains. *J. Orthop. Sci.* 6, 295–301.
8. McCreadie, B.R. and Hollister, S.J. (1997). Strain concentrations surrounding an ellipsoid model of lacunae and osteocytes. *Comput. Methods Biomech. Biomed. Eng.* 1, 61–68.
9. Mullender, M.G. and Huiskes, R. (1997). Osteocytes and bone lining cells: which are the best candidates for mechano-sensors in cancellous bone?. *Bone* 20, 527–532.
10. Thomson, R.C., Yaszemski, M.J., Powers, J.M. and Mikos, A.G. (1995). Fabrication of biodegradable polymer scaffolds to engineer trabecular bone. *J. Biomater. Sci. Polym. Ed.* 7(1), 23–38
11. Mikos, A.G., Thorsen, A.J., Czerwonka, L.A., Bao, Y., Langer, R., Winslow, D.N. and Vacanti, J.P. (1994). Preparation and characterization of poly(l-lactic acid) foams. *Polymer* 35, 1068–1077.
12. Schugens, C., Maquet, V., Grandfils, C., Jerome, R. and Teyssie, P. (1996). Polylactide macroporous biodegradable implants for cell transplantation. 2. Preparation of polylactide foams by liquid–liquid phase separation. *J. Biomed. Mater. Res.* 30, 449–461.
13. Nam, Y.S. and Park, T.G. (1999). Porous biodegradable polymeric scaffolds prepared by thermally induced phase separation. *J. Biomed. Mater. Res.* 47(1), 8–17.
14. Hollister, S.J. and Kikuchi, N. (1994). Homogenization theory and digital imaging: a basis for studying the mechanics and design principles of bone tissue. *Biotech. Bioeng.* 43, 586–596.
15. Hornung, U., Ed. (1997). *Homogenization and Porous Media.* Springer-Verlag, New York.
16. Hollister, S.J., Maddox, R.D. and Taboas, J.M. (2002). Optimal design and fabrication of scaffolds to mimic tissue properties and satisfy biological constraints. *Biomaterials* 23, 4095–4103.
17. Sigmund, O. (1994). Materials with prescribed constitutive parameters — an inverse homogenization problem. *Int. J. Solid Struct.* 31, 2513–2329.
18. Lin, C.Y., Kikuchi, N. and Hollister, S.J. (2003). Scaffold internal architecture design for elastic properties and porosity by a topology microstructure optimization method. *Proc. Soc. Biomater.* 29, 354.
19. Svanberg, K. (1987). The method of moving asymptotes — a new method for structural optimization. *Int. J. Numer. Methods Eng.* 24, 359.

20. Dillow, A.K., Lowman, A.M., and Hudgins, K.A., eds., (2002). *Biomimetic Materials and Design: Biointerfacial Strategies, Tissue Engineering, and Targeted Drug Delivery*. Marcel-Dekker, Inc., New York.
21. Cooper, K.G. (2001). *Rapid Prototyping Technology: Selection and Application.* Marcel-Dekker, Inc., New York.
22. Zein, I., Hutmacher, D.W., Tan, K.C. and Teoh, S.H. (2002). Fused deposition modeling of novel scaffold architectures for tissue engineering applications. *Biomaterials* 23, 1169–1185.
23. Goldstein, S.A. (1987). The mechanical properties of trabecular bone: dependence on anatomic location and function. *J. Biomech.* 20, 1055–1061.
24. Sherwood, J.K., Riley, S.L., Palazzolo, R., Brown, S.C., Monkhouse, D.C., Coates, M., Griffith, L.G., Landeen, L.K. and Ratcliffe, A. (2002). A three-dimensional osteochondral composite scaffold for articular cartilage repair. *Biomaterials* 23, 4739–4751.
25. Fisher, J.P., Vehof, J.W.M., Dean, D., van der Waerden, J.P.C.M., Holland, T.A., Mikos, A.G. and Jansen, J.A. (2002). Soft and hard tissue response to photocrosslinked poly(propylene fumarate) scaffolds in a rabbit model. *J. Biomed. Mater. Res.* 59, 547–556.
26. Fisher, J.P., Dean, D. and Mikos, A.G. (2002). Photocrosslinking characteristics and mechanical properties of diethyl fumarate/poly(propylene fumarate) biomaterials. *Biomaterials* 23, 4333–4343.
27. Cooke, M.N., Fisher, J.P., Dean, D., Rimnac, C. and Mikos, A.G. (2002). Use of stereolithography to manufacture critical-sized 3D biodegradable scaffolds for bone ingrowth. *J. Biomed. Mater. Res. Part B: Appl. Biomater.* 64B, 65–69.
28. Das, S., Hollister, S.J., Flanagan, C., Adewunmi, A., Bark, K., Chen, C., Ramaswamy, K., Rose, D. and Widjaja, E. (2003). Freeform fabrication of Nylon-6 tissue engineering scaffolds, *Rapid Prototyping J.* 9, 43–49.
29. Chu, T.M.G., Halloran, J.W., Hollister, S.J. and Feinberg, S.E. (2001). Hydroxyapatite implants with designed internal architecture. *J. Mater. Sci. Mater. Med.* 12, 471–478.
30. Chu, T.M.G., Orton, D.G., Hollister, S.J., Feinberg, S.E. and Halloran, J.W. (2002). Mechanical and *in vivo* performance of hydroxyapatite implants with controlled architectures. *Biomaterials* 23, 1283–1293.
31. Taboas, J.M., Maddox, R.D., Krebsbach, P.H. and Hollister, S.J. (2003). Indirect solid free form fabrication of local and global porous, biomimetic and composite 3D polymer-ceramic scaffolds. *Biomaterials* 24, 181–194.
32. Charriere, E., Terrazzoni, S., Pittet, C., Mordasini, Ph., Dutoit, M., Lemaitre, J. and Zysset, Ph. (2001). Mechanical characterization of brushite and hydroxyapatite cements. *Biomaterials* 22, 2937–2945.
33. Charriere, E., Lemaitre, J. and Zysset, Ph. (2003). Hydroxyapatite cement scaffolds with controlled macroporosity: fabrication protocol and mechanical properties. *Biomaterials* 24, 809–817.

SECTION III

Applied Principles of Bone Tissue Engineering

8 Study Design and Statistical Analysis

Robert T. Rubin and Jeffrey O. Hollinger

Contents

INTRODUCTION

Study design and statistical analysis are research components of immense practical consequence. Careful attention to study design, at the very beginning of one's thinking about a research project, will insure that questions and hypotheses are formulated that can be adequately addressed through data collection. It also insures that all potentially relevant measures are collected in such a way that their importance can be determined, and it insures that extraneous factors are excluded, so that they do not contaminate the data.

Statistical analysis is one element of study design and should be considered at the very beginning of project planning. Before a study is undertaken, the structure of the data to be collected and the statistical tests to be used should be clearly delineated. Statistical analysis addresses the relative importance of each of the measures (statistical significance), and it provides an indication of the overall definitiveness of the study (statistical power). Also, statistical techniques provide a way to combine separate studies that address the same research question in order to reach an overarching, more definitive conclusion than was provided by any of the studies separately (so-called meta-analysis).

The experienced investigator understands the practical necessity of specifying the design of a study, including the statistical tests to be used at the very beginning of research project planning. The history of scientific research has many examples

0-8493-1621-9/05/$0.00+$1.50
© 2005 by CRC Press LLC

of time-consuming and expensive research projects that were flawed, either because important measures were omitted, inappropriate comparison groups were used, or very few subjects were included. It cannot be stressed too strongly that "an ounce of prevention is worth a pound of cure." Unfortunately, once completed, there may be no cure at all for a badly designed study. This introduction will set forth only one rule: the statistician should be a part of the investigative team from the very beginning of study design through data analysis and report preparation, and his/her consultative services should be used early and often.

Herein, we highlight aspects of study design and data analysis that we consider to be fundamental considerations for any research project, including both basic and clinical bone tissue-engineering studies. There are many excellent reference texts targeted for specific kinds of research — clinical drug trials, cancer research, endocrinology, behavioral research, epidemiological studies, etc. There also are many excellent and detailed texts available on specific techniques of statistical analysis, for example, analysis of variance, factor analysis, and distribution-free ("nonparametric") statistics. A number of these texts will be referenced, so that the reader will have available the next level of discourse to aid in research planning and execution. Also, the Internet is an excellent source of statistical information. Using a search engine to address specific statistical issues can yield a wealth of useful information and often results in hits on even the most arcane search terms.

STUDY DESIGN

CATEGORIES OF STUDY DESIGN

A study can be *descriptive* or *quantitative*, *basic* or *clinical*, or *retrospective* or *prospective*. Both descriptive and quantitative studies may be valuable, depending on their purpose. For example, one might wish to characterize the types of fractures (femoral neck, vertebral compression, etc.) occurring in osteoporotic, postmenopausal women, the treatments (hip replacement, spine fixation, etc.) used 30 years ago vs. the present time, and the associated costs per case (hospitalization, physician fees, post-hospitalization rehabilitation, etc.). Such a descriptive study could be very useful for planning medical and social services, establishing in-home safety programs, and engineering tissues for improved medical treatments. Quantitative studies, in which variables are measured along some comparative scale, encompass the majority of research projects. Many descriptive studies have some quantitative data, so that the boundary between these two types of studies is often not clear.

The boundary between basic and clinical studies similarly can be vague, for example, is a study of a basic physiological process in healthy or sick human subjects basic or clinical research? Some investigators narrowly apply the term *clinical* to mean applied research, for example, drug and device trials in patients. The term *translational* ("from bench to bedside") is sometimes used to define clinical research. Most of the time, however, these descriptors are extraneous to the importance of a well-designed project.

The collection of past (historical) information represents a retrospective study design. A common example is the review of case reports or patient charts to gather data relevant to a particular problem. The future collection of data represents a prospective study design. A clinical example relevant to bone physiology would be a trial in which an osteoanabolic treatment is administered to one group of postmenopausal women who have osteoporosis, an inactive treatment (placebo) is administered to a comparable subject group, bone mineral content and bone density are measured pre- and post-therapy in both groups, and the data are compared statistically to determine the effectiveness of the treatment. A basic example relevant to bone physiology would be a study of the cellular effects of the nuclear transcription factors Runx2 and osterix, both of which are downstream osteoblast differentiation regulators.

Elements of Study Development and Design

Several key elements are necessary for every scientific study. These include setting the stage for the study (background for and significance of the proposed research) and the specifics of study design (defining hypotheses, variables, the study population, sampling methodology and sample sizes needed, anticipated structure of database, anticipated statistical tests needed).[1] While this phase may seem tedious, it sharpens the focus of one's project planning because gaps in the existing literature can be identified. Occasionally, one learns that the study being contemplated has already been done —an unpleasant but time-saving discovery. Concerning our example of osteoporosis therapy, the review should include a succinct but comprehensive discussion of osteoporosis, and the limitations and comparative advantages of currently available treatments. For our basic example, the review should cover essential aspects of osteoblast differentiation, signaling pathways, and key gene expression markers, and what is already known about Runx2 and osterix. This introductory information underpins the next element: the need for the study.

Need and Significance: One must clearly state the need for the study; otherwise, it may appear to be an intellectual exercise with little scientific importance and not worth the time and expense to do it. With regard to our example of osteoanabolic therapy for osteoporosis, the key question is: is there a need for the therapy? For our basic example, the key question might be: will overexpression of Runx2 and osterix downregulate type I and/or type II bone morphogenetic protein (BMP) receptors? The practical importance of the question should be indicated for every study being proposed, providing the investigator(s), readers of the study report(s), and grant review panels (if the project is being submitted for funding) a clear view of how the proposed study will develop knowledge critical to the advancement of the field.

Questions and Hypotheses: When thinking about a problem of interest, one usually first asks a question: Will this new osteoanabolic therapy really improve bone strength and fracture resistance in postmenopausal women? If mice are bred to overexpress Runx2 and/or osterix, will there be an effect on osteogenesis through downregulation of BMP receptors? In order to develop the appropriate study design to answer these questions, they are usually recast as specific hypotheses: This osteoanabolic therapy will significantly increase bone mineral content and bone density. Overexpression of Runx2 and osterix will significantly downregulate BMP receptors. Setting up the questions as hypotheses helps to focus the question and determine the types of measures needed for comparison.

As a final step, for statistical purposes hypotheses are often recast in the negative: This osteoanabolic therapy will have no significant effect on bone mineral content and bone density. Overexpression of Runx2 and osterix will not significantly downregulate BMP receptors. These are called null hypotheses[1,2] and while seemingly convoluted extrapolations from the original questions, they in fact have an important rationale: From a statistical standpoint, one does not accept a hypothesis as true, because it implies certainty. Rather, one rejects or fails to reject a null hypothesis with a specific degree of probability. Thus, if the osteoanabolic treatment turns out to be a wonder drug, one still cannot say that the clinical trial proved its efficacy, but one can state that the null hypothesis of no difference between inactive (placebo) and drug-treatment groups was rejected at a particular level of significance (to be discussed later). This is the same probability with which one can accept the alternative hypothesis of a treatment effect, thereby preventing one's conclusions from overreaching the data and forcing recognition that scientific "truths" are ultimately a matter of probabilities. As stated by one statistician, "The logic goes in a direction that seems intuitively backwards...calculations of [statistical significance] start with an assumption about the population (the null hypothesis) and determine the probability of randomly selecting samples with as large a difference as [that] observed."[3, p. 96]

This highlights a fundamental maxim of scientific inquiry: the enduring worth of any research finding is its replication by others, that is, its durability in the real world. The greater the statistical significance of rejection of the null hypothesis in a particular study, the greater the likelihood others will be able to replicate the study with samples drawn from the same population, so that the finding will hold. A corollary of this is "do not multiply hypotheses beyond necessity"[4, p.11]; that is, use the simplest model that adequately covers the study.

Accuracy and Precision of Measurements: Data constitute the "meat" of statistical analysis, and they should be collected as accurately and precisely as possible. *Accuracy* is the degree to which a measurement method collects the information it is supposed to collect. For example, does a test for bone density in a living subject really measure bone density only, or bone density plus some other factor that compromises the true reflection of bone density compared to, say, an *in vitro* technique? Does an assay for BMP quantitate all the BMP in a sample, yet only the BMP and not other, interfering proteins? A useful way to determine the accuracy of a test is to calibrate it with external standards, for example, a set of standards made with increasing concentrations of pure BMP in the same carrier solution as that used in the BMP assay.

Precision refers to the repeatability of each measurement. If the same BMP standard is assayed 100 times under the same assay conditions, how close are the values? These repeated measurements should be Gaussian-distributed; that is, forming a symmetrical, bell-shaped curve. Accuracy and precision are not necessarily related —a technique could be accurate but imprecise, yielding unacceptably differing values for repeated measurements of the same substance, or it could be precise but inaccurate, quantitating the wrong substance with exquisite replicability. Analytical tools used in research must be both accurate and precise.

Independent and Dependent Variables: Independent variables are measures of the specific conditions of the study as set up by the investigator, and dependent variables are the measures used to test the hypothesis. In our clinical example, different doses of an osteoanabolic agent and a placebo dose would be independent variables, and bone mineral content and bone density would be dependent variables. If groups of subjects in addition to postmenopausal women were included, for example, young and elderly men, then age and sex would be additional independent variables, and the dependent variables could be compared on the basis of age and sex of the subjects, as well as on drug dose vs. placebo. Independent and dependent variables are sometimes called predictor and outcome variables, respectively, for example, an increase in bone density (the dependent or outcome variable) is "predicted" by (dependent on) the amount of osteoanabolic agent administered (the independent or predictor variable).

There are three main types of variables to keep in mind. Categorical variables are not intrinsically ordered, for example, sex (male, female) and race (Caucasian, African-American, Asian). The order in which they are included in the data analysis makes no difference. Ordinal variables are categorical in name, but they are intrinsically ordered, for example, developmental stages (child, adolescent, young adult, elderly) or age deciles (0–10, 11–20, 21–30 years). Their order of inclusion does affect the study results: imagine a graph of height in relation to developmental stage or age decile in which the independent variable categories were out of order. Finally, there are continuous or interval variables, for example, bone mineral content, bone density, or receptor activity, which are measured on a continuous scale with a large number of possible values.

The types of independent and dependent variables dictate to a large extent the type of statistical analysis that should be applied to the data. In statistical testing, it is better to use continuous dependent variables to the fullest extent of their measurability, rather than collapsing them into ordinal scales. For example, let us assume that bone mineral content, as a ratio of total bone mass, can be confidently measured to the nearest percentage. It is better to analyze this as a continuous variable that can range between 0 and 100% than to, say, group the data into deciles (0–10%, 11–20%, 21–30%, etc.). Because the latter reduces the dependent variable to only 10 possible values instead of 100, it disregards the precision of the measurements, thereby sacrificing information content.

In addition to the independent and dependent variables necessary to structure a study, there may be confounding variables — the researcher's nemesis. Confounding variables have a potential influence on the results but frequently are not considered or controlled for in the initial study design; thus, their effects cannot be determined. A confounding variable can render a study worthless. For example, let us assume we have completed our study of the effect of osteoanabolic treatment on bone mineral content and bone density in two groups of postmenopausal women: one given the active treatment and the other given an inactive treatment (placebo). Later, we learn from other sources that estrogen treatment also has osteoanabolic effects. How much estrogen the postmenopausal women in our study might have been taking was either not determined during the study, or it was determined that significantly more of the women in the group receiving the active osteoanabolic treatment were taking estrogen than were the women in the group receiving placebo.

In this case, estrogen replacement therapy is a confounding variable whose influence cannot be readily teased apart from the ostensible positive effect of the new treatment, thus calling into question the validity of the results. It is extremely disconcerting to experience this in one's research career, especially when a manuscript reporting a study has been submitted for publication and a reviewer highlights an irrevocable confound as rendering the study fatally flawed. Such a potential disaster underscores the immense importance of adequately preparing for a study by learning everything available about the research question before beginning, and by considering all possible extraneous influences on study outcome and how they will be controlled for in the study design.

Constructing the Sample: Proper development of the sample upon which the experiment is to be conducted depends on understanding the underlying population of interest from which the sample is to be drawn. For statistical testing, some guidelines should be followed to insure that the sample is as representative of the population (the total entity under consideration) as possible. For example, our study of osteoanabolic treatment for osteoporosis in postmenopausal women should be done on a sample that includes women of a broad postmenopausal age range, in order for the results to be reasonably generalizable to all postmenopausal women. In addition to having the general characteristics of the population of interest, each person in the sample should be independent from the others and randomly drawn from the population, in order for the sample to be a representative subset. "By randomization, systematic effects are turned into error."[4, p. 130]

If, for our study, we want 100 women between the ages of 50 and 90 years in each group (active treatment vs. placebo), we need to select 200 unrelated women at random from this age range, so that each woman can be counted as an independent unit in the statistical analysis. If we were to select 100 sets of identical twins, there clearly would not be independence of responses to treatment within the twin pairs, and this redundancy reduces the number of independent units in the sample. (For those thinking ahead, one could specifically assign one member of each twin pair to active treatment and the other to placebo; this would be a paired, case–control study with only 100 independent "units," but it likely will be powerful, because extraneous between-group differences in physiological response are reduced by the pairing. (More on this later.)

Power: The power of a statistical test is its ability to reject the null hypothesis when it is indeed false, that is, the ability of the test to detect a real difference between groups. Power depends on four related elements. First is the size of the sample in terms of independent units, for example, the number of postmenopausal women in each group — the larger the sample, the greater the power of the test. Second is the magnitude of the difference between groups — finding large differences requires fewer subjects in each group than finding small differences. Third is the variability of the data in each group — if there is a large spread of individual values in each group, even though the group averages are not the same, the "smear" of the data may obscure the difference in averages. Fourth is the desired statistical significance of the results. A convention among statisticians is to set the significance level (so-called α) at 0.05 or less, that is, a probability of 5% or less that the null hypothesis is being rejected by chance (a significant difference indicated when in fact there is none: a Type I error). Thus, a statistical test will have the greatest power when the sample size of each group is large, the treatment effect (difference between groups) is large, the variability of the individual values within each group is small, and the α used is 5% rather than a smaller percentage.

Another convention among statisticians is to set the probability that the null hypothesis fails to be rejected when in fact it should be rejected (a Type II error) at 0.20 or less (so-called β). In other words, studies should be designed to achieve an analytic power ($1 - \beta$) of 0.80 or greater, that is, an 80% or greater probability of correctly identifying a significant difference. Determining the sample sizes needed to achieve a power of 80% or greater in statistical testing should be done as part of every study design. Unfortunately, underpowered studies, in which an apparently important real-world difference is reported as statistically nonsignificant, can be found in the scientific literature.

Tables for determining sample sizes for a specified significance level and power require the determination of the effect size (d), a dimensionless estimate of the magnitude of the difference.[5] As a simple example, the d for an independent-samples t-test (see Statistical Hypothesis Testing section for description of t-test) is determined as the absolute difference between the means of the two samples divided by the standard deviation of either sample (because the standard deviations are assumed to be equal).

One can determine d in two ways. Previous, similar studies can be used to calculate d from their data, and an average of the calculated d's can be used. Or, a small, medium, or large d can be assumed. (By convention, effect sizes can be considered small, medium, or large; e.g., for the t test, a small $d = 0.2$, a medium $d = 0.5$, and a large $d = 0.8$.) Once d is estimated, the needed sample sizes to achieve significance at a specified α level can be determined from published power tables.[5] For an independent-samples t-test, for example, given a large effect size ($d = 0.8$) and a desired power of 0.80, to achieve significance at the 0.05 level 26 subjects will be required in each group.[5, p. 36] Not only will power analysis help avoid the fecklessness of an underpowered study, funding agencies often require power analysis for each proposed experiment as part of submitted research grant applications.

As mentioned above, the enduring worth of any research is its replication by others. Special attention should be given to sample sizes in replication studies; these often need to be larger than the sample size of the original study, because the significance level (α) should be set lower. For example, one would intuitively conclude that, if the original study was significant at, say, the 0.05 level of significance with a particular sample size, there would be a 95% chance of a replication study achieving the same significance level with the same sample size. In fact, the odds of achieving the same significance level would be about 50%, that is, a 50/50 chance of α being 0.05 or lower vs. 0.06 or higher. In the latter case, the replication study would be considered nonsignificant and contrary to the findings of the original study, even though the effect sizes in the two studies might be very close. So, replication studies should be more stringent, that is, a significance level set at 0.01 and power analysis for sample sizes done accordingly, especially if the significance of the original study is just at the 0.05 level.

Control groups: Implicit in the foregoing discussion is that the need for a reference measurement against which the condition of interest, e.g., a new treatment, is compared. This may be a predetermined value, or it may be a non-treatment group; i.e., a control group, which is as close to the experimental group as possible in all respects, except that it does not receive the treatment under consideration. Often in clinical trials this is called a placebo control group,because subjects in this group undergo the same contacts with investigators and apparent experimental manipulations, only the manipulations are therapeutically inactive. A common placebo example is a pill identical in shape, color, markings, etc. to a new medication being tested, but containing an inert substance. Because the medication of interest may have side effects, "active placebos" are sometimes used, which contain compounds that mimic the major side effects of the new medication but have no therapeutic activity.

STATISTICAL HYPOTHESIS TESTING

This section will provide an introduction to this topic, with some comments on aspects of statistical testing that are fundamental to any experimental situation. Only a few specific tests will be presented, as examples of tests to use on particular types of data. Statistical theory will be avoided as much as possible, but should be recognized as the underlying rationale for all of the statistical testing used in the real world. Formulas for specific tests will be presented only in the detail necessary to convey the concept of the test. Full formulas can be found in many reference sources, both print and electronic (Internet), and there are many statistical software packages that incorporate these and many other tests.

Statistical inference involves using real-life *samples* that one collects to estimate characteristics of a *population* of interest from which the samples are drawn. Unless the sample equals the population in size, there will always be an inherent error in drawing conclusions about the population from the analysis of the sample. The more the samples that are randomly drawn from the population for analysis and the larger the size of each sample, the higher the probability that the findings are an accurate reflection of the characteristics of the population. This "great truth" underlies the rest of this section.

One frequently sees the terms *parametric* and *nonparametric* in reference to statistical analysis. A parameter is any measurable characteristic of the population, such as the mean and variance for a Gaussian distribution.[6, p. 39; 7, p. 84] For samples, the mean, median, variance, standard deviation, etc. are statistics and are used as estimators of population parameters. Parametric statistical tests, therefore, use particular statistics to estimate and compare these parameters across different groups of subjects. In nonparametric statistical analysis (sometimes called "distribution-free"), no parameters are estimated, and there are no assumptions made about the underlying distribution(s) of the data. It should be emphasized that both independent, random sampling and random assignment to groups are essential, no matter which type of statistical test is used.

Data Cleaning: Before any statistical analysis is undertaken, a mundane but extremely important task is to insure the accuracy of the data set being analyzed. Laboratory errors, transcription errors, coding errors, etc. can plague every data set. The raw data should be inspected to insure that all data entries are correct. Are there missing data, and, if so, how should this problem be handled? The inspection process includes spot-checking data entries, complete checking of all entries if the data set is not too large, plotting of data to ascertain possible outliers, checking of the source of the data to determine if the suspected outlier was correctly transcribed, and calculation of summary statistics to determine the characteristics of the data set. The importance of this step cannot be overemphasized; if the data set is inaccurate, its analysis is correspondingly compromised.

Data Distributions: As indicated in the section on measurement precision, replicate measurements on the same item will be distributed along a symmetrical, bell-shaped curve, the Gaussian distribution, also called the normal distribution. Also, many continuous-variable measures on independently drawn samples in nature are approximately Gaussian-distributed. There are several advantageous characteristics of Gaussian-distributed data: they are unimodal, the mode being the most probable value. The mode coincides with two other measures of central tendency: the mean (the average of all the data points) and the median (the value above and below which 50% of the data points lie). The variance and standard deviation (SD), other parameters to be discussed below, describe the spread of the data around the mean. So, for a Gaussian distribution, the mean, variance, and SD convey a great deal of information about the underlying data.

In addition to the need for independent, random sampling, as discussed above, the statistical tests commonly known as parametric tests require the assumptions of an underlying Gaussian data distribution and equality of variances, that is, spread of the data about the mean. Deviations from these assumptions can degrade the power of parametric tests, requiring either mathematical data transformation to approximate a Gaussian distribution or the use of nonparametric tests.

In biological samples, the data often are distorted versions of a Gaussian distribution, usually being "skewed to the right," in that some values are very high. This produces a long right tail to the distribution, makes the median higher than the mean, and spreads out the SD. The degree of skewness can be calculated mathematically and can usually be corrected to a large extent by data transformation. With data that show positive skewness, converting the values to logarithms often compresses the long right tail into an acceptable approximation of a Gaussian distribution (a "log-normal" distribution). The results of parametric tests conducted on transformed data are, strictly speaking, referable only to the transformed data, but they are generally used to draw inferences about the untransformed (raw) data.

Examples of basic parametric tests, to be presented below, are the Student's *t*-tests for independent and paired samples, the analysis of variance, the Pearson correlation coefficient, and linear regression. Examples of basic nonparametric tests will be the Mann–Whitney *U*-test, the Wilcoxon signed-rank test, the Kruskal–Wallis analysis of variance, the Spearman rank correlation coefficient, and the chi-squared test. Before highlighting these statistical tests, some further definition of terms is in order.

As mentioned above, for a data set the *mode* is the most frequently occurring value, the *median* is the value above and below which half the data points lie, and the *mean* is the average of all the values (sum, or Σ, of the individual values, $X_i \ldots X_N$, divided by the sample size, N):

$$\bar{X} = \sum_{i=1}^{N} (X_i \ldots X_N) \Big/ N.$$

In the following equations, we will dispense with the formal notation of $\Sigma_{i=1}^{N}$ and X_i and simply refer to Σ and X.

The *variance* is a basic measure of the spread of the data around the mean:

$$S^2 = \frac{\Sigma(X-\bar{X})^2}{N-1}.$$

There are some aspects of this equation that require comment. The numerator is the sum (Σ) of the squares of each of the individual data points minus their mean ($X-\bar{X}$). These values are squared so that each will be positive; otherwise, their sum would be 0. The denominator is the number of data points (sample size) minus the number of parameters estimated in the equation. In this case, there is one estimated parameter: the mean ($\bar{X}$). Subtracting the number of estimated parameters from the sample size yields the *degrees of freedom*, an important concept in statistical testing.[3, p. 27] For example, as will be detailed below, the degrees of freedom associated with a two-sample Student's *t*-test are N_1+N_2-2, the sum of the two sample sizes minus the number of parameters estimated (the means of the two samples).

The *standard deviation* (SD), which is the square root of the variance ($\sqrt{S^2}$), is a very useful parameter, because for a Gaussian distribution, 66.6% of the data points lie within the area of the mean $\pm$1SD, 95% lie within the mean $\pm$2SD, and 99% lie within the mean $\pm$3SD. Individual data points that lie beyond $\pm$3SD may be outliers, especially if they are far away from the rest of the data set. Outlying data points are at times discarded, especially if they can be traced to some kind of measurement error. An important use of the SD is to convert data to *z* scores by dividing each data point by its SD, thereby standardizing its distance from the group mean. In this way, variables having different measurement units can be compared, as in the correlation coefficient discussed below.

Whereas the SD describes the spread of individual data points in a sample distribution, the *standard error of the mean* (SEM) describes the spread of repeatedly determined means from randomly selected samples of size *N* from the underlying population. The shape of the population distribution is irrelevant. "If the distribution of the [population] has a finite variance, the sampling distribution of the means of random samples will be approximately [Gaussian] *if the sample size is sufficiently large* (emphasis added)."[6, p. 59] This is illustrated in Figure 8.1, in which means of repeated samples of the indicated sizes, randomly drawn from a rectangular population (shown at the top of the figure), are plotted. The larger the size of the samples, the more tightly clustered the distribution is around the true mean.

The SEM is calculated from the standard deviation:

$$\text{SEM} = \frac{\text{SD}}{\sqrt{N}}$$

The importance of sample size is again highlighted in this equation: not only does the SD decrease with increasing *N*, the SEM denominator increases with increasing *N*, the effect of the denominator being especially influential when sample sizes are relatively small. The SEM, therefore, provides a sense of confidence of the accuracy of the sample mean as it relates to the population mean. If, for example, the means and SEMs of two independently drawn samples with approximately equal variances are calculated, the approximate 95% confidence interval of each mean can be calculated as the mean $\pm$2 SEM. If each mean lies outside the 95% confidence interval of

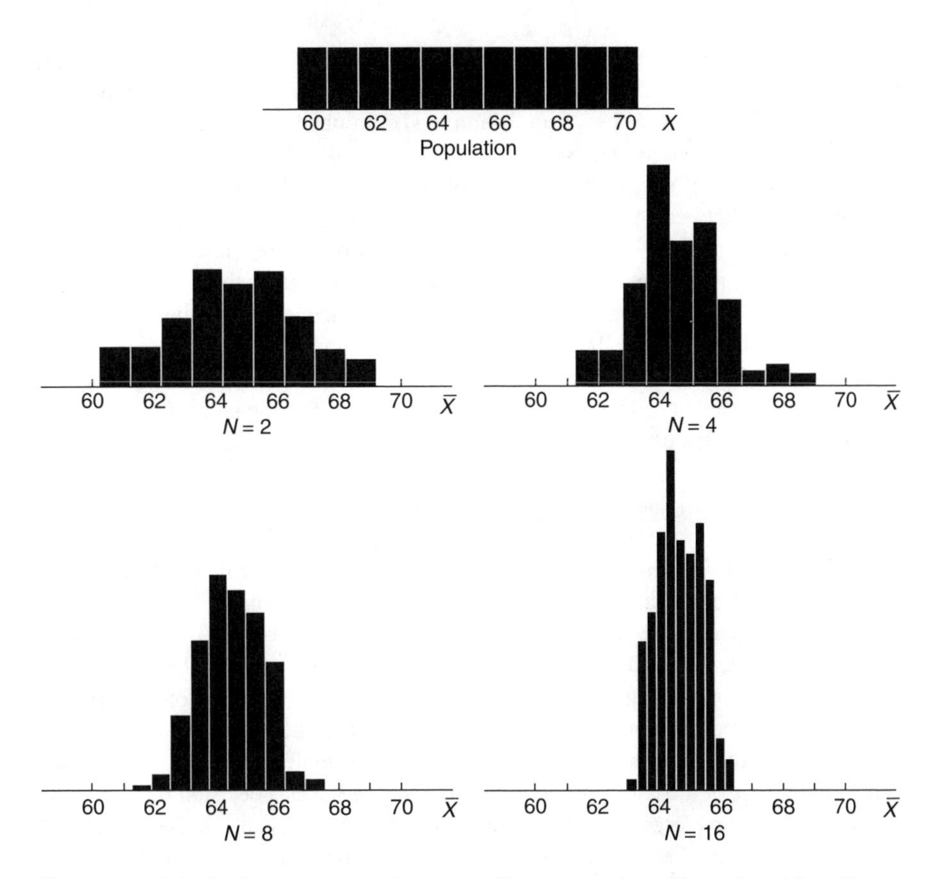

Figure 8.1 Distributions of means of samples of increasing sizes. (Reproduced from Dixon, W. J. and Massey, F. J. (1969). *Introduction to Statistical Analysis, 3rd ed.*, McGraw-Hill, New York. With permission.)

the other mean, the means can be considered different at the 0.05 level of significance. If each mean lies outside of the 99% confidence interval of the other mean (mean ±3 SEM), then the means are different at the 0.01 level of significance. This concept of confidence intervals forms the basis of most statistical testing. (It should be noted, however, that overlapping confidence intervals do not necessarily imply lack of a significant difference.[4, p.39])

Parametric Statistical Tests

Student's t-Test for Independent Samples: This tests the null hypothesis of no significant difference between the means of two independently drawn samples, for example, two groups of postmenopausal women, one given an osteoanabolic treatment and the other given a placebo. Let $\bar{X}_A$ = the mean of the first group and $\bar{X}_B$ = the mean of the second group:

$$t = \frac{\bar{X}_A - \bar{X}_B}{\text{SE of difference}}.$$

Calculation of the SE of the difference depends on whether the two sample sizes (N_A and N_B) are equal or unequal; the formulas can be found in many texts. The degrees of freedom (df) $= N_A + N_B - 2$ (total sample size minus the number of parameters estimated, that is, the two means).

Student's t-Test for Paired Samples: This tests the null hypothesis of no significant difference between the means of two conditions imposed on a single, independently drawn sample, for example, one group of postmenopausal women given an osteoanabolic treatment at one time and a placebo at another time. From a design standpoint, the order of treatments should be randomized across subjects, and the treatment lengths should be the same. Because each woman serves as her own control, the difference in outcome between active treatment and placebo yields a single score for each subject, and the null hypothesis is that the mean of the difference scores is not significantly different from 0:

$$t = \frac{\bar{X}}{\text{SEM}}.$$

Here, df $= N-1$, because only one group of women was studied, and only one parameter (mean of the individual difference scores) was estimated.

There are two considerations as to whether an independent-sample or paired-sample *t*-test should be used. The first is that a paired-sample study design, that is, two conditions imposed on the same subjects, should eliminate the inter-subject variance, giving a more precise estimate of the difference between the two conditions. On the other hand, the df is only half that of an independent-sample *t*-test. An empirical test of the relatedness of the two conditions has been suggested as a decision rule for using an independent-sample vs. a paired-sample *t*-test: If the across-subjects correlation of the outcome variable between the two conditions is >0.50, then the paired-sample *t*-test gives a more precise estimate of the significance of the between-condition difference.[4, p. 61] (See below for a discussion of the correlation coefficient.)

With these tests, we have two values: *t* and df. To determine the significance level of *t*, we use tables that list α's for *t* values at various df. One consideration is whether the α for a given *t* and df is one-tailed or two-tailed. For a one-tailed test, the direction of the alternative hypothesis must have been specified as part of the study design, for example, a null hypothesis of no significant difference between means vs. the alternative hypothesis of a specified mean being significantly greater than the other mean. If, in fact, it turns out that the specified mean is significantly smaller than the other, this cannot be considered an experimental outcome, because it was not proposed as an *a priori* alternative hypothesis. In this case, the null hypothesis simply fails to be rejected. Because a one-tailed α is easier to achieve than a two-tailed α, some investigators try to squeeze statistical significance out of a data set by rationalizing a one-sided alternative hypothesis after the study has been started. This is not permissible, because it biases the study toward significance. Some statisticians therefore advise researchers to always use two-sided alternative hypotheses, for which a statistically significant difference in means in either direction can be a legitimate outcome.

Analysis of Variance (ANOVA): The *t*-test is applicable to one or two sets of data, that is, paired or independent samples. For more than two data sets, the ANOVA is often used. Here, the null hypothesis is that the means of all the groups are not significantly different. ANOVAs can incorporate both independent and repeated measures, have complex designs such as nested and Latin squares, determine significance across more than one group of means, have more than one dependent variable (multivariate ANOVA), and control for the influence of a potentially confounding variable (analysis of covariance: ANCOVA). For illustrative purposes, we will discuss the simplest design: a one-way ANOVA on three independently drawn samples (no repeated measures on the same subjects).

The procedure is to partition the total variance of the data set into two components: that arising from differences among the group means and that arising within each group. Recall the numerator of the equation for variance above — the sum of squares: $\Sigma(X-\bar{X})^2$. For a one-way ANOVA, three sums of squares (SS) are calculated: The total SS is calculated on the differences of all individual values in the data set from the overall, grand mean. The between-group SS is calculated from the differences of the three group means from the grand mean. And, the within-group SS is calculated from the differences of the individual values within each group from that particular group's mean. The sum of the between-group and within-group SS = the total SS.

Next, the mean square (MS) for each is calculated by dividing the SS by its df. For the between-group SS, df $= k - 1$, where k = the number of groups (in this case three, so that df = 2). For the within-group SS, df $= N - k$, where N = the total sample size. Finally, the between-groups MS is divided by the within-groups MS to yield the F ratio. For clarity, ANOVA results are usually presented in tabular form:

	Sum of Squares (SS)	DF	Mean /square (MS)	F Ratio
Between groups	of group-mean differences from grand mean	k–1	SS_{BG}/df_{BG}	MS_{BG}/MS_{WG}
Within groups	of individual-value differences from group means	N–k	SS_{WG}/df_{WG}	
Total	of individual-value differences from grand mean	N–1		

BG=Between Groups; Wg=Within Groups

The premise is that, if the null hypothesis is indeed true, the variance due to between-group differences will be small in comparison to the variance within groups, and the latter will be close to the total variance. Therefore, to reject the null hypothesis, the F ratio must be larger than a critical value. To determine this, tables are used that set out the critical F ratios for different α's. The numerator df (that for the between-groups SS) and the denominator df (that for the within-groups SS) are both needed to find the significance level of a given F ratio.

It should be noted that F-ratio distributions are one-tailed. It is of interest that an F ratio calculated from just two groups of independent data will equal the square of t calculated on the same data. Also, the F ratio, with numerator and denominator df's of $k - 1 = 1$ and $N - k = N - 2$, respectively, will have the same level of significance as t with df $= N - 2$. Thus, the ANOVA is a mathematical extension of the t test to three or more groups.

The ANOVA is an omnibus test, that is, a significant F ratio only indicates that there is at least one significant difference between means somewhere in the data set.

To determine the location of the significance, multiple comparison tests designed for this purpose can be used. The more multiple comparisons done on a data set, the greater the chance of a Type I error, that is, the chance indication of a statistically significant comparison when in fact there is none. Correction of α levels is considered in many of the multiple comparison tests, such as the Scheffé, Tukey, Neuman–Keuls, and Duncan. Multiple t-tests also can be used, with correction of the α level needed to indicate a significant difference by the formula $1-(1-\alpha)^m$, where m represents the number of t-tests performed. For example, for three t tests used to perform pairwise comparisons of the three means in the ANOVA example, a nominal α of < 0.02 would be required to achieve a true α of 0.05. This, however, may be an overly conservative correction for α, and one of the aforementioned tests for *a posteriori* comparisons is usually preferable.

In a one-way ANOVA, there is only one main effect, that between groups. ANOVAs, however, can be multi-way, in which the data are represented along several dimensions. For example, consider two groups of randomly sampled postmenopausal women receiving osteoanabolic treatment, one on estrogen hormone replacement and one not, and two more groups of postmenopausal women receiving placebo, again one on estrogen hormone replacement and one not. Here, two main effects can be discerned: (1) treatment vs. placebo, including all the women in each group, both hormone-replaced and not replaced, and (2) hormone replacement vs. no hormone replacement, again including all the women in each group, those given both osteoanabolic treatment and placebo.

In a two-way ANOVA, in addition to F ratios for the two main effects, there is a third F ratio that can be calculated: the interaction of the main effects. For example, if the osteoanabolic treatment increases bone density to the same degree in both the estrogen-replaced and nonreplaced women, the effects of osteoanabolic treatment and estrogen are additive and there is no significant interaction. On the other hand, if osteoanabolic treatment increases bone density three times as much in the hormone-replaced women as it does in the nonreplaced women, there is an interaction between the osteoanabolic treatment and hormone replacement, the significance of which is determined from the interaction F ratio.

Pearson Product–Moment Correlation Coefficient: This parametric test is a measure of association. It indicates how well sets of paired data relate to each other, for example, measures of bone mineral content and bone density following osteoanabolic treatment, or measurement of one of these outcomes by two different techniques. For this parametric test, each set of data should be Gaussian distributed, the so-called bivariate normal distribution. If the values for each subject are plotted on graph paper, with, for example, bone mineral content on one axis and bone density on the other, the data points will form a "cloud," with the greatest density in the middle of the cloud corresponding to the values around the means, and decreasing density toward the edges of the cloud corresponding to the upper and lower tails of the Gaussian distributions. This is illustrated in Figure 8.2, in which the heights of 356 men are plotted against their weight. Collapsing all the heights down to the X-axis would reveal their Gaussian distribution, as would collapsing all the weights against the Y-axis.

The shape of the cloud provides a qualitative, visual indication of how closely one variable is related to the other: A circular cloud indicates little or no correlation;

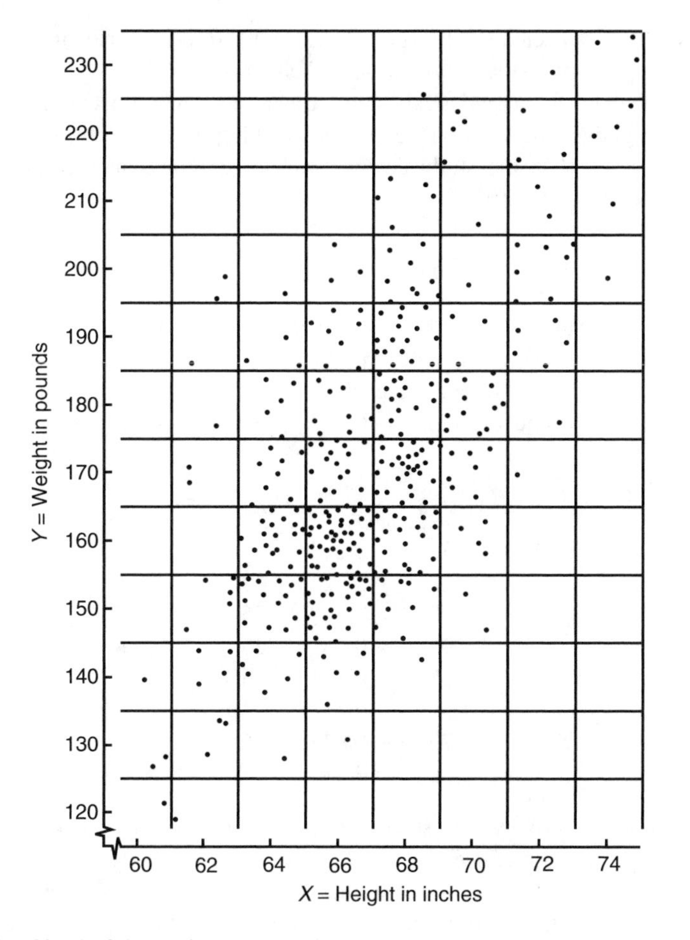

Figure 8.2 Cloud of data points representing a bivariate normal distribution. (Reproduced from Dixon, W. J. and Massey, F. J. (1969). *Introduction to Statistical Analysis, 3rd ed.*, McGraw-Hill, New York. With permission.)

a cigar-shaped cloud indicates some degree of association (the longer and thinner the shape, the greater the association); and the direction of the axis of the cloud indicates a positive or negative relationship. In Figure 8.2, the cigar-shaped cloud increasing from bottom left to top right indicates a positive correlation between the two measures. In contrast, one decreasing from top left to bottom right would indicate a negative correlation.

To calculate the correlation coefficient (r), for each data pair let X= the first variable, Y = the second variable, and N = the number of data pairs:

$$r = \frac{\Sigma\lfloor (X-\overline{X})/SD_X\,(Y-\overline{Y})/SD_Y \rfloor}{N-1}$$

Note that dividing each value-minus-its-group-mean $(X-\overline{X})$ by its group SD standardizes these values as z scores, thereby canceling out any differing measurement units between the variables.

Correlation coefficients can vary between −1.0 (a perfect negative correlation) and +1.0 (a perfect positive correlation). Their df = N − 2, and their significance can be determined from statistical tables. Because correlation coefficients can be positive or negative, unless a one-sided alternative hypothesis for the direction of the correlation is stated in the study design, two-tailed significance levels should be used. t can be calculated from r:[3, p. 163]

$$t_{(df = N - 2)} = r\sqrt{\frac{N - 2}{1 - r^2}}$$

so that the significance of r can be determined from t-test tables.

A useful concept is r^2, which describes the shared variance between the two sets of data. Consider the following table of r's needed to achieve a given amount of shared variance (r^2) and the sample size (N) needed to achieve a two-tailed α of < 0.05:[6, p. 569]

r^2(%)	$\pm r$	$N(\alpha = .05)$
5	0.22	80
10	0.32	40
15	0.39	26
20	0.45	20
25	0.50	16
50	0.71	8
70	0.84	6
90	0.95	<5

Note that there is relatively little shared variance until r exceeds ±0.4. Also note that the N required for significance of $r = \pm 0.22$ is 80, for $r = \pm 0.39$ is 26, and that the N decreases sharply for r's of increasing magnitude. Thus, studies often report low$r$'s that are statistically significant because of large sample sizes. These low r's are relatively meaningless, however, because the shared variance is negligible, indicating considerable independence of the variables that are reported to be significantly correlated.

Two final points about correlational analysis: First, it is symmetrical, Variables X and Y correlate with each other to the same degree, and one cannot infer causality, that is, which variable influences which. Second, differences in measurement techniques cannot be determined from correlational analysis. For example, two measures of bone density may correlate perfectly across a group of subjects, that is, subject 1 has the highest values for both measures, subject 2 the second highest values for both, etc. However, in bone-density units, the two measures may be significantly different, and one or both may be highly inaccurate compared to a known standard. Recall that in correlational analysis, the values for each measure are reduced to a common unit (z scores). This allows their relationship to be determined but obviates comparing them in their original units.

Linear Regression: This is also a measure of association; but unlike the correlation coefficient, it does not require a bivariate normal distribution. For example, if one were treating groups of postmenopausal women with increasing doses of an osteoanabolic drug and measuring bone density as the outcome variable, the independent variable, drug dose, would be assumed to be fixed, with no variance. That is, each woman receiving a given drug dose is assumed to be receiving exactly that dose. The outcome variable, on the other hand, is assumed to be Gaussian-distributed across the women in the group receiving each specific drug dose.

In linear regression analysis, the slope of a straight line that best fits the data and the intercept of that line (value on the Y, or dependent-variable axis when the X, or independent-variable axis $= 0$) are calculated. The significance of the slope can be determined by standard formulas. Some researchers confuse regression and correlation models and report a correlation coefficient from a regression model. A regression model is not symmetrical, however, because the independent variable is fixed, not random. An advantage of linear regression is that one can infer some causality or "prediction" of the dependent variable, based on the independent variable, as in the example given above, that is, increasing doses of osteoanabolic drug lead to increasing bone density.

Multivariate Analysis: So far, we have considered data sets in which there is one outcome variable and, in the case of regression, one independent variable. There are studies in which more than one variable of each type are measured, and there are multivariate statistical tests that can handle these situations.[7] For example, multivariate ANOVA could handle both bone mineral content and bone density as dependent variables, and multiple regression could handle osteoanabolic drug dose, whether or not concurrent estrogen replacement is being given age, etc, as independent variables. The computational algorithms for multivariate analysis include a covariance matrix that indicates the relationships of all the variables among each other. Those that are highly correlated convey redundant information. For example, if a stepwise multiple regression is performed, the most influential independent variable is included in the regression equation first, the covariance matrix is calculated, the variables highly correlated with the first are set aside, and the most influential variable given the presence of the first is included next. The process is iterated and stops when a specified significance level for inclusion of subsequent variables fails to be reached.

Even categorical data can be included as independent variables in multiple regression, for example, inclusion of postmenopausal women of different racial backgrounds in an osteoanabolic treatment trial. This can be accomplished by the use of "dummy" variables. This has led some statisticians to advocate multiple regression as a general data analytic procedure, having more flexibility than ANOVA.[8]

An important consideration in multivariate analysis is the ratio of independent variables to sample size. An N of 50–100 is often considered the minimum sample size, and 5–10 subjects per independent variable are considered the minimum ratio. The greater the total N and the fewer the independent variables, the more stable, that is, replicable with similar samples, the multivariate analysis.

Nonparametric Statistical Tests

Examples of basic nonparametric tests to be presented are the Mann–Whitney U-test, the Wilcoxon signed-rank test, the Kruskal–Wallis analysis of variance, the

Spearman rank correlation coefficient, and the chi-squared test. These are analogous to several of the parametric tests indicated above. They rely on rank-ordering of continuous variables that may be severely non-Gaussian distributed, or on counts of categorical variables.

Mann–Whitney U-Test: This test, also known as the Wilcoxon rank-sum test, is analogous to the *t* test for independent samples. For the U test, the data points in both samples are rank-ordered as if they were a single sample. Ties are ranked as the average of the two or more ranks for which they are tied, for example, the data series 10, 13, 14, 17, 17, 23, 30 would be ranked 1, 2, 3, 4.5, 4.5, 6, 7. The ranks in each separate sample are then summed, and tables are used to determine the significance of the difference in sums by entering the table with the sum of ranks of the smaller sample at the appropriate point for the two sample sizes.[6, p.545; 9, p.757] Under the null hypothesis, the difference between the sums of ranks for the two samples = 0.

Wilcoxon Signed-Rank Test: This is the nonparametric equivalent of the *t* test for paired samples. For the signed-rank test, the difference score for each data pair is recorded, and the absolute numbers are ranked. Ties are ranked as the average of the two or more ranks for which they are tied. The ranks of the positive difference scores are summed, as are the ranks of the negative difference scores. Tables are used to determine the significance of the difference in sums by entering the table with the smaller sum of ranks at the appropriate point for the sample size.[6, p.543; 9, p.756] Under the null hypothesis, the difference between the sums of ranks for the positive and negative difference scores = 0.

Kruskal–Wallis Analysis of Variance: This is analogous to the one-way parametric ANOVA. As in the *U*-test, the data points in all samples are rank-ordered as if they were a single sample. Ties are ranked as the average of the two or more ranks for which they are tied. The calculations are carried out on the ranks of each sample. Sums of squares, mean squares, and *F* ratios are generated as in the parametric ANOVA. The Friedman two-way ANOVA extends this nonparametric model to more than one main effect and can be used for repeated-measures data.

Spearman Rank Correlation Coefficient(r_S): This is the nonparametric equivalent of the Pearson product–moment correlation. The data in each group are ranked, and r_S is calculated in the same way as *r*, above.

Chi-Square (χ^2) Test: This test is performed on counts of categorical variables and compares two or more proportions. The 2 × 2 table is the simplest, but tables with more cells also can be analyzed by χ^2. For example, assume that 40 postmenopausal women were given osteoanabolic treatment and 40 were given placebo treatment. Twenty-four women given active drug developed a skin rash, whereas only eight given placebo developed a rash. Is this a significant difference? The χ^2 test compares the actual proportions against the expected proportions based on the sample sizes in each group. The 2 × 2 table is constructed as follows:

	Drug	*Placebo*	*Totals*
Rash+	24(16)	8(16)	32
Rash−	16(24)	32(24)	48
Totals	40	40	80

For each of the four cells, the expected value, shown in parentheses, is calculated as the row total for that cell × the column total for that cell divided by the grand total, for example, for the Drug/Rash+ and Placebo/Rash+ cells, the expected value is $32 \cdot 40/80 = 16$. Then, for each cell the expected value is subtracted from the observed value, and the difference is squared and divided by the expected value. For example, for the Drug/Rash+ cell, $8^2/16 = 4$. χ^2 = the sum of these across all the cells. Its degrees of freedom are $(N_{rows}-1)(N_{columns}-1)$, and its significance is determined from a χ^2 distribution table. For our example, $\chi^2 = 4 + 4 + 2.67 + 2.67 = 13.34$, with df = 1. This is significant at the 0.001 level.

For χ^2 to be accurate, the number of observations in each cell should be ≥ 5, and the total number of observations should be ≥ 20. For smaller sample sizes, the Fisher exact test can be used. For 2 × 2 χ^2 tables, the Yates correction is generally applied, whereby the absolute observed-minus-expected value is reduced by 0.5 before it is squared. Some statisticians, however, consider this correction to be too conservative.

The following table summarizes the tests reviewed above:

Independent samples	Correlated samples
Parametric	
Two groups	Two groups
Student's *t*-test — independent	Student's *t*-test — paired
	Pearson's correlation coefficient
	Linear regression
Three or more groups	Three or more groups
Analysis of variance (ANOVA); one-way, multi-way	Repeated-measures analysis of variance (ANOVA); one-way, multi-way
Analysis of covariance (ANCOVA)	Repeated-measures analysis of covariance
ANOVA and ANCOVA followed by multiple comparison tests	ANOVA and ANCOVA followed by multiple comparison tests
	Multivariate analyses
Nonparametric	
Two groups	Two groups
Mann–Whitney U test	Wilcoxon signed-rank test
	Spearman correlation coefficient
Chi-square contingency table	
Three or More Groups	Three or More Groups
Kruskal–Wallis ANOVA	
Friedman two-way ANOVA	Repeated-measures Friedman ANOVA
Chi-square contingency table	

CONCLUDING REMARKS

Herein, we have set forth some principles of study design and examples of basic statistical techniques that are relevant to bone tissue engineering. Of course, they are much more widely applicable, being fundamental scientific concepts and techniques. The references listed are primarily general statistics textbooks that provide, considerably more rigorously than we have done, the theoretical framework behind the statistical tests. This theory is not esoteric; it delimits the applicability of statistical testing and guards against erroneous usage of tests in real-life circumstances. Statisticians are more conversant with statistical theory than are most scientists, and their expertise is invaluable to the practicing researcher who has opportunities to go astray, from the beginnings of the study design to the final data analytic steps.

Even when a research study is carefully designed, with meticulous methodology and sophisticated statistical analysis, the researcher is always the last word on the importance of the findings. He or she is the most knowledgeable about the work and the best person to both report it and put it into perspective for the scientific community. The study should be presented in a sufficiently complete manner such that another investigator will have the information necessary to perform a replication study.

One final comment: some years ago, the first author of this chapter was traveling to a professional meeting to present some new research findings. Near the airport, there was a billboard over a parking lot that espoused a truth we all should remember (see Figure 8.3).

Figure 8.3 © 2003, Robert T. Rubin.

Acknowledgment

Joseph F. Lucke, Ph.D. provided valuable statistical consultation and advice on the preparation of this chapter.

References

1. Haley, R. W. (1994). Designing clinical research. In *Techniques of Patient-Oriented Research*, Pak, C. Y. C. and Adams, P. M., Eds., Raven Press, New York, pp. 47–80.
2. Swinscow, T. D. V. and Campbell, M. J. (1997). *Statistics at Square One, 9th ed.*, BMJ Publishing Group, London. http://bmj.com/collections/statsbk/index.shtml
3. Motulsky, H. (1995). *Intuitive Biostatistics.* Oxford University Press, New York.
4. van Belle, G. (2002). *Statistical Rules of Thumb.* Wiley-Interscience, New York.
5. Cohen, J. (1988). *Statistical Power Analysis for the Behavioral Sciences, Second Edition.* Lawrence Erlbaum Associates, Mahwah, NJ.
6. Dixon, W. J. and Massey, F. J. (1969). *Introduction to Statistical Analysis, 3rd ed.*, McGraw-Hill, New York.
7. Armitage, P., Berry, G. and Matthews, J. N. S. (2002). *Statistical Methods in Medical Research, 4th ed.*, Blackwell Science, Oxford.
8. Afifi, A. A. and Clark, V. (1996). *Computer-Aided Multivariate Analysis, 3rd ed.*, CRC Press, Boca Raton, FL.
9 Cohen, J. and Cohen, P. (1984). *Applied Multiple Regression/Correlation Analysis for the Behavioral Sciences.* Lawrence Erlbaum Associates, Mahwah, NJ.

9 Animal Models for Bone Tissue Engineering of Critical-sized Defects (CSDs), Bone Pathologies, and Orthopedic Disease States

Mark P. Mooney and Michael I. Siegel

Contents

INTRODUCTION: THE GENERAL UTILITY OF ANIMAL MODELS FOR BONE TISSUE ENGINEERING

Experimental animal models have proved instrumental in the preclinical evaluation of the functional efficacy and safety of new bone tissue engineering methodologies.[1, 2]

0-8493-1621-9/05/$0.00+$1.50
© 2005 by CRC Press LLC

However, determining an appropriate animal model is critical to successful experimental design and extrapolation to the clinical setting.[1,3–13] It is also a complicated procedure and often a heated topic of debate.[6,13–22]

In this chapter we will: (1) discuss a number of practical criteria to facilitate animal model choice for bone tissue engineering studies;[1,7,8,12] (2) review the advantages and disadvantages of various animal models for basic bone biology and the testing of bone replacement materials in critical-sized defects; and (3) identify abnormal animal models with clinically relevant pathologies that may be used to develop specific bone tissue engineering paradigms and strategies. This discussion is intended as an aid to less experienced researchers and students designing experimental studies, and to clinicians who must critically evaluate the results obtained from published animal studies for validity and potential extrapolation to the human clinical condition.

LEVELS OF HYPOTHESIS TESTING AND APPROPRIATE ANIMAL MODEL CHOICE

A number of general factors should be considered when choosing an animal model for bone tissue engineering studies.[1, 7–9, 23, 24] These include: (1) animal model appropriateness; (2) potential and expected extrapolation to the clinical setting; (3) genetic homogeneity of the specific animal model; (4) available data concerning skeletal anatomy, bone physiology and biomechanical properties, and osseous wound healing; (5) cost and availability of the model; (6) generalizability of the results across species; (7) ease and adaptability of the model to experimental and laboratory manipulations; (8) ecological considerations; and (9) ethical and societal implications.

Although nonhuman primates are phylogenetically closer to humans than other mammalian groups, which may fulfill many of the considerations listed above, and make the extrapolation of findings to humans theoretically "better," not all bone tissue engineering experimental manipulations require nonhuman primate models. As we have suggested previously for craniofacial biology studies,[7, 12, 25–27] the choice of an appropriate animal model for bone tissue engineering should be based, in part, on criteria initially suggested by Smith,[28] Reynolds,[29] and Goldsmith and Moor-Jankowski[30] for determining appropriate animal models in toxicologic studies. Smith,[28] based on comparative studies of drug disposition across a large number of mammalian taxa, reported great variation in drug metabolism within the various primate groups. He suggested that the choice of the best animal model should be based on similar physiologic pathways between the model and the human condition, and not necessarily phyletic affinity. Smith[28] also suggested that the level of hypothesis testing and the expected extrapolation of the results to humans should be of major concern in choosing the best animal model. Based on these practical criteria, we propose that the choice of an appropriate model for bone tissue engineering studies should also be linked to the level of hypothesis testing, the expected extrapolation to specific human clinical condition under study, and the expected commercial market of the tissue-engineered construct.[1]

The three main strategies taken to engineer bony tissue involve the following: (1) osteoconduction, which utilizes a scaffold as a mechanical barrier to exclude

Table 9.1
Levels of Hypothesis Testing in Bone Tissue Engineering Paradigms

	Bone Tissue Engineering Paradigms		
Level of Hypothesis Testing	**Osteoconduction**	**Osteoinduction**	**Osteogenic Cell Transplantation**
Basic bone cell biology models	*In vitro* or *in vivo* manipulation of cell or tissue response to trauma using mechanical or environmental factors.	*In vitro* or *in vivo* manipulation of cell or tissue response to trauma using growth factors, cytokines, or genes.	*In vitro* or *in vivo* manipulation of autologous or isohistogenic donor osteogenic cell response to host bone trauma.
General clinical models	Repair of critical-sized defects using a passive mechanical scaffold.	Repair of critical-sized defects using resorbable scaffolds seeded with growth factors, cytokines, or genes.	Repair of critical-sized defects using resorbable scaffolds seeded with osteoblasts or stem cells.
Specific clinical models	Repair of bony defects in specific clinical conditions (e.g., periodontitis, diabetes, cleft palate, osteoporosis, etc.) using a passive mechanical scaffold.	Repair of bony defects in specific clinical conditions (e.g., periodontitis, diabetes, cleft palate, osteoporosis, etc.) using resorbable scaffolds seeded with growth factors, cytokines, or genes.	Repair of bony defects in specific clinical conditions (e.g., periodontitis, diabetes, cleft palate, osteoporosis, etc.) using resorbable scaffolds seeded with osteoblasts or stem cells.

osteoinhibitory tissues and protect osteogenic tissues in a bony defect; (2) osteoinduction, which utilizes the release of bioactive molecules that bind only to specific host cells with receptors for the molecules and stimulate cell migration and osteogenesis to repair a bony defect; and (3) osteogenic cell transplantation, which involves the seeding of osteogenic cells from a donor source on a synthetic bone construct that is subsequently implanted into the defect (Table 9.1).[31–33] Animal model "goodness-of-fit" will vary across these paradigms as the need for extrapolation to the human condition increases and can be divided into three levels of hypothesis testing: (1) "generic" animal models of basic bone-cell biology, (2) phylogenetically "closer" models with comparable anatomy and bony wound healing sequelae, and (3) "fitting" the appropriate animal model to various clinical conditions.

"Generic" Animal Models and Basic Bone-cell Biology

Studies involving *in vitro* manipulations of genes, molecules, cells, and tissue are viewed as the most basic. Humans share primitive developmental and regulatory genetic mechanisms that produce homologous cortical and trabecular bony structures, similar bone physiology and wound healing sequelae, and even comparable bone pathophysiology and disease pathogenesis, and progression with many other taxonomic orders.[2, 10,32, 34–36, 38–41] In general, osteogenic cells from normal and genetically engineered rodents and lagomorphs have been used extensively to elucidate a basic understanding of cell–gene, cell–factor, and cell–cell interactions (see other chapters in this volume) as an aid to designing bone tissue engineering strategies (Table 9.1)[1, 2, 33, 41–44] The data derived from studies using these "generic" animal model are as valid as those derived from carnivore or primate models, given that no attempt is made to grossly extrapolate beyond this level of hypothesis testing.

PHYLOGENETICALLY "CLOSER" ANIMAL MODELS AND GENERAL CLINICAL ISSUES

In vivo bone tissue engineering studies at this level of hypothesis testing typically attempt to model clinical problems and complications associated with general skeletal trauma and postoperative bony wound healing. These studies typically involve manipulations of general clinical skeletal models (e.g., long bone, vertebral body, and craniofacial fractures, oncologic resections, elective craniofacial and maxillofacial procedures, etc.) and present more difficult problems for choosing an appropriate animal model. Depending on the level of extrapolation to the human condition, not all animal models may be appropriate at this level of hypothesis testing. The commonly used "generic" animal models (such as rodents and rabbits) differ somewhat in their basic bone microarchitecture and biomechanics, functional gross skeletal anatomy, and healing of critical-sized defects compared to phylogentically closer models (such as some carnivores and nonhuman primates).[7–10, 12, 24, 34, 35, 37, 45–48] Such interspecific differences should be taken into account before choosing an appropriate model at this level of hypothesis testing.

Basic Bone Microarchitecture and Biomechanics

Appropriate animal models at this level of hypothesis testing should mimic normal human bone microanatomy and systemic physiological processes. The human skeleton shows significant osteoporotic changes during immobilization, normal aging, and/or menopause as a consequence of systemic and environmental factors.[10, 27, 35] In young individuals, the central cancellous bone network is a highly connected, very strong, plate–strut network. In older individuals, osteoclastic activity changes this network to a more fragile, disconnected, strut–strut network that is relatively weak.[35] Due to these endocrine-mediated, age-related changes, 40–45% of the cancellous bone is lost in the vertebral bodies, long bones, and femoral neck with a concomitant loss of approximately 80–90% strength and a reduced vertebral cortical shell thickness of approximately 200–400 μm. Peak cortical bone mass, connectivity, and thickness are also reduced due to both cortico-endosteal and intracortical (Haversian) bone remodeling.[10, 35] These changes are accelerated and more pronounced during the development of osteoporosis (age-, menopausal-, or immobilization-related) and result in strength changes of the human femur during compression and tension which range from 90 to 143 and 90 to 167 (MPa), respectively. Age-related changes in Young's modulus are also observed, which range from approximately 5 to 15.5 during compression and 4 to 17 (GPa) during tension.[37] Such age- or disease-related changes in cancellous bone loss, cortical bone thinning, and decreased bone strength and Young's modulus are responsible, in part, for the clinical problems associated with long bone, femoral neck, and vertebral body fractures. Such clinical complications have stimulated an impressive bone tissue engineering initiative to address these issues.[1, 2]

Rodents and rabbits are probably the most commonly used animal models at this level of hypothesis testing.[24, 35, 9, 37, 10] The advantages of these models include the following: (1) standardization of experimental conditions; (2) genetically specific or mutant strains can be acquired; (3) relatively inexpensive to house and maintain; (4) relatively shorter life spans and fast bone turnover rates; (5) extensive documentation of bone metabolism and skeletal effect of diet; (6) similar lamellar bone architecture; (7) similar cancellous bone thinning and fragility, as well as remodeling rates and sites; and (8) ovarectomy mimics human skeletal ageing and menopause.[24, 35, 9, 32, 10] However, there are also a number of disadvantages of the rodent and rabbit models, which should be taken into account before designing bone tissue engineering studies utilizing these models. These include: (1) different skeletal loading patterns; (2) open epiphyses at various growth plates upto the age of 12–24 months; (3) minimal intra-cortical remodeling; (4) rodents lack Haversian canal systems; (5) rodents also have hemopoietic bone marrow at most skeletal sites, which increases bone turnover rates compared to primates; (6) they have a smaller proportion of cancellous bone to total bone mass; (7) they do not experience a natural menopause and do not show impaired osteoblast function during late stages of estrogen deficiency; and (8) their relatively small size for the testing of prosthetic devices, repeated bone biopsies, or blood testing.[24, 35, 9, 10] The rodent model also shows age-related changes in femoral head strength during compression, which ranges from approximately 1 to 5 (MPa), and in Young's modulus during compression, which ranges from approximately 12 to 405 (GPa).[37] If one can keep the disadvantages of the small animal models in mind during data interpretation, the rodent and rabbit models are excellent for the initial testing of novel bone tissue engineering paradigms (Table 9.1).

In contrast, larger animal models (nonhuman primates, dogs, cats, sheep, goats, and swine) show microarchitecture, bone physiology, and biomechanical properties more similar to humans than do the rodent and rabbit models, especially with regard to intracortical bone remodeling.[24, 35, 9, 10, 49] While each large animal model has its own advantages and disadvantages,[24, 35, 9, 10] nonhuman primates are thought to be the model of choice at this level of hypothesis testing, based on very similar anatomical, gastrointestinal/dietary, endocrine, bone metabolism, and circadian rhythm factors.[9,10,49–51] However, cost, availability, and genetic heterogeneity of nonhuman primates can be problematic, which necessitates the utilization of other large animal models. In general, the advantages of large animal models over small animal models include: (1) well-developed Haversian and trabecular bone remodeling; (2) greater skeletal surface areas and volumes; (3) similar skeletal disuse atrophy results; (4) highly localized bone fragility associated with stress shielding by orthopedic implants; and (5) similar postmenopausal related osteopenia (with the exception of the dog model). Depending on the large animal model, there are also a number of disadvantages, which include: (1) relatively high cost and maintenance expense; (2) USDA housing and space requirements; (3) relatively long life spans; (4) varying gastrointestinal and dietary (trace mineral) requirements, which can affect skeletal calcium:phosphorus ratios; (5) problems with generating large, homogenous samples for statistical testing; (6) potential reservoirs for a host of zoonotic diseases; (7) emotional attachment; and (8) varying ethical concerns.[24, 35, 9, 10] It is also difficult to create the disproportionate loss of bone strength (characteristic for humans) in many of the large animal models, with some large animal models having a vertebral cortical shell 1500–3000 μm thick.[35] The large animal models also show age-related changes in femur strength during compression and tension, which range from approximately 136 to 195 and 93 to 172 (MPa), respectively. They also showed age-related changes in Young's modulus during compression and tension, which range from approximately 19 to 27 and 5 to 25 (GPa), respectively.[37]

Gross Anatomical, Functional, and Growth Considerations

There are also a number of gross anatomical differences in the long bones, vertebrae, and craniofacial complex of different animal models (Figures 9.1–9.5) that need to be considered before determining animal model appropriateness. Such gross skeletal differences are due, in part, to differing endocrine, biomechanical and functional (locomotor, postural, and masticatory), and dietary influences.[49–51,24, 9,35,37,10] These factors can govern the regulation of bone mass and three-dimensional structure, as well as modulate the "set point" for bone adaptation and bony wound healing.[24]

While only humans and some birds (Emus[40]) are habitually bipedal, most nonhuman primates and rodents maintain upright postures during feeding and grooming.[45, 52] Such comparable functional loading produces similarly elongated femoral necks and heads (compare Figures 9.1a and d), similarly shaped vertebral bodies, vertebral laminar thicknesses, and oriented spinous and mammillary processes of the lumbar vertebrae (compare Figures 9.2a and b to Figures 9.2g and h). In contrast, more habitually quadrupedal species (rabbits and dogs in this case) show more robust femoral shafts, thicker and shorter femoral necks (Figures 9.1b and c), vertically compressed vertebral bodies, relatively thinner vertebral laminae, and differently oriented bony vertebral processes (Figures 9.2c–f) compared to monkeys and rats. Thus, rats may be an appropriate, less expensive model to nonprimates for modeling human femoral or vertebral body morphology if postural loading on tissue-engineered constructs is of primary concern.

Craniofacial skeletal tissue loss due to congenital defects, disease, and injury is a major clinical problem.[33] The human craniofacial skeleton is a complex region to model and postural, masticatory, dental, visceral, and neural factors all contribute to its unique adult shape.[9, 53, 12] The more common "generic" animal models (mice and rats) are monophyodonts, exhibiting only a single set of the molar and incisor classes,[45, 46, 52, 54] while other generic models (rabbits and guinea pigs, for example) have succedaneous dentition but maintain continually erupting incisors.[48, 55, 5] These "generic" animal models also have large maxillary and mandibular diastema between incisors and premolars/molars with limited alveolar bone in this area (Figure 9.3). They are also dolichocephalic (i.e., long, narrow headed) from limited brain cerebralization[45–47] with

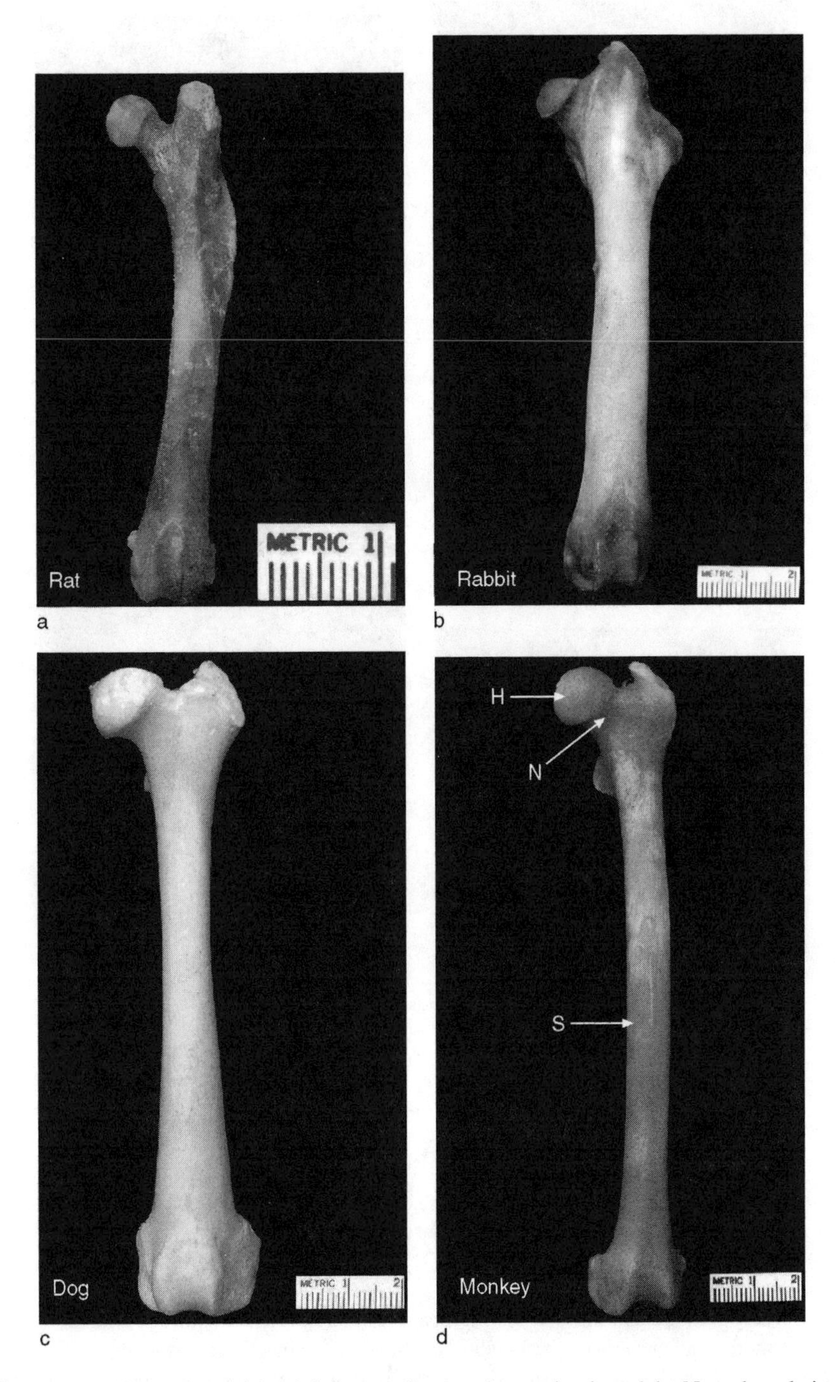

Figure 9.1 Cleaned and dried left femora from various animal models. Note the relatively longer femoral necks in the rat and monkey compared to the Rabbit and Dog, which reflect a more habitual upright posture. (H = femoral head; N = femoral neck; S = femoral shaft)

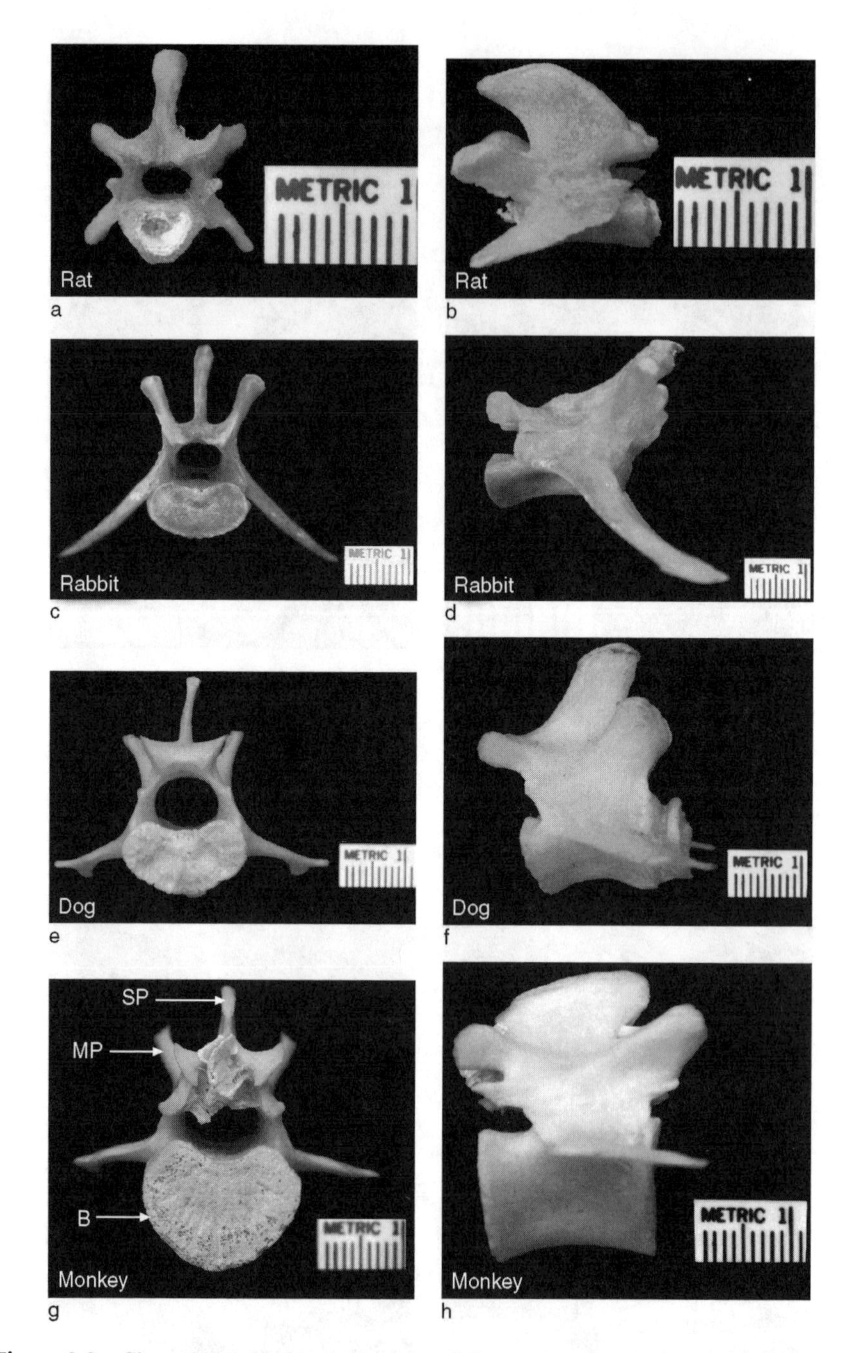

Figure 9.2 Cleaned and dried second lumbar (L2) vertebrae from various animal models. Note the more oval-shaped vertebral bodies and broader spinous processes in the rat and monkey compared to the rabbit and dog, which reflect a more habitual upright posture and vertebral loading. (B = vertebral body; MP = mammillary process; SP = spinous process)

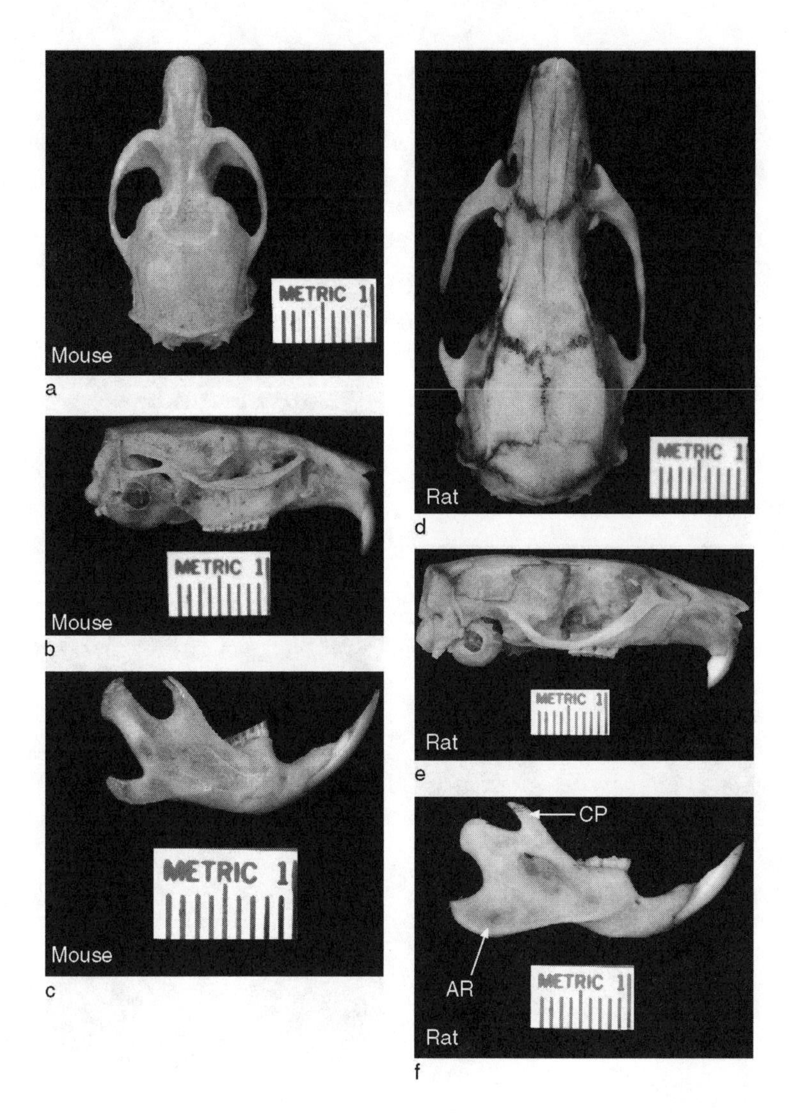

Figure 9.3 Cleaned and dried skulls and mandibles from the small "generic" animal models. Note the reduced coronoid processes in the guinea pig and rabbit compared to the mouse and rat, which reflect differing masticatory functions. (AR = ascending ramus; CP = coronoid process).

relatively long, narrow, and flat cranial vault bones (Figure 9.3) compared to the human condition. Rats and mice are omnivorous while rabbits and guinea pigs are herbivorous. These functional differences are reflected in the enlarged attachments sites of the masseter m. on the ascending ramus and the reduced coronoid processes seen in the mandibles of rabbits and guinea pigs (Figures 9.3i and l). These models are also relatively small, which may affect the design and feasibility of testing synthetic constructs and orthopedic devices for implantation and osteointegration.

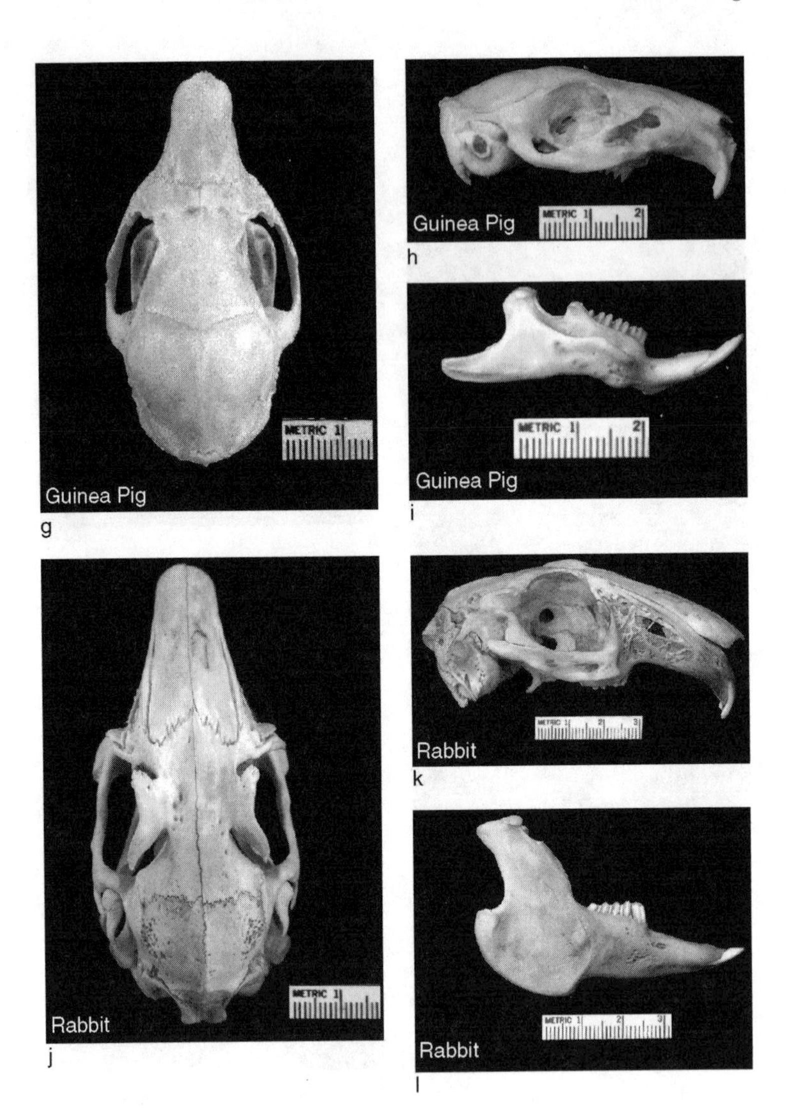

Figure 9.3 (Continued)

Larger, generic animal models have been utilized to overcome some of the disadvantages of the smaller models discussed above. These models include sheep, goats, and swine (both large and miniature versions) and have been well described as models for osteoporosis.[9–11, 24, 34, 35, 37] All three models have varying dental formulas, and goat and sheep have maxillary and mandibular diastemas (Figure 9.4).[57] They are also dolichocephalic (i.e., long, narrow headed) from anteroposterior brain growth and limited superior cerebralization.[45–47] Their cranial bones are relatively long, narrow, and flat, but they are still much larger than those of rodents and lagomorphs (Figure 9.4). The goat also has horns, which limits the accessibility and surface area of the cranial vault bones for surgical manipulations (Figures 9.4a and b).[58, 59] Sheep and goats are herbivorous and ruminants, while pigs are omnivorous. These functional differences are reflected in the enlarged attachment sites of the masseter m. on the ascending mandibular ramus and zygomatic arch, the reduced temporal fossae for the attachment site of the temporalis m. seen in the skulls and jaws of sheep and goats compared to swine (Figure 9.4). Although omnivorous, the pig also has a very strong masticatory apparatus with an enlarged ascending mandibular ramus for masseter m. attachment. These larger, "generic" animal models are more suitable to studies that need a greater bony surface area, more alveolar bone, or increased jaw basal bone for testing compared to the smaller, "generic" animal models. However, these models are much more expensive to purchase and house, bone turnover rate and osseous healing are relatively slow, and they are less genetically homogenous compared to the smaller, "generic" animal models, although the minipig model overcomes some of these disadvantages.[111]

In contrast, the phylogentically "closer" models (carnivores and nonhuman primates) exhibit craniofacial structures more similar to humans than do the "generic" models (Figure 9.5). They possess succedaneous dentition with varying eruption times of both the deciduous and permanent teeth,[45, 46, 57, 7, 60–63] some with very large canine teeth. They are relatively more brachycephalic (i.e., short, wide headed) from increased brain corticalization and gyrification.[45, 56, 47, 53, 64–66, 12] Their cranial bones are relatively shorter, wider, and more curved (Figure 9.5) compared to the "generic" animal models, which makes them more similar to the human condition. While cats and dogs are carnivores, laboratory-reared species usually have a much softer, standardized diet than their wild-reared counterparts. Cats and dogs have enlarged coronoid processes for temporalis m. attachment and more inferiorly located mandibular condyles compared to primates, which affords them strong lever arms, reduced horizontal excursion, and increased vertical power. In contrast, some nonhuman primates are omnivorous, while others are herbivorous, although laboratory-reared nonhuman primates also have a standardized, relatively soft diet. Their dental formulae and masticatory apparatus are also closest to humans (Figure 9.5). Based on these striking similarities in systemic and oral skeletal physiology and anatomy, nonhuman primates have often become the model of choice of many investigators.[49–51, 24, 9, 35,] However, the decreasing availability of wild-caught and laboratory-reared primates, their increased cost, and the social and ethical concerns of using nonhuman primates necessitates the development and utilization of alternative models.[6–8, 13–21, 23, 67]

Early surgical intervention of craniofacial and dentofacial deformities also recently entailed the development of strategies to engineer bone and soft-tissue constructs that grow along with the individual.[33, 44] The testing of these growing constructs requires animal models that show similar human craniofacial growth patterns.[7, 25, 26, 12] Appropriate animal models at this level of hypothesis testing would be those that exhibit similar regional growth patterns to humans. In general, human regional craniofacial growth patterns are characterized by a high degree of cranial base flexion (kyphosis), nasomaxillary reduction, anteroinferior displacement of the midface, and significant vertical and transverse increases in both the neurocranium and upper face.[45, 68, 48, 53] The human craniofacial growth pattern differs from the primitive adult mammalian condition of a long, slender dolichocephalic craniofacial skeleton to a broad, vertically flattened brachycephalic cranium without a prominent snout (Figures 9.3–9.5) (for an excellent review, see Enlow[53]). Thus, experimental studies designed to manipulate regional craniofacial

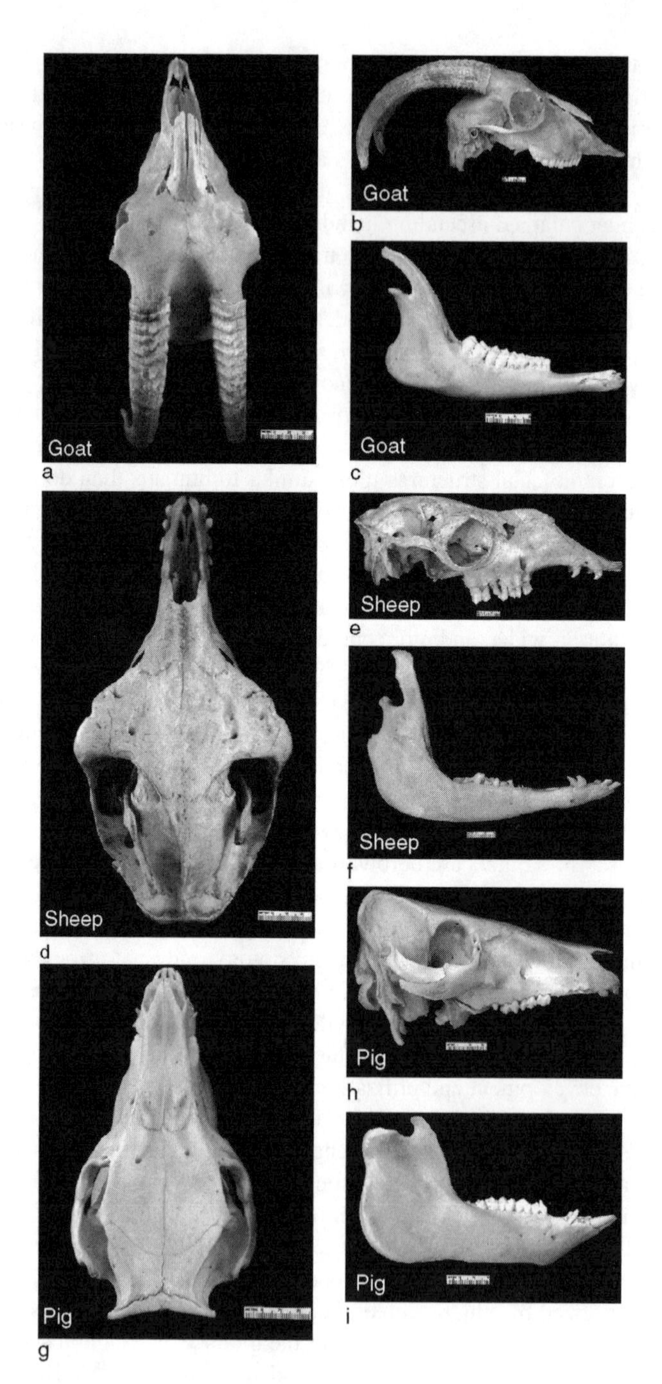

Figure 9.4 Cleaned and dried skulls and mandibles from the large "generic" animal models. Note the different cranial vaults, dentition, and the enlarged coronoid processes in the goat and sheep compared to the pig, which reflect differing brain morphology and masticatory functions.

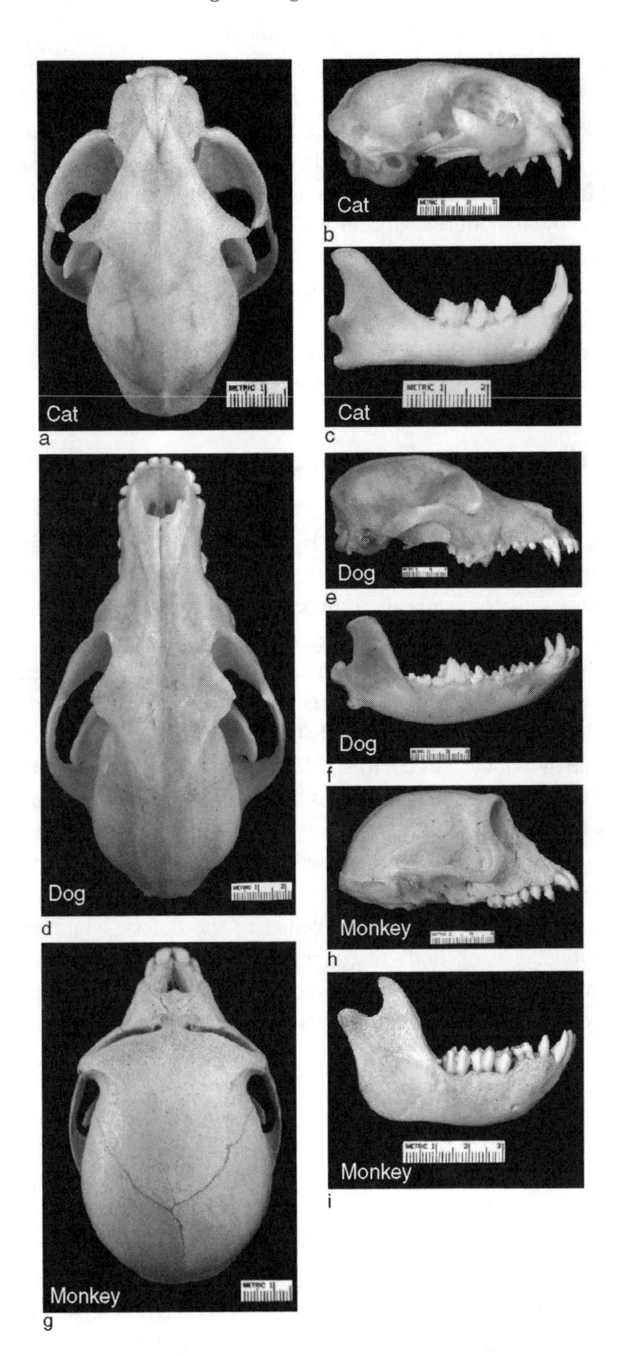

Figure 9.5 Cleaned and dried skulls and mandibles from the phylogenetically "closer" animal models. Note overall differences in the shape of the skull, dentition, and mandibles compared to the "generic" animal models, which reflect further changes in brain morphology and differing masticatory functions.

growth patterns should choose an animal model based on similar regional growth vectors and patterns, and not necessarily only on phyletic affinity.

In general, under conditions of modeling specific human craniofacial growth patterns, a single animal model may not be appropriate for all regions. For example, short-faced animals (felids and short-faced nonhuman primates) show midfacial and nasal capsule growth patterns similar to humans,[9,26] while juvenile canids show mandibular growth patterns similar to humans.[25] Thus, if the objectives of the study are to improve the human condition, the use of a primate model simply on the basis of phyletic affinity is not justified and that an understanding of the relationship between the growth patterns in humans and the animal under consideration is a prerequisite to appropriate animal model choice.[9, 25–27]

Bony Wound Healing and Critical-Sized Defects

The most common replacement therapy for bony tissue loss, especially in the craniofacial complex, is autologous bone grafting. However, limited supply and donor site morbidity have stimulated the search for bone replacement materials (BRMs).[69–75] Bone tissue engineering aims to restore the function and/or replace the damaged or diseased bony tissue through the principles of osteogenesis, osteoinduction, and osteoconduction (Table 9.1). As new BRMs are being designed and developed, a consistent, standardized testing paradigm using critical-sized defects (CSDs) has been proposed.[76, 8]

A CSD is defined as a defect that heals by less than 10% bony regeneration during the lifetime of the individual.[8] In contrast, fibrous nonunions can form through a combination of fibrous and bony healing that does not return bony continuity and function, but may eventually heal. CSDs are thought to result from a number of factors, including: (1) an uneven gradient of soluble osseous growth factors in the wound, which fail to reach the center of the defect and uniformly promote osteogenesis; (2) uneven biomechanical stress and loading, especially toward the center of the defect, which reduces uniform bone cell migration, adhesion, and subsequent osteoid and bone matrix formation; and/or (3) the differential and more rapid migration of fibroblasts (compared to osteoblasts) into the wound site during healing, which results in a higher prevalence of fibrous tissue in the center of the defect.[8, 76–79] Thus, the CSD model fails to heal only because it exceeds the body's ability to regenerate adequate amounts of bone fast enough.

CSDs make an excellent model to test bone tissue engineering paradigms.[76,8] However, there are a number of confounding variables that can influence the bony wound healing of CSDs and should be taken into account before choosing an appropriate animal model. These include: (1) species-specific differences in bony wound healing; (2) skeletal age of the individual; and (3) anatomic location of the CSD.[80, 76, 8]

Species-specific differences in bony wound healing have been reported.[80, 76, 8] Spontaneous osseous wound healing was observed to decrease in higher order species.[80, 8] As noted by Hollinger and Kleinschmidt[8] (p. 61), "The defect size to volume ratio in the rat is 1.96, while in the human the ratio is 0.12. If the human could regenerate as much bone as the rat it could regenerate bone to fill a defect 23 cm in diameter." This nonallometric relationship is also seen in Table 9.2. CSD size in the rodent calvaria ranges from 5 to 8 mm in diameter while in larger animal models, the CSD size ranges from 15 to 25 mm in diameter.

There are also significant age-related changes in the healing of CSDs, which also vary across taxonomic levels. Bony defects in skeletally immature animals of most species heal at a faster rate than skeletally mature individuals. If BRMs are tested in immature animals, falsely high expectations of the osteoconductive properties of the material could be generated.[81, 8] Hollinger and Kleinschmidt[8] suggest that biological age, as determined by radiographic confirmation of epiphyseal closure, be used in conjunction with chronological age and body weight to ascertain skeletal maturity, prior to CSD creation. Rats are the exception to this rule because they grow constantly throughout their life cycle.[52, 8] While in most animals, the initial size of the CSD decreases with age, the initial size of the CSD in rats remains constant, regardless of age.[8] However, there has recently been a push to engineer bone and soft-tissue constructs that grow along with the individual.[33, 44] The testing of these growing constructs may require the use of CSDs in growing animal models; information regarding this is currently sparse in the literature.

Anatomic location of the CSD can also affect healing rate and interpretation. The craniofacial complex is much more vascular and more richly innervated than the long bones and vertebral column. Even within species, CSD size varies across these anatomic locations, and CSD sizes in the skull and mandible are relatively smaller than those produced elsewhere (Table 9.2). Thus, anatomic location and CSD size need to be taken into consideration before choosing an appropriate animal model.

Table 9.2 reviews studies of selected CSD sizes by anatomic location and species. As can be seen in the table, the majority of the CSD studies have been performed on the calvaria, mandible, and femur (Figures 9.1–9.5) across a wide variety of animal species. These regions are easily accessible for surgery and represent biomechanically loaded (mandible and femur) or unloaded (calvaria) regions to simulate varying clinical conditions. Animal model choice will vary depending on the paradigm being tested (i.e., size of scaffold, duration of healing, length of construct bioactivity, etc.).

"Fitting" the Appropriate Animal Model to Specific Human Clinical Conditions

As tissue engineering paradigms become more reproducible and sophisticated, we find bone tissue engineers trying to tackle osseous wound healing problems in specific, clinically compromised, subpopulations of patients (i.e., diabetic, osteoporotic, arthritic, geriatric, congenital anomalies, etc.). Addressing osseous wound healing problems in these clinical situations will require specific animal models of the pathology or disease state. Table 9.3 presents a selected list of animal models of specific bone pathologies and bone metabolic disease states. A number of these models have been developed for specific purposes (i.e., to study the onset and treatment of postmenopausal osteoporosis, inflammatory arthritis, periodontitis-induced bone loss, etc.); but these models can also be adapted to study the various bone tissue engineering paradigms. Osteoconduction, osteoinduction, or cell transplant may be significantly altered in these human clinical groups and animals that model these specific clinical conditions will be instrumental in developing new treatment strategies to regenerate new bone and improve their quality of life.

Table 9.2
Selected Critical-Sized Defect (CSD) Studies by Animal Model and Anatomic Region

	Animal Model									
Anatomic Region	**Mouse**	**Rat**	**Rabbit**	**Cat**	**Dog**	**Sheep**	**Pig**	**Horse**	**Nonhuman primates**	**Goat**
Calvaria	5 mm round[82]	8 mm round[83–85]	15 mm round[80, 86–88]	25 mm round[89]	20 mm round[81, 90]	20 mm round[91]			15 mm round (rhesus[8] baboons[92]) 25 mm round (baboons[93])	14 mm round[58, 59]
Palate			5 mm seg.[94, 95]	5 mm[96]	5 mm seg.[97, 98]					
Alveolus	3 mm seg.[99]	4 mm seg.[100]	5 mm seg[101, 102, 95, 79] 10 mm seg.[103]						10–15 mm[104, 8]	
Mandible	4 mm round[105]	4 mm seg.[106, 107]	5 mm seg.[108]		17–45 mm seg.[109, 8]	35 mm seg[110]	15 mm×15 mm seg[111]		15 mm[8]	
Zygomatic Arch			5 mm seg.[77, 78]							
Nasal Bones			20 mm×8 mm seg.[112]							

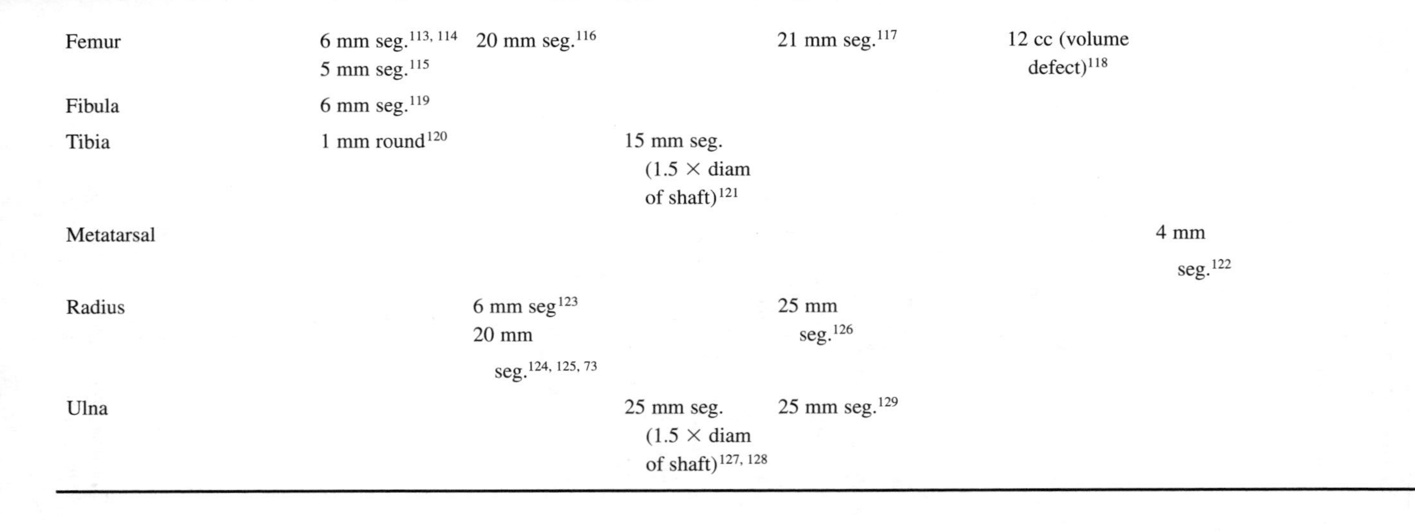

Femur	6 mm seg.[113, 114] 5 mm seg.[115]	20 mm seg.[116]		21 mm seg.[117]	12 cc (volume defect)[118]	
Fibula	6 mm seg.[119]					
Tibia	1 mm round[120]		15 mm seg. (1.5 × diam of shaft)[121]			
Metatarsal						4 mm seg.[122]
Radius		6 mm seg[123] 20 mm seg.[124, 125, 73]		25 mm seg.[126]		
Ulna			25 mm seg. (1.5 × diam of shaft)[127, 128]	25 mm seg.[129]		

Table 9.3
Selected Animal Model Studies of Bone Pathologies and Bone Metabolic Disease States

Condition	Technique	Species	Comments	Reference
		Inflammatory arthritis models		
Osteoarthritis	Naturally occurring	Various species	Good comparative review of pathogenesis and models	130
Type II collagen Arthritis	Immune response to Injection of type II Bovine collagen	DBA/LACJ mouse	Acute inflammatory response in extremity joints and loss of bone mineral density	131, 132
Spondyloarthopathies	Immune response to injection of bacterial flora	HLA-B27 transgenic rat	Arthritis, loss of bone mineral density, and tarsal joint ankylosis	133
Rheumatoid arthritis	Autoimmune diseases	Rat, mouse	Inflammatory response and loss of bone mineral density	134
		Osteonecrotic and iatrogenic models		
Femoral head osteonecrosis	Surgically created via ischemic and/or cryogenic insults	Emu, dog, sheep	Developed early and late stage human pathology. Emu good bipedal loading model	40, 135
Osteomyelitis	Injections with various pathogens (*S. aureus*, *B. fragilis*,*E. coli*, ...)	Rat, rabbit, dog, guinea pig, chicken	Decreased bone mineral density, increased risk of fracture, and increased cortical and trabecular bone loss	36, 136, 137, 138
Disuse atrophy and underloading-induced osteopenia	Limb immobilization, denervation, micro-gravity	Rat, dog, monkey	Trabecular and cortical bone loss, decreases in bone formation, Increases in bone resorption	34, 139
Age-induced osteopenia	Natural aging	Mouse, rat, monkey	Reduction in skeletal mass and osteopenia	34
	Natural Aging Postmenopausal	Monkey	Age-related decline in skeletal bone mineral content in males and females	140, 49
	Orchidectomy	Rat	Cancellous bone osteopenia and reduced bone turnover	34
Seasonal-induced osteopenia	Antler formation, diet changes, hibernation	Reindeer, deer caribou, bats	Experience transient mineral deficit and reduced bone turnover	34

Table 9.3 Continued

Condition	Technique	Species	Comments	Reference
Reproductive cycle-inducedosteopenia	Pregnancy, lactation, egg laying	Sheep, rat, dog, chicken, turtle	Altered calcium storage and release with subsequent osteopenia	34
Diet-induced osteopenia	Calcium and phosphorus dietary manipulations	Cat	Develop bone disease and decreased bone mineral density	141, 9
	Vitamin C and D deficiency	Rat, rabbit	Osteomalacia, rickets, osteopenia	34
Alcohol-induced osteopenia	Pre- and postnatal alcohol exposure	Rat	Osteopenia and reduced skeletal growth	142, 143
Glucocorticoid-induced osteopenia	Glucocorticoid exposure	Rat, rabbit, dog, sheep	Varies from species to species but decreases bone mass	34
Bone metastasis-induced osteopenia	Injection of a breast cancer cell line	Mouse	Causes osteolytic bone metastasis and osteoclasis	144
	Spontaneous mammary tumor model	Mouse	Causes osteolytic bone metastasis and osteoclasis	144
		Oral Bone Loss Models		
Periodontitis	Naturally occurring in laboratory-reared models	Baboon, dog, monkey, marmoset	Isolated vertical bone loss in maxillary and mandibular alveolar and basal bone	145, 146, 50, 51, 147, 148, 149, 150
	Oral infection by *P. gingivalis*	Mouse, rat, rhesus monkey, baboon, dog, sheep	Reliably produced maxillary and mandibular alveolar and basal bone loss	151
		Experimentally Induced, Estrogen-Deficient Bone Loss Models		
Post-menopausal osteoporosis and osteopenia	Various procedures	Various models	Good discussions and critiques of various techniques and and models	34, 139, 35, 82, 152
	Ovariectomy	Sheep	Bone loss in mandible, proximal femur, and vertebral body	11, 153, 9
	Ovariectomy	Nonhuman primates	Decreased bone mineral density and reduced vertebral cancellous bone volume	154, 9, 24
	Ovariectomy	Ferret	Increased skeletal bone loss	155

Table 9.3 Continued

Condition	Technique	Species	Comments	Reference
	Ovariectomy	Dog	Similar bone microarchitecture but does not develop osteoporosis	156, 157, 9
	Ovariectomy	Rat	Developed cortical thinning and significant bone loss	158, 159, 160, 9
	Ovariectomy	Guinea pig	Showed no consequence on bone	161, 9
	Ovariectomy and calcium-restricted diet	Sinclair S-1 minipig	Showed bone removal and loss comparable to humans	162, 9
		Perturbed Bone Metabolism Models		
Genetic models of various metabolic bone diseases	Knockout and overexpression models	Transgenic mice	Skeletal phenotypes range from osteopetrosis to osteopenia	10, 38
Diabetic osteopenia	Various experimental and transgenic insulin deficiency models	Mouse, rat	Decreased bone mass and increased prevalence of skeletal fractures	163, 164, 165
Growth hormone resistant dwarfism (Laron syndrome)	Congenital models of GH receptor mutation and GH-resistance	Sex-linked dwarf chicken, guinea pig, miniature poodle, cattle, and pigs	Severe skeletal growth retardation and dysmorphic facial features	166, 167
Calvarial hyperostosis and craniosynostosis	Various congenital, experimental, and transgenic models	Various species	Features clinical manifestations ranging from microcephaly to craniofacial dysostoses with digital involvement	12
Facial clefting	Various congenital, experimental, and transgenic models	Various species	Features clinical manifestations ranging from minimal alveolar and palatal clefts to median and oblique facial clefts	168

Animal model choice at this level of hypothesis testing should not necessarily be based only on phyletic affinity but should take into account the etiology of the particular bone pathology or disease state. Although the expectation is that phyletic affinity and similarity of etiology, disease progression, and pathophysiology are reasonably related,[50, 51, 34, 82, 10, 38] phyletic affinity should not be the exclusive criterion, but should be used in conjunction with similar biomechanical, anatomical, and wound healing factors.

CONCLUSIONS

From the preceding review, it should be evident that many animal models have been developed and utilized to test various bone tissue engineering paradigms. It should be understood that each animal model has its own utility as well as limitations for the study of human bone tissue engineering. In this chapter, we have proposed a number of practical criteria for choosing and evaluating various animal models, and have suggested that appropriate animal model choice should also be linked to levels of hypothesis testing and/or expected extrapolation to the human clinical condition. The continued intelligent and conservative use of animal models should be based on such objective criteria and not necessarily on phyletic affinity alone.

ACKNOWLEDGMENTS

The authors would like to thank Ms. Ronal Mitchell for her help with the figures and Ms. M. Elizabeth Hommel for her help with library research. This work was supported in part by a Comprehensive Oral Health Research Center of Discovery (COHRCD) program/project grant from NIH/NIDCR (P60 DE13078) to the Center for Craniofacial Development and Disorders, Johns Hopkins University, Baltimore MD.

References

1. Goldstein, S.A. (2002). Tissue engineering: functional assessment and clinical outcome. *Ann. NY Acad. Sci.* 961, 183–92.
2. Vacanti, C.A. and Bonassar, L.J. (1999). An overview of tissue engineered bone. *Clin. Orthop.* 367 (Suppl), S375–S381.
3. Schmidt-Nielsen, B. (1961). Choice of experimental animals for research. *Fed. Proc.* 20, 902–906.
4. Brown, A.M. (1963). Matching the animal with the experiment. In *Animals for Research*, Lane-Petter, W., Ed. Academic Press, New York, 261–280.
5. Navia, J.M. (1977). *Animal Models in Dental Research.* University Alabama Press, Birmingham.
6. Prichard, R.W. (1978). The need for new animal models — a philosophic approach. *JAVMA* 173, 1208–1209.
7. Siegel, M.I. and Mooney, M.P. (1990). Appropriate animal models for craniofacial biology. *Cleft Pal.-Craniofac. J.* 27, 18–25.
8. Hollinger, J.O. and Kleimnschmidt, J.C. (1990). The critical size defect as an experimental model to test bone repair methods. *J. Craniofac. Surg.* 1(1), 60–68.

9. Newman, E., Turner, A.S., and Wark, J.D. (1995). The potential of sheep for the study of osteopenia: current status and comparison with other animal models. *Bone* 16(4 Suppl), 277S–284S.
10. Turner, R.T. et al. (2001). Animal models for osteoporosis. *Rev. Endocr. Metab. Disord.* 2(1), 117–127.
11. Turner, A.S. (2002). The sheep as a model for osteoporosis in humans. *Vet. J.* 163(3), 232–239.
12. Mooney, M.P., Siegel, M.I. and Opperman, L.A. (2002). Animal models of craniosynostosis: experimental, congenital, and transgenic. In *Understanding Craniofacial Anomalies: The Etiopathogenesis of Craniosynostosis and Facial Clefting,* Chapter 9, Mooney, M.P. and Siegel, M.I., Eds. John Wiley and Sons, New York, pp. 209–250.
13. Murray, R. (2002). Animal models for orthopaedic disease — who benefits?. *Vet. J.* 163(3), 230–231.
14. Prince, A.M. et al. (1988). Chimpanzees and AIDS research. *Nature* 333, 513–514.
15. Rowsey, J.J. (1988). Responsibilities in animal experimentation. *Opthalm. Surg.* 19, 161–162.
16. Fernandes, D. (1989). Animal experimentation: necessary or not?. *Cleft Pal.-Craniofac. J.* 26, 258–259.
17. Loeb, J.M. et al. (1989). Human vs. animal rights: in defense of animal research. *J. Am. Med. Assor.*, 262, 2716–2720.
18. Roach, H.I., Shearer, J.R. and Archer, C. (1989). The choice of an experimental model: a guide for research workers. *J. Bone Joint Surg.*, 71-B, 549–553.
19. English, D.C. (1989). Using animals for the training of physicians and surgeons. *Theor. Med.* 10, 43–52.
20. Taylor, I. et al. (1991). Dilemmas facing surgical research in the '90s. *Ann. R. Coll. Surg. England* 73, 70–72.
21. Cramer, M. (1998). Experiments using animals. *Plast. Reconstr. Surg.* 102, 926–927.
22. An, Y.H. and Friedman, R.J. (1999). Animal selections in orthopaedic research. In *Animal models in Orthopaedic Research*, An, Y.H. and Friedman, R.J., Eds. CRC Press, London, pp. 39–57.
23. Davidson, M.K., Lindsey, J.R. and Davis, J.K. (1987). Requirements and selection of an animal model. *Isr. J. Med. Sci.* 23, 551–555.
24. Rodgers, J.B., Monier-Faugere, M.C. and Malluche, H. (1993). Animal models for the study of bone loss after cessation of ovarian function. *Bone* 14 (3): 369–377.
25. Losken, A., Mooney, M.P. and Siegel, M.I. (1992). Comparative analysis of mandibular growth patterns in seven animal models. *J. Oral Maxillofac. Surg.* 50, 490–495.
26. Losken, A., Mooney, M.P. and Siegel, M.I. (1994). Comparative cephalometric study of nasal cavity growth patterns in seven animal models. *Cleft Pal.-Craniofac. J.* 31, 17–23.
27. Mooney, M.P. and Siegel, M.I. (1993). Appropriate animal models for craniofacial biology II: their applications to clinical investigations. Paper presented at the annual meeting of the American Cleft Palate — Craniofacial Association, Pittsburgh, PA. April.
28. Smith, G.C. (1969). Value of nonhuman primates in predicting disposition of drugs in man. *Ann. NY Acad. Sci.* 162, 600–603.
29. Reynolds, H.H. (1969). Nonhuman primates in the study of toxicological effects on the central nervous system: a review. *Ann. NY Acad. Sci.* 162, 604–609.
30. Goldsmith, E.I. and Moor-Jankowski, J. (1969). Experimental medicine and surgery in primates, *Ann. NY Acad. Sci.* 162, 1–704.
31. Langer, R. and Vacanti, J.P. (1993). Tissue engineering. *Science* 260, 920–926.
32. Putnam, A.J. and Mooney, D.J. (1996). Tissue engineering using synthetic extracellular matrices. *Nat. Med.* 2, 824–826.

33. Alsberg, E., Hill, E.E. and Mooney, D.J. (2001). Craniofacial tissue engineering. *Crit. Rev. Oral Biol. Med.* 12(1), 64–75.
34. Miller, S.C., Bowman, B.M. and Jee, W.S. (1995). Available animal models of osteopenia—small and large. *Bone* 17(4 Suppl),117S–123S.
35. Mosekilde, L. (1995). Assessing bone quality—animal models in preclinical osteoporosis research. *Bone* 17(4 Suppl), 343S–352S.
36. Cremieux, A.C. and Carbon, C. (1997). Experimental models of bone and prosthetic joint infections. *Clin. Infect. Dis.* 25(6), 1295–1302.
37. Athanasiou, K.A. et al. (2000). Fundamentals of biomechanics in tissue engineering of bone. *Tissue Eng.* 6(4), 361–381.
38. McCauley, L.K. (2001). Transgenic mouse models of metabolic bone disease. *Curr. Opin. Rheumatol.* 13(4), 316–325.
39. Green, D., Walsh, D., Mann, S., Oreffo, R.O. (2002). The potential of biomimesis in bone tissue engineering: lessons from the design and synthesis of invertebrate skeletons. *Bone* 30(6), 810–815.
40. Conzemius, M.G. et al. (2002). A new animal model of femoral head osteonecrosis: one that progresses to human-like mechanical failure. *J. Orthop. Res.* 20(2), 303–309.
41. Krebsbach, P.H. and Gehron-Robey, P. (2002). Dental and skeletal stem cells: potential cellular therapeutics for craniofacial regeneration. *J. Dent. Ed.* 66:766–773.
42. Boyan, B.D. et al. (1999). Bone and cartilage tissue engineering. *Clin. Plast. Surg.* 26(4), 629–645.
43. Goldstein, S.A., Patil P.V. and Moalli M.R. (1999). Perspectives on tissue engineering of bone. *Clin. Orthop.* 367 (Suppl),S419–S423.
44. Alsberg, E. et al. (2002). Engineering growing tissues. *Proc. Natl. Acad. Sci. USA*, 99(19), 12025–12030.
45. Gregory, W.K. (1929). *Our Face from Fish to Man.* Hafner Publ. Co., New York.
46. De Beer, G.R. (1937). *The Development of the Vertebrate Skull.* Oxford University Press, London.
47. Romer, A.S. and Parsons, T.S. (1986). *The Vertebrate Body,* 5th ed. Saunders, Philadelphia.
48. Young, J.Z. (1975). *The Life of Mammals*, 3rd ed. Clarendon Press, Oxford.
49. Jerome, C.P. and Peterson P.E. (2001). Nonhuman primate models in skeletal research. *Bone* 29(1), 1–6.
50. Aufdemorte, T.B. et al. (1993). Diagnostic tools and biologic markers: animal models in the study of osteoporosis and oral bone loss. *J. Bone. Miner. Res.* 8 (Suppl 2), S529–S534.
51. Aufdemorte, T.B. et al. (1993). A non-human primate model for the study of osteoporosis and oral bone loss. *Bone* 14, 581–586.
52. Farris, E.F. and Griffith, J.Q. (1949). *The Rat in Laboratory Investigation*, Hafner Press, New York.
53. Enlow, D.H. (1990). *Handbook of Facial Growth.* Saunders, New York.
54. Greene, E.C. (1968). *Anatomy of the Rat.* Hafner Publ. Co., New York.
55. Cooper, G. and Schiller, S. (1975). *Anatomy of the Guinea Pig.* Harvard University Press, Cambridge, MA.
56. Kier, E.L. (1976). Phylogenetic and ontogenetic changes of the brain relevant to the evolution of the skull. In *Development of the Basicranium*, Bosma, J., Ed., DHEW/NIH Publ #76–989. Bethesda, MD, 468–499.
57. Getty, R. (1975). *Sissons and Grossman's The Anatomy of the Domestic Animals.* WB Saunders, Philadelphia.
58. Merkx, M.A. et al. (1999a). Incorporation of three types of bone block implants in the facial skeleton. *Biomaterials* 20(7), 639–645.

59. Merkx, M.A. et al. (1999b). Incorporation of particulated bone implants in the facial skeleton. *Biomaterials* 20(21), 2029–2035.
60. Hartman, C.F. and Strauss, W.L. (1933). *The Anatomy of the Rhesus Monkey*. Hafner Publ. Co., New York.
61. Elliot, R. (1963). *Reighard and Jennings' Anatomy of the Cat*. Holt, Rinehart, and Winston, New York.
62. Swindler, D.R. and Wood, C.D. (1982). *An Atlas of Primate Gross Anatomy: Baboon, Chimpanzee, and Man.* Robert E. Krieger Publ., Malabar, FL.
63. Sirianni, J. (1985). Nonhuman primates as models for human craniofacial growth. In *Nonhuman Primate Models for Human Growth and Development*, Watts, E.S. Ed. A.R. Liss, New York, pp. 95–124.
64. Enlow, D.H. and McNamara, J. (1973). The neurocranial basis for facial form and pattern. *Angle Orthod.* 43, 256–271.
65. Sirianni, J.E. and Swindler, D.R. (1979). A review of the postnatal craniofacial growth in old world monkeys and apes. *Yrbk. Phys. Anth.* 22, 80–104.
66. Siebert, JR and Swindler, DR. (2002). Evolutionary changes in the midface and mandible: establishing the primate form. In *Understanding Craniofacial Anomalies: The Etiopathogenesis of Craniosynostosis and Facial Clefting*, Mooney, M.P. and Siegel, M.I., Eds. John Wiley and Sons, New York, 345–378.
67. Geddes, A.D. (1996). Animal models of bone disease. In *Principles of Bone Biology*, Bilezikians, J.P., Raisz, L.G., and Rodan, G.A., Eds. Academic Press, San Diego, 1343–1354.
68. Enlow, D.H. and Azuma, M. (1975). Functional growth boundaries in the human and mammalian face. *Birth Defect* 11, 217–230.
69. Whang, K. et al. (1999). Engineering bone regeneration with bioabsorbable scaffolds with novel microarchitecture. *Tissue Eng*. 5, 35–51.
70. Service, R.F. (2000). Bone remodeling and repair. *Science* 289, 1421–1640.
71. Murphy, W.l. et al. (2000). Sustained release of vascular endothelial growth factor from mineralized poly lactide-co-glycolic scaffolds for tissue engineering. *Biomaterials* 21, 2521–2527.
72. Shea, L.D. et al. (2000). Engineered bone development from pre-osteoblast cell line on three-dimensional scaffolds. *Tissue Eng*. 6, 605–617.
73. Mackenzie, D.J. et al. (2001). Recombinant human acidic fibroblast growth factor and fibrin carrier regenerates bone. *Plast. Reconstr. Surg.* 1 107(4), 989–996.
74. Hutmacher, D.W. (2001). Scaffold design and fabrication technologies for engineering tissues—state of the art future perspectives. *J. Biomater. Sci. Polym. Ed.* 12, 102–124.
75. Einhorn, T. and Lee, C.A. (2001). Bone regeneration: new findings and potential clinical applications. *J. Am. Acad. Orthop. Surg.* 9,157–165.
76. Schmitz, J.P. and Hollinger, J.O. (1986). The critical size defect as an experimental model for craniomandibulofacial nonunions. *Clin. Orthop.* 205, 299–308.
77. Mundell, R. et al. (1993). Osseous guided tissue regeneration using a collagen barrier membrane. *J. Oral Maxillofac. Surg.* 51, 1004 – 1012.
78. Mooney, M.P. et al. (1996). The effects of guided tissue regeneration and fixation technique on osseous wound healing in rabbit zygomatic arch osteotomies. *J. Craniofac. Surg*. 7(1), 46–53.
79. Stetzer, K.M. et al. (2002). The effects of fixation type and guided tissue regeneration on maxillary osteotomy healing in rabbits. *J. Oral Maxillofac. Surg*. 60, 427–436.
80. Frame, J.W. (1980). A convenient animal model for testing bone substitution materials. *J. Oral Surg.* 38, 176–180.

81. Prolo, D.J. et al. (1982). Superior osteogenesis in transplanted allogenic canine skull following chemical sterilization. *Clin. Orthop*. 108, 203.
82. Thompson, D.D. et al. (1995). FDA Guidelines and animal models for osteoporosis. *Bone* 17 (4 Suppl), 125S–133S.
82. Lee, J.Y. et al. (2001). Effect of bone morphogenetic protein-2-expressing muscle-derived cells on healing of critical-sized bone defects in mice. *J. Bone Joint Surg. Am.* 83-A, 1032–1039.
83. Takagi, K. and Urist, M.R. (1982). The role of bone marrow induced repair of femoral massive diaphyseal defects. *Clin. Orthop*. 171, 224–231.
84. Hollinger, J.O. et al. (1988). A synthetic polypentapeptide of elastin for initiating calcification. *Calcif. Tissue Int.* 42, 231–236.
85. Winn, S.R. et al. (1999). Tissue-engineered bone biomimetic to regenerate calvarial critical-sized defects in athymic rats. *J. Biomed. Mater. Res.* 45(4), 414–421.
86. Clokie, C.M. et al. (2002). Closure of critical sized defects with allogenic and alloplastic bone substitutes. *J. Craniofac. Surg.* 13(1), 111–121.
87. Mooney, M.P. et al. (2001). Correction of coronal suture synostosis using suture and dura mater allografts in rabbits with familial craniosynostosis. *Cleft Pal.Craniofac. J.* 38, 72–91.
88. Vesala, A.L. et al. (2002). Bone tissue engineering: treatment of cranial bone defects in rabbits using self-reinforced poly-L,D-lactide 96/4 sheets, *J. Craniofac. Surg.*, 13(5), 607–613.
89. Costantino, P.D. et al. (1992). Experimental hydroxyapatite cement cranioplasty, *Plast. Reconstr. Surg.* 90(2), 174–185.
90. Urist, M.R. (1984). New advanced in bone research. *West. J. Med.* 141,71.
91. Shang, Q., Wang, Z., Liu, W., Shi, Y., Cui, L., Cao, Y. (2001). Tissue-engineered bone repair of sheep cranial defects with autologous bone marrow stromal cells. *J. Craniofac. Surg.* 12(6), 586–593.
92. Hollinger, J.O. et al. (1989). Calvarial bone regeneration using osteogenin. *J. Oral Maxillofac. Surg.* 47,1182–1186.
93. Ripamonti, U. et al. (2001). Bone induction by BMPs/OPs and related family members in primates. *J. Bone Joint Surg. Am.* 83-A (Suppl 1)(Part 2), S116–S127.
94. Bardach, J., Roberts, D.M. and Klausner, E.C. (1979). Influence of two-flap palatoplasty on facial growth in rabbits. *Cleft Pal. J.* 16, 402–411.
95. Bardach, J. et al. (1980). The influence of simultaneous cleft lip and palate repair on facial growth in rabbits. *Cleft Pal. J.* 17(4), 309–318.
96. Freng, A. (1979). Transversal maxillary growth in experimental submucous mid-palatal clefts. A roentgen-cephalometric study in the cat. *Scand. J. Plast. Reconstr. Surg*. 13(3), 409–416.
97. Bardach, J., Mooney, M. and Bardach, E. (1982). The influence of two-flap palatoplasty on facial growth in beagles. *Plast. Reconstr. Surg*. 69(6), 927–936.
98. Bardach, J. et al. (1985). Bone formation in the canine palate following partial resection. In *Normal and Abnormal Bone Growth: Basic and Clinical Research,* Dixon, AD., Sarnat, B.V., and Hoyte P.A.N., Eds. CRC Press, Boca Raton, FL, 365–377.
99. Kawata, T. et al. (2001). Transplantation of new autologous biomaterials into jaw cleft, *J. Int. Med. Res.* 29, 287–291.
100. Takano-Yamamoto, T., Kawakami, M. and Sakuda, M. (1993). Defects of the rat premaxilla as a model of alveolar clefts for testing bone-inductive agents. *J. Oral Maxillofac. Surg.* 51, 887–891.
101. Verwoerd, C.D.A., Verwoerd-Verhoef, H.L. and Urbanis, N.A.M. (1976). Skulls with facial clefts. Experimental surgery on the facial skeleton. *Acta Otolaryngol.* 81,249–256.

102. Eisbach, K.J., Bardach, J. and Klausner, E.C. (1978). The influence of primary unilateral cleft lip repair on facial growth, Part II: direct cephalometry of the skull. *Cleft Pal. J.* 15, 109–117.
103. Verschueren, D. et al. (2003). Le Fort I osteotomy healing in rabbits with the use of guided tissue regeneration (GTR). *J. Dent. Res.* 82, 1574.
104. El Deeb, M., Horswell, B. and Waite, D. (1985). A primate model for producing experimental alveolar cleft defects. *J. Oral Maxillofac. Surg.* 43, 523–527.
105. Alden, T.D., Jane Jr., J.A., Hudson, S.B., Helm, G.A. (2000). The use of bone morphogenetic protein gene therapy in craniofacial bone repair. *J. Craniofac. Surg.* 11(1):24–30.
106. Kaban, L.B., Glowacki, J. and Murray, J.E. (1979). Repair of experimental bony defects in rats. *Surg. Forum* 30, 519–521.
107. Saadeh, P.B. et al. (2001). Repair of a critical size defect in the rat mandible using allogenic type I collagen. *J. Craniofac. Surg.* 12(6), 573–579.
108. Kahnberg, K. (1979). Restoration of mandibular jaw defects in the rabbit by subperiosteally implanted Teflon® mantle leaf. *Int. J. Oral Surg.* 8, 449–456.
109. Hollinger, J.O. and Schmitz, J.P. (1987). Restoration of bone discontinuities in dogs using a biodegradable implant. *J. Maxillofac. Surg.* 45, 594–600.
110. Schliephake, H. et al. (2001). Use of cultivated osteoprogenitor cells to increase bone formation in segmental mandibular defects: an experimental pilot study in sheep. *Int. J. Oral Maxillofac. Surg.* 30(6):531–537.
111. Henkel, K.O. et al. (2002). Stimulating regeneration of bone defects by implantation of bioceramics and autologous osteoblast transplantation. *Mund Kiefer Gesichtschir.* 6(2):59–65.
112. Lindsey, W.H. et al. (1998). A nasal critical-size defect: an experimental model for the evaluation of facial osseous repair techniques. *Arch. Otolaryngol. Head Neck Surg.* 124(8), 912–915.
113. Einhorn, T.A. et al. (1984). The healing of segmental bone defects induced by demineralized bone matrix. *J. Bone Joint Surg.* 66A, 274–279.
114. Lieberman, J.R. et al. (1999). The effect of regional gene therapy with bone morphogenetic protein-2-producing bone-marrow cells on the repair of segmental femoral defects in rats. *J. Bone Joint Surg. Am.* 81(7), 905–917.
115. Lane, J.M. et al. (1999). Bone marrow and recombinant human bone morphogenetic protein-2 in osseous repair. *Clin Orthop.* 361, 216–227.
116. Wheeler, D.L. et al. (2000). Assessment of abosorbable resorbable bioactive material for grafting of critical-size cancellous defects. *J. Orthop Res.* 18(1), 140–148.
117. Bruder, S.P. et al. (1998). The effect of implants loaded with autologous mesenchymal stem cells on the healing of canine segmental bone defects. *J. Bone Joint Surg. Am.* 80(7), 985–996.
118. Lange, T.A. et al. (1986). Granular tricalcium phosphate in large cancellous defects. *Ann. Clin. Lab. Sci.* 16, 467–472.
119. Narang, R. and Laskin, D.M. (1976). Experimental osteogenesis at fracture sites and gaps. *J. Oral Surg.* 34, 225–231.
120. Landry, P.S. et al. (1996). Bone injury response. An animal model for testing theories of regulation *Clin. Orthop.* 332, 260–273.
121. Toombe, J.P. et al. (1985). Evaluation of Key's hypothesis in the feline tibia: an experimental model for augmented bone healing studies. *Am. J. Vet. Res.* 46, 513–518.
122. Collier, M.A. et al. (1985). Direct current stimulation of bone production in the horse: preliminary study with a "gap healing" model. *Am. J. Vet. Res.* 46, 612–621.
123. Ben-fu, C. and Xue-ming, T. (1986). Ultrastructural investigation of experimental non-union of fractures. *Chinese Med. J.* 99, 207–214.

124. Wheeler, D.L. et al. (1998). Radiomorphometry and biomechanical assessment of recombinant human bone morphogenetic protein 2 and polymer in rabbit radius ostectomy model. *J. Biomed. Mater. Res.* 43(4), 365–373.
125. Hollinger, J.O. et al. (1998). Recombinant human bone morphogenetic protein-2 and collagen for bone regeneration. *J. Biomed. Mater. Res.* 43(4), 356–364.
126. Sciadini, M.F., Dawson, J.M. and Johnson, K.D. (1997). Evaluation of bovine-derived bone protein with a natural coral carrier as a bone-graft substitute in a canine segmental defect model. *J. Orthop. Res.* 15(6), 844–857.
127. Key, J.A. (1934). The effects of local calcium depot on osteogenesis and healing of fractures. *J. Bone Joint Surg. (Am.)* 16, 176–184.
128. Nilsson, O. et al. (1986). Bone repair induced by bone morphogenic protein in ulnar defects in dogs. *J. Bone Joint Surg.* 63B, 635–642.
129. Salkeld, S.L. et al. (2001). The effect of osteogenic protein-1 on the healing of segmental bone defects treated with autograft or allograft bone. *J. Bone Joint Surg. Am.* 83-A(6), 803–816.
130. Oegema, T.R. and Visco, D. (1999). Animal models of osteoarthritis. In *Animal Models in Orthopaedic Research*, An, Y.H. and Friedman, R.J., Eds. CRC Press, London, pp. 349–367.
131. Badger A.M. et al. (1996). Pharmacological profile of SB 203580, a selective inhibitor of cytokine suppressive binding protein/p38 kinase, in animal models of arthritis, bone resorption, endotoxin shock and immune function. *J. Pharmacol. Exp. Ther.* 279(3), 1453–1461.
132. Wooley, P.H. (1988). Collagen-induced arthritis in the mouse. *Methods Enzymol.* 162, 361–373.
133. Breban, M. (1998). Animal models and *in vitro* models for the study of aetiopathogenesis of spondyloarthropathies. *Baillieres Clin. Rheumatol.* 12(4), 611–626.
134. Joe, B. et al. (1999). Animal models of rheumatoid arthritis and related inflammation. *Curr. Rheumatol. Rep.* 1(2), 139–148.
135. Phillips, T.W., Johnston, G. and Wood, P. (1987). Selection of an animal model for resurfacing hip arthroplasty. *J. Arthroplasty* 2(2), 111–117.
136. Mader, J.P. (1985). Animal models of osteomyelitis. *Am. J. Med.* 78(Suppl 6B), 213–217.
137. Norden, C.W. (1988). Lessons learned from animal models of osteomyelitis. *Rev. Infect. Dis.* 10, 103–110.
138. Rissing, J.P. (1990). Animal models of osteomyelitis: knowledge, hypothesis, and speculation. *Infect. Dis. Clin. North Am.* 4(3), 377–390.
139. Cesnjaj, M., Stavljenic, A. and Vukicevic, S. (1991). In vivo models in the study of osteopenias. *Eur. J. Clin. Chem. Clin. Biochem.* 29(4), 211–219.
140. Black, A. et al. (2001). A nonhuman primate model of age-related bone loss: a longitudinal study in male and premenopausal female rhesus monkeys. *Bone* 28(3), 295–302.
141. Draper, H.H. (1985). *Advances in Nutritional Research*. Plenum Press, New York, pp. 172–186.
142. Riesenfeld, A. et al. (1991). The effects of perinatal alcohol exposure and dietary calcium supplements on skeletal and dental growth in rats. *Acta Anatomica* 140, 1–7.
143. Turner, R.T. et al. (1988). Chronic alcohol treatment results in disturbed vitamin D metabolism and skeletal abnormalities in rates. *Alcohol Clin. Exp. Res.* 12, 159–162.
144. Yoneda, T. et al. (1999). Use of bisphosphonates for the treatment of bone metastasis in experimental animal models. *Cancer Treat Rev.* 25(5), 293–299.
145. Miller, D.R. et al. (1995). Periodontitis in the baboon: a potential model for human disease. *J. Periodontal Res.* 30(6), 404–409.

146. Haney, J.M., Zimmerman, G.J. and Wikesjo, U.M. (1995). Periodontal repair in dogs: Evaluation of the natural disease model. *J. Clin. Periodontol.* 22(3), 208–213.
147. Levy, B.M. et al. (1971). Primates in dental research. In *Proceedings of the Third Conference on Experimental Medicine and Surgery in Primates, Medical Primatology,* Goldsmith, E.I., and Moor-Jankowski, J., Eds. Karger Press, Basel, pp. 859–869.
148. Levy, B.M. (1976). Animal model of human disease: chronic destructive periodontitis (periodontal disease, pyorrhea alveolaris, pyorrhea). *Am. J. Pathol.* 83, 637–640.
149. Levy, B.M. (1980). Animal analogues for the study of dental and oral diseases. *Dev. Biol. Stand.* 45, 51–59.
150. Dreizen, S. and Levy, B.M. (1977). Monkey models in dental research. *J. Med. Primatol.* 6, 133–144.
151. Genco, C.A., Van Dyke, T. and Amar, S. (1998). Animal models for Porphyromonas gingivalis-mediated periodontal disease. *Trends Microbiol.* 6(11), 444–449.
152. Sietsema, W.K. (1995). Animal models of cortical porosity. *Bone* 17(4 Suppl), 297S–305S.
153. Lamghari, M. et al. (1999). A model for evaluating injectable bone replacements in the vertebrae of sheep: radiological and histological study. *Biomaterials* 20, 2107–2114.
154. Miller, L.C. et al. (1986). Effects of ovariectomy on vertebral bone in the cynomolgus monkey (Macaca fascicularis). *Calcif. Tissue Int.* 38, 62–65.
155. Mackey, M.S. et al. (1995). The ferret as a small animal model with bmu-based remodeling for skeletal research. *Bone* 17(4), 191S–196S.
156. Kimmel, D.B. (1991). The oophorectomized beagle as an experimental model for estrogen-depletion bone loss in the adult human. *Cells Mater.* (Suppl.), 75–84.
157. Kimmel, D.B. (1996). Animal models for in vivo experimentation in osteoporosis research, In Osteoporosis, Marcus, Feldman, Kelsey, Eds. Academic Press, San Diego, pp. 671–690.
158. Wronski, T.J. and Yen, C.F. (1991). The ovariectomized rat as an animal model for postmenopausal bone loss. *Cells Mater.* (Suppl. 1), 69–74.
159. Frost, H.M. and Jee, W.S.S. (1992). On the rat model of human osteopenia and osteoporosis. *Bone Miner.* 18, 227–236.
160. Kalu, D.N. (1991). The ovariectomized rat model of postmenopausal bone loss. *Bone Miner.* 15, 175–192.
161. Vanderschueren, D. et al. (1992). Bone and mineral metabolism in the adult guinea pig: Long-term effects of estrogen and androgen deficiency. *J. Bone Miner. Res.* 7, 1407–1415.
162. Mosekilde, L. et al. (1993). Calcium-restricted ovariectomized Sinclair S-1 minipigs: An animal model of osteopenia and trabecular plate perforation. *Bone*, 14, 379–382.
163. Kagel, E.M., Majeska, R.J. and Einhorn, T.A. (1995). Effects of diabetes and steroids on fracture healing. *Curr. Opin. Orthop.* 6(5), 7–13.
164. Hough, F.S. (1987). Alterations of bone and mineral metabolism in diabetes mellitus, Part I, An overview. *S. Afr. Med. J.* 18, 72(2), 116–119.
165. Zapf, J. (1998). Growth promotion by insulin-like growth factor I in hypophysectomized and diabetic rats. *Mol. Cell Endocrinol.* 25, 140(1–2), 143–149.
166. Daughaday, W.H. (1992). Animal models of abnormal GH receptor binding and post-binding mechanisms. *Pediatr. Adolesc. Endocrinol.* 24, 282–287.
167. Hull, K.L. and Harvey, S. (1999). Growth hormone resistance: clinical states and animal models. *J. Endocrinol.* 163(2), 165–172.
168. Diewert, V.M. and Lozanoff, S. (2002). Animal Models of Facial Clefting — Experimental, Congenital, and Transgenic. In *Understanding Craniofacial Anomalies: The Etiopathogenesis of Craniosynostosis and Facial Clefting*, Mooney, M.P. and Siegel, M.I., Eds. John Wiley and Sons, New York, pp. 251–272.

10 Biomechanics

Dennis M. Cullinane and Kristy T. Salisbury

Contents

0-8493-1621-9/05/$0.00+$1.50
© 2005 by CRC Press LLC

INTRODUCTION

WHAT IS BIOMECHANICS?

Biomechanics is the application of engineering principles to biological structures and systems. It includes the characterization and quantification of load application, load response, and the various physiological responses to load. Biomechanics is indispensable if one is to understand how organisms exist in a mechanical environment, and how the structures and systems of those organisms are adapted to, and function in that environment. Biomechanics is also essential in understanding and evaluating the etiology of skeletal pathologies and trauma, as well as evaluating the various medical modalities applied to prevent or heal them. Biomechanical evaluation can take the form of noninvasive image analyses, finite element models, gait analyses, strain gauge analyses, or direct mechanical tests. This chapter has been written to serve as both a review of basic skeletal biomechanics, and as an overview of the potential applications of biomechanics in evaluating the efficacy of tissue engineering as an intervention process in musculoskeletal medicine.

HOW IS BIOMECHANICS RELEVANT TO MUSCULOSKELETAL MEDICINE?

Biomechanics is especially relevant to musculoskeletal medicine, from the etiology of a disease or trauma, to the application of a modality such as fracture fixation or joint replacement, and finally to the local mechanical environment that will direct tissue repair during healing. A skeletal injury event can be broken down into three basic mechanical components (Figure 10.1). First are the external forces acting on the skeletal structure (e.g., during a fall, the impact on the greater trochanter of the femur, resulting in a femoral neck fracture). This impact can be further broken down into a reaction force from the femoral head and acetabulum, and the internally generated shear stresses resulting from the two juxtaposed loads. Second is the existing mechanical integrity of the structure (in this example, the combination of bone

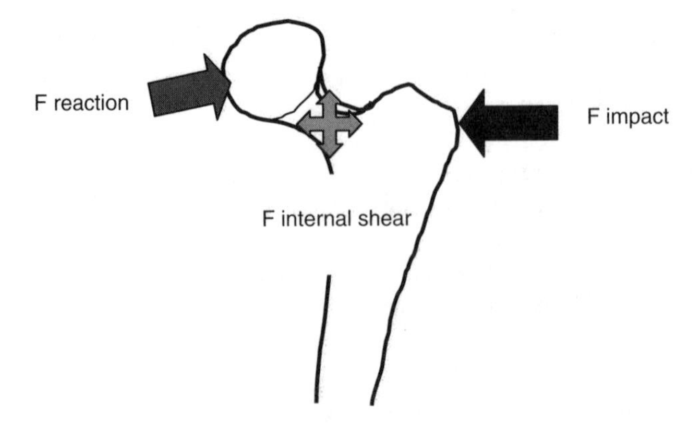

Figure 10.1 The initial impact on the greater trochanter of the femur causes a reaction force on the femoral head by the acetabulum. These juxtaposed forces generate internal shear forces within the hip, and lead to a fracture.

density and morphological configuration of the proximal femur). Third is the response of the structure to the external load, in this case a fall on the trochanter causing a fracture due to shear stresses. This is further complicated by the neuromuscular response to falling: when an individual falls, they contract their muscles and the tension induced within the musculature stiffens the joints, concentrating stress. Finally, biomechanics plays a role during the healing process of this fracture, where the local mechanical environment will influence the type of tissues that form (fibrous tissue, cartilage, bone), as well as their molecular organization[1] and resulting mechanical properties.[2] This process of mechanical influence on tissue formation is also evident systemically in cases of reduced loading under microgravity, where even transient hypophysiological loads lead to bone resorption.[3]

How does Biomechanics Factor in Tissue Engineering?

Tissue engineering is comprised chiefly of cellular-level therapies conducted through the transfer of genetic information via some vector. In orthopedic medicine, this process of tissue engineering typically focuses on the augmentation of skeletal mechanical parameters such as bone strength or stiffness. The function of the skeleton is to act as an articulated system of mechanically optimized load-bearing elements. Any pathological aberration or trauma would reduce the load-carrying capacity of the skeleton, and so it would be advantageous to identify treatments to counteract these pathologies. Tissue engineering is the manipulation of the genetic parameters controlling either skeletal configuration (shape and size) or skeletal composition (mineral density or collagen configuration, for example), and thus skeletal mechanics. These necessary processes of mechanical alteration and mechanical evaluation create a role for biomechanics in tissue engineering.

Why is Biomechanics Important to Skeletal Tissue Engineering?

The vast majority of tissue-engineered skeletal models involve factors that affect the mechanical properties of bone. The primary focus of these models is the enhancement of mechanical integrity, or prevention of any pathological reduction in mechanical integrity. In order to accomplish their goals, tissue engineering models must be evaluated for their pertinence to skeletal mechanical integrity, and biomechanics is essential in evaluating such relationships. Using the principles of biomechanics, one can quantify parameters such as strength, stiffness, Young's modulus, and Poisson's ratio; mechanical descriptors that characterize a specimen's mechanical integrity. Biomechanical evaluations can take the form of standard biomechanical tests that utilize actual biological samples in direct tests of mechanical integrity. These typically take the form of compression, tension, torsion, bending, shear, or noninvasive analyses such as a simple radiograph or dual-energy X-ray absorptiometry (DEXA). Noninvasive tests are surrogates of actual mechanical tests, and whose measures are meaningful only because they can be correlated to actual mechanical values. Biomechanical analyses then can include a spectrum of tests from invasive, direct measures of mechanical integrity, to noninvasive imaging and modeling procedures.

The role of biomechanics in skeletal tissue engineering is threefold. First, biomechanics is used in characterizing and quantifying the need for modification of skeletal mechanical parameters such as density, size, or geometry. These are based on evaluations of skeletal mechanical performance in cases of disease, general pathology, trauma, or during the repair process. This establishes a baseline set of values by which inadequacies in mechanical integrity are identified. Second, biomechanics is essential in evaluating the efficacy of any tissue engineering model. A model of mouse fracture repair, for example, must be evaluated for its capacity to enhance the mechanical properties of a healing callus, and the only way to absolutely quantify material properties is through direct mechanical testing. Any image-based evaluation, no matter how high the resolution, will have to be correlated to actual mechanical test results. Finally, mechanobiology is a subfield of biomechanics that describes how the local mechanical environment can direct the differentiation of tissues during development and healing. This relationship can be used to predict not only the type of tissue that will form but also the macro and microarchitecture and resulting mechanical properties of those tissues.

ANATOMY AND BIOMECHANICS

BONE AS TISSUE AND ORGAN

Bone is both a tissue (material) and an organ (structure), and the difference between these two categories must be kept in mind when designing a biomechanical test. Bone as a tissue constitutes the bulk of the skeleton. It is composed of a collagen matrix impregnated with a hydroxyapatite crystalline lattice. The collagen portion of bone tissue is arranged in specific configurations based on the type of bone (trabecular), the specific bone within the body (femur), the locale within the particular bone (proximal metaphysis), and the daily loading conditions to which the bone is exposed. Bone tissue throughout the body can vary in terms of its organic and mineral composition, making site specificity a fundamental factor in mechanical properties.[4–9] Finally, mechanical tests of bone as a tissue are fundamentally different from tests on bones as structures, and although the results can be correlated, they are not synonymous.

TYPES OF BONES

The basic types of bones are long bones, round bones, irregular-shaped bones, and the dermal or flat bones of the skull.[10] *Long bones* constitute the bulk of the appendicular skeleton and include the humerus, radius, ulna, femur, tibia, and fibula. The metacarpals and metatarsals can also be mechanically considered long bones, in addition to the phalanges. *Round bones* typically comprise the bending portion of complex joints like the wrist. These are the carpals and tarsals. The vertebrae are commonly thought of as *irregular bones*; however, some other irregularly shaped bones include the calcaneous, the pelvis, the scapula, and the mandible. The skull is composed of *dermal* or *flat bones* like the frontal and occipital.

Different types of bones are tested using specific and appropriate mechanical models. They can be modeled as beams, columns, rods, cylinders, or any other standard engineering geometric. These models are based on assumptions of how the bones are loaded *in vivo* during everyday activities. The purpose of any mechanical test is to mimic or somehow reflect normal *in vivo* loading conditions, in terms of both character and magnitude. Long bones like the femur, tibia, and humerus are typically tested in torsion or bending, whereas irregular bones like the vertebrae are typically tested in axial compression or shear. The type of test used for any particular bone or bone sample should be related to the way the bone is loaded *in vivo*.

Trabecular vs. Cortical Bone

Through examination of a sagittal or coronal section of the radius, for example, we can see that it is composed of two basic bone types: cortical and trabecular (Figure 10.2). The dense outer shell is composed of a solid bone tissue known as *cortical* or *compact bone*, while the inner framework of the head, neck, and epiphyses is composed of a sponge-like trabecular or cancellous bone. *Trabecular bone* is composed of mini struts of bone called *trabeculae*, which as a whole take on a mean preferred orientation that can be correlated to the principal lines of loading *in vivo*. In the proximal femur, this orientation has been correlated to principal stress lines resulting from weight bearing.

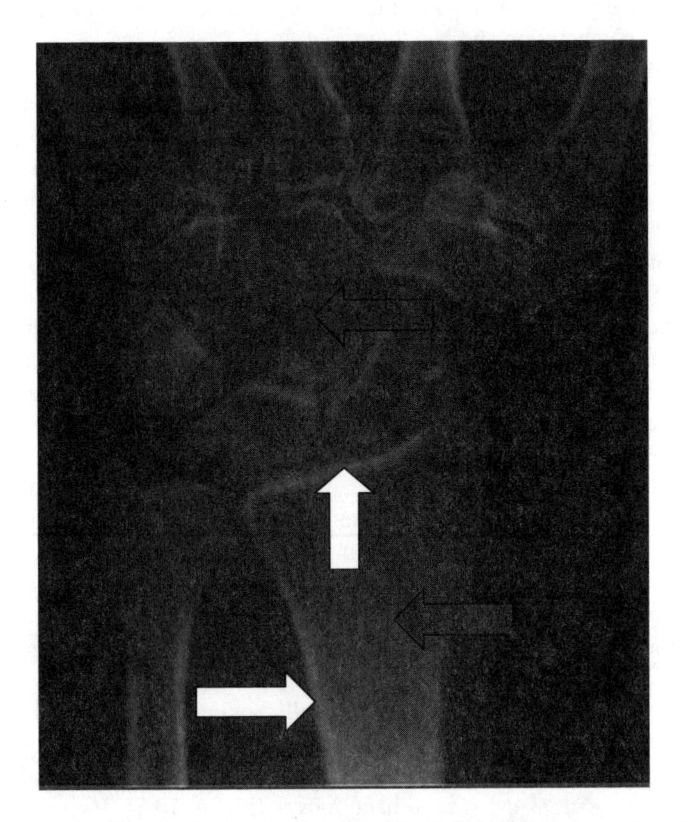

Figure 10.2 A radiograph of the wrist and palm illustrating the presence of cortical (white arrows) and trabecular bone (red arrows). The cortical bone creates the shell of a bone, while the trabecular bone acts as a supportive internal strut-like material.

BASIC MECHANICS

TERMINOLOGY

A mechanical test (compression, indentation, tension, torsion, bending, or shear) consists of a sample and an applied stress (tension, compression, or shear). Each mechanical test can be resolved into primary and secondary stresses acting simultaneously on a specimen (bending generates tension, compression, and shear, while torsion generates rotational shear and tension). Pure *tension* is generated when two equal but divergently directed forces act on a specimen along a common line, tending to stretch a structure or sample (Figure 10.3a). Conversely, pure *compression* is generated when two equal and convergently directed forces act along a common line, tending to compress a structure or sample (Figure 10.3b). Pure *shear* is generated when two forces act in convergent and parallel directions, but along slightly offset lines or planes (Figure 10.3c). Linear shear can be thought of as an off-axis tension or compression resulting in a splitting of a structure or specimen, whereas torsional shear results in twisting of the structure or specimen, one end rotating relative to the other. Beyond these basic methods, there are more complicated mechanical tests, which are actually combinations of the basic applications, an example of which is simultaneous compression and bending.

Thus, there are five different ways in which one can physically test a bone (neglecting combination testing such as simultaneous compression and torsion). These are tension, compression (including indentation), bending, shear, and torsion (a specialized kind of shear). For the sake of simplicity, and unless otherwise stated, we will refer to these loads throughout the chapter as if they were applied either to a structure like the femur (bending or torsion), a vertebra (compression or shear), a cube (compression or shear), milled cylinder (tension or torsion), or a beam (bending) of bone tissue. The four most common methods of mechanical tests for bone tissue and whole bones are tension, compression, bending, and torsion. *Tension* tests involve a milled uniform sample of bone with a reduced diameter at its midpoint to concentrate stress and assure failure within the specimen, rather than at the specimen-grip transition. Tension units are typically Newtons/area. *Compression* tests involve either a milled uniform sample or an intact bone surface, and use a compressive plate or indentor to apply the load. Compression tests are sometimes performed in chambers that confine the material to deformation in one axis; these are known as confined compression tests. Compression units are typically Pascals (pressure) or Newtons/area. *Bending* tests involve rigidly holding the structure or specimen and applying a load either as a cantilever (from one end only, the other stationary) or on the middle portion (with both ends stationary), resulting in a deformation of the specimen along its long axis (Figure 10.4). Bending units depend on the type of bending being used, but it is typically applied in Newtons. A *torsion* test involves a twist being applied to a specimen, with one long end rotating relative to the other. It results in angular reconfiguration of the specimen long axis, like the drive shaft of a clock as the second hand moves (Figure 10.5). Torsion, sometimes described as a moment, has units in Newtons $\times$ length (of force application from the specimens' neutral axis). In daily life, the skeleton is typically loaded in combinations

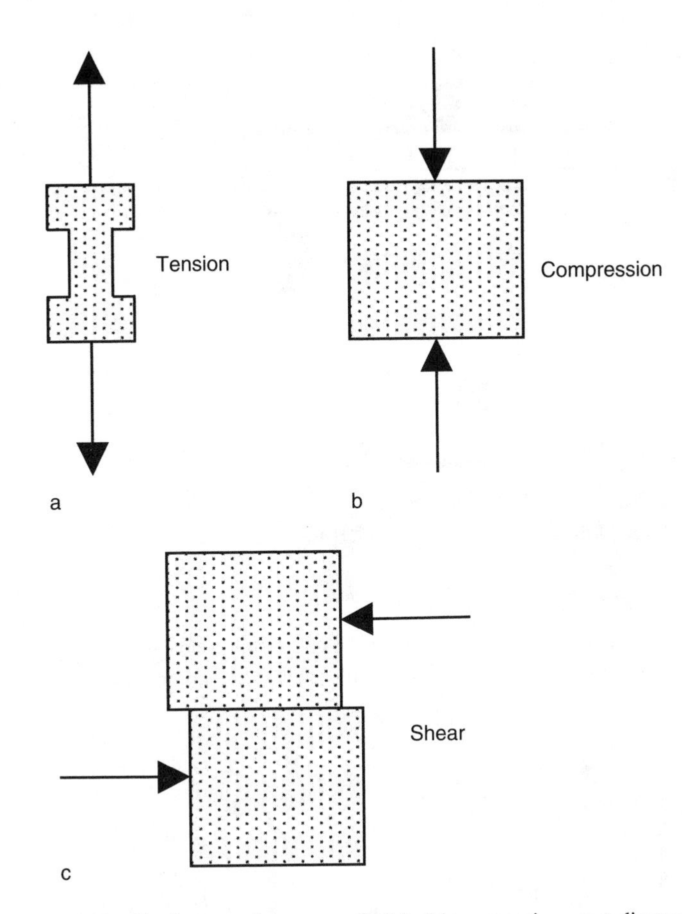

Figure 10.3 (a) Tensile forces acting on a cylindrical bone specimen act divergently along a single line. The middle portion of the specimen is milled down to concentrate stress and assure failure within the specimen rather than at the specimen–fixation junction. (b) Compression vectors on a cube of bone act convergently in a single line. (c) Shear is created by off-axis convergent forces. It is a common load experienced by vertebral bodies, in addition to compression.

of these simple load types, like the femur under compression, bending, and torsion, for example; but mechanical tests are typically simplified to one load type.

When a load is applied to a bone, the bone must respond mechanically or it will experience failure (a fracture). This effort by the bone to resist an applied load is called the stress. *Stress* is directly proportional to the applied load/area of application and is usually reported as the applied load per area, but it is technically a *response* by the specimen to the applied load/area. The amount of deformation that a specimen undergoes as a consequence of the applied load and resulting stress is called strain. *Strain* is measured as the change in specimen configuration (length, angle, or volume), divided by the original configuration (length, angle, or volume). If one torques a long bone like the femur along its long axis, it will start to deform rotationally, the strain being measured by the angle of deflection. Conversely, if a vertebra is

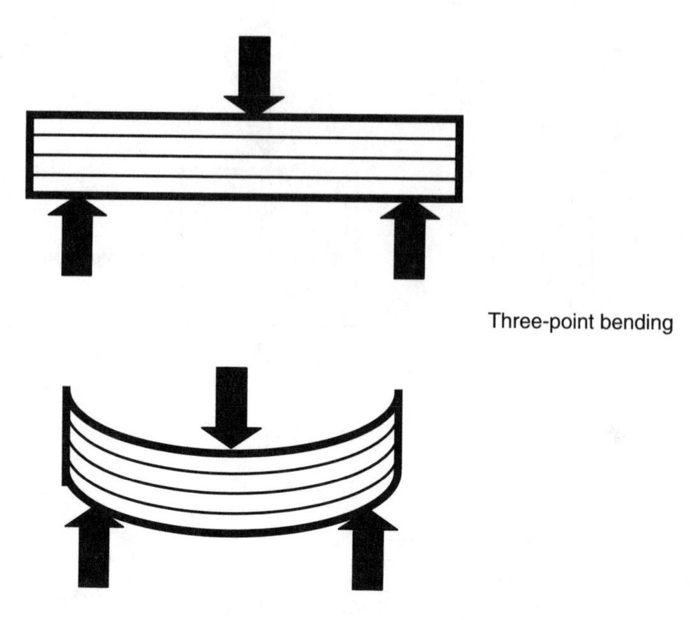

Figure 10.4 A graphic representation of three-point bending and its effects on a beam. Strain is measured in linear displacement and can be reported as a fraction of the width dimension.

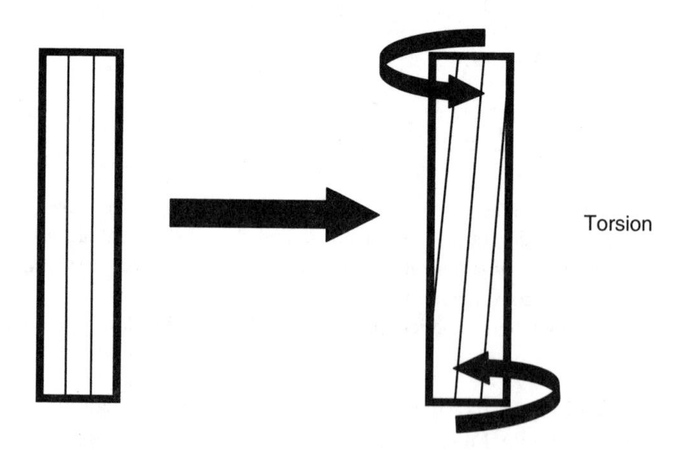

Figure 10.5 Torsion as experienced by a column. Strain is measured in axial angular displacement.

compressed from top to bottom (superior to inferior), the vertebra will shorten in its height in some proportion to the applied load (Figure 10.6). This change in angular configuration, or ratio of the post-test height to original height is the strain (in addition to shortening, the vertebra will likely bulge laterally in response to the compressive strain; see the section *Biomechanical descriptors*). Strain is typically proportional to stress, and that precise relationship (the shape of the resulting stress–strain curve) defines the mechanical properties of a structure or tissue.

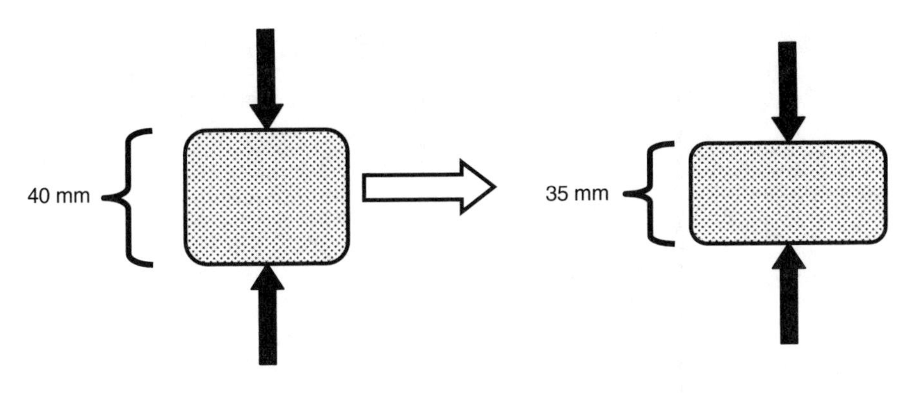

Figure 10.6 Idealized vertebral compression acts to reduce the height of the vertebral body. Strain is measured by the change in height over the original height.

Anisotropy

Bone is a highly organized tissue whose architecture varies depending on its location in the skeleton, and thus its function. The orientation of a specimen is an important issue when designing a mechanical test because the molecular- and tissue-level architectures of bone vary from location to location within the skeleton, as well as by the orientation of a single specimen. Materials that have uniform structural properties and equal mechanical properties regardless of the orientation and direction of load are known as *isotropic*: examples are steel or methylmethacrylate. More complex materials like those composing biological tissues have molecular- and tissue-level differences in structure that vary depending on the orientation of the specimen. These materials are said to be *anisotropic*. Bone is an example of an anisotropic material with different mechanical properties in different directions of loading.

Load Response: The Stress–Strain Curve

The curve from the bivariate plot of stress and strain (stress by convention is on the Y-axis) is called the *stress–strain curve*. It should be noted here that the stress–strain curve is similar but not synonymous with the load–displacement curve, and so special attention must be paid to which curve is being referenced (see below). The relationship of stress with strain is crucial in biomechanics and relates how a tissue or object responds to a given load, illustrating in two dimensions the mechanical properties of the tissue or structure. The stress–strain curve can be broken down into different regions and points along the curve, which are then used to identify the mechanical properties of the test subject (Figure 10.7).

The curve typically has an initial linear portion, known as the *elastic region*, and this is followed by a nonlinear or *plastic region*. Over the elastic region of the curve, the structure or tissue is temporarily deforming and storing the energy elastically with no permanent damage. Ideally, if the stress is returned to zero, the strain also returns to zero, with no permanent damage to the anatomic/molecular structure of the tissue. Under plastic deformation, and thus past the elastic region, the material is permanently deformed (damaged) as a result of loading. The point where the curve

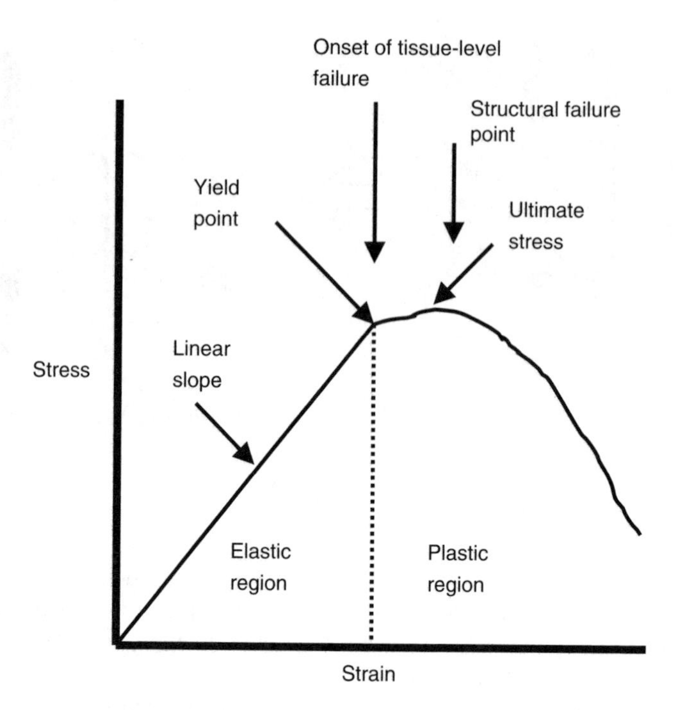

Figure 10.7 An idealized stress–strain curve showing an elastic region, plastic region, failure point, and ultimate stress. Note that the ultimate stress need not correspond exactly with the point of failure.

changes from elastic to plastic is known as the *yield* point. Yield involves plastic flow within the material so during and past the point of yield, the tissue or structure is said to yield over the range of plastic deformation. At the onset of yield, the specimen experiences a *tissue-level failure*, and at the termination of the range of plastic deformation the sample experiences a *structural-level failure*. Typically, the structural-level failure of the specimen is reported as the *failure point*. A bone fracture is a structural-level failure while a stress fracture is a tissue-level failure. The peak level of stress achieved on the stress–strain curve over the entire duration of the test is known as the *ultimate stress*. The ultimate stress is typically found close to the transition zone from elastic to plastic loading and thus near the yield point. A good analogy of elastic vs. plastic loading is an archery bow being pulled back to shoot an arrow. If the bow is pulled back short of its breaking point, then the arrow is shot using almost the entire muscle energy that was stored elastically in the bow (some miniscule fraction of energy is always lost to heat). If the arrow is released after the bow has been pulled back beyond its failure point (when the bow begins to crack), then some portion of the applied muscle energy is lost under plastic deformation of the bow, and that portion of energy is not stored elastically and not imparted to the arrow upon release. When a material is said to be elastic, the reference does not imply a rubber band-like behavior. Rather, it implies that the material stores energy linearly and produces a straight-line slope on the stress–strain curve.

An alternative bivariate plot to the stress–strain curve is a simple load vs. displacement curve. Although they closely parallel one another, there are important differences in the mechanical properties described by the load–displacement curve vs. the stress–strain curve. The first is stiffness. *Stiffness* is the slope of the linear portion of the load–displacement curve, whereas the elastic modulus is described by the slope of the linear portion of the stress–strain curve. Stiffness is related to elastic modulus by the following equation:

$$\text{Stiffness } (k) = \text{Elastic Modulus } (E) * \text{Area } (A)/\text{initial Length } (l_0).$$

The failure load corresponds to the ultimate strength at failure and they are related as follows:

$$\text{Failure load } (U) = \text{Stress } (\sigma) * \text{Area } (A).$$

The difference lies in the fact that the stress–strain curve describes the material behavior of the tissue comprising the specimen, whereas the load–displacement curve describes the structural properties of the specimen. The structural properties reflect the geometry and material properties of a tissue. Thus, the mechanical properties of any particular tissue specimen do not entirely reflect the mechanical properties of the donor structure or tissue, but those of that specific sample and its particular geometry. However, the material properties obtained from a stress–strain curve reflect the whole tissue, regardless of location and geometry (disregarding microarchitectural differences).

Biomechanical Descriptors

The slope of the linear portion of the curve (the stress over the strain at any point along the curve) describes the *modulus* of the specimen. Modulus is calculated by dividing the stress by the strain, from any point along the linear portion of the curve. It may seem counterintuitive, but every point on the linear elastic region of any particular stress–strain curve will generate an identical stiffness value. A structure or tissue with an initially linear stress–strain curve is said to be *linearly elastic*. Linear elasticity implies the ability to reversibly store energy. The *modulus* of a tissue is calculated as the slope of the elastic region of the stress–strain curve, normalized for geometry like cross-sectional area, and can refer to stiffness in tension, compression, torsion, etc. Modulus then, is typically a tissue-level descriptor while stiffness is more of a structural-level one. Probably the most commonly used descriptor in biomechanics is *Young's modulus* (E), which describes the slope of the stress–strain curve for any structure or specimen, and typically in tension. *Shear modulus* (G) describes the behavior of material in torsion or linear shear, and *Bulk modulus* (K) describes mechanical behavior under compressive pressure. The three moduli are related to one another via the equation:

$$E = 3G/(1 + G/3K)$$

Poisson's ratio (ν) describes the strain of a specimen induced in a secondary dimension (typically negative for tension and positive for compression), relative to the strain induced in the primary dimension of the load application. Poisson's ratio varies between zero for an ideal rigid solid (diamond comes close) and 0.5 for an ideal liquid (water comes close). As an example, if one applies tension to a ligament, the ligament will strain in the direction of tension and becomes longer, but it will also simultaneously strain in the perpendicular dimension, or across its diameter, becoming thinner with increasing tensile strain. Specimens often exhibit complex behaviors under stress and it is these behaviors that are used to characterize the relative mechanical traits of a specimen.

When a tissue or structure is mechanically loaded to a given stress level, it will strain until it reaches a tangential equilibrium state, dependent on its material and structural properties, and the specifications of the applied load. If the stress is maintained constantly over time and the material or structure continues to strain, it is said to exhibit creep. *Creep* is the continuation of strain in a specimen over time under constant load application and following the initial strain response equilibrium (Figure 10.8a). Creep is a common phenomenon in bone and especially trabecular bone, where the smaller trabecular elements can individually fail over time.[11] Conversely, a constant strain applied to a specimen tends to allow an event somewhat related to creep, but resulting in a diminution of stress over time.

If a predetermined strain level is applied to a biological specimen and then held constant, a phenomenon known as *stress relaxation* can occur, in which the resulting stress level diminishes over time, despite a statically maintained level of strain (Figure 10.8b). Stress relaxation is common in soft connective tissues such as tendons and ligaments, but also occurs in bone and cartilage. Stress relaxation results from a relaxation of the constituent materials within a specimen, as well as the flow of the liquid phase within the material.

Biological tissues are also known to exhibit *viscoelasticity*, which is a tendency to mechanically behave like a hybrid material, somewhat between a solid and a liquid. This phenomenon derives from the fact that the tissues constituting a specimen contain fluid, and thus their response to mechanical load is, like that of a fluid, rate-dependent (Figure 10.9). Thus, the mechanical properties of viscoelastic materials are sensitive to the rate of load application, exhibiting greater stiffness values with higher load application rates. As an example, if one takes their hand and slowly dips it into a tub of water, there is no perceptible resistance to the hand. However, if one slaps their hand into the same tub of water, there is a great deal of resistance encountered, the only difference being the rates of loading. The mechanical properties of water are rate-dependent, so, if not given enough time, the water cannot move out of the way of the incoming hand, resisting the applied load. This same phenomenon occurs in biological structures. A fresh bone specimen that is loaded at a low rate of bending will eventually crack in what is known as a green stick fracture, splintering slowly like a fresh green stick. If the same bone is loaded by an extremely high velocity impact (extremely high loading rate), then the bone shatters more like a brittle solid. The rate of load application then, is an important factor to bear in mind when designing and interpreting the results of a mechanical test.

Another important concept to consider involves tissue fatigue under cyclic loading. Thus far, all discussions in this chapter have considered only static loading. But, *in vivo*, bone usually undergoes dynamic or cyclic loading. *Cyclic loading* is a repeated loading cycle with an initial load, a reduction in that applied load, and a

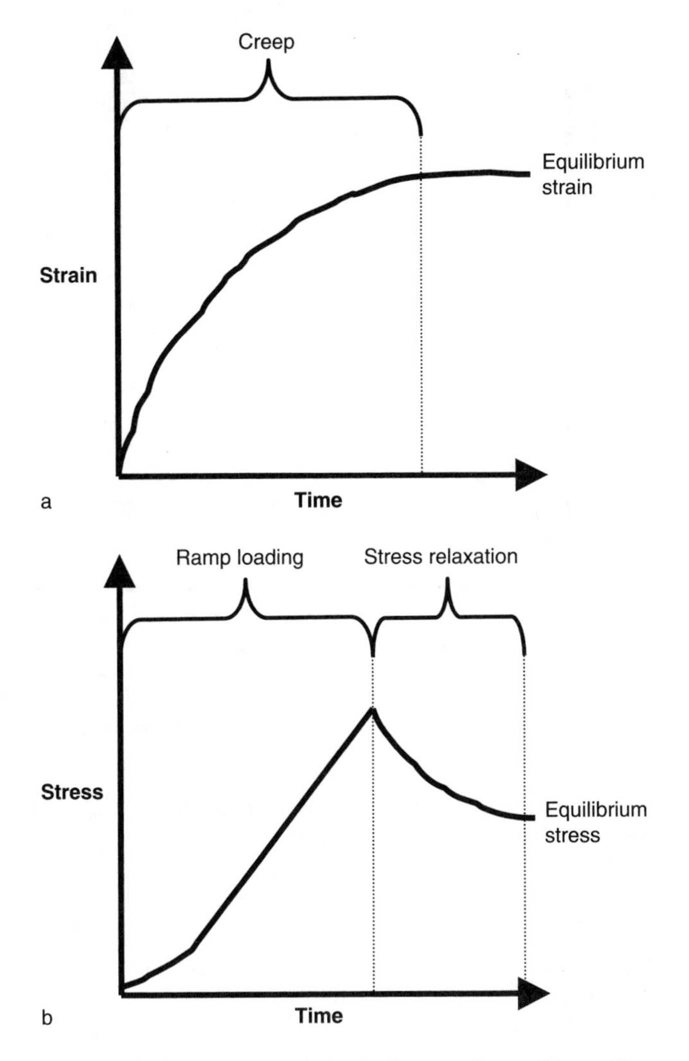

Figure 10.8 (a) An illustration of creep behavior in a specimen of bone. The specimen has already reached equilibrium under a static stress; but the net strain continues to increase over time. A final equilibrium strain is reached at the termination of the creep behavior. (b) An illustration of stress relaxation immediately following ramp loading. As the strain level is increased and then held static, the stress level declines as the material relaxes.

repeat of the initial load, usually for multiple cycles. It can cause failure in a specimen even though the level of applied stress does not exceed the specimen's ultimate strength. This is demonstrated in the stress–strain curve as *hysteresis*, and represents a loss in the applied energy during each cycle of cyclic loading. Although the specimen may appear to go back to prestress conditions macroscopically, there may be microscopic damage that accounts for the loss in energy. A portion of the energy loss due to hysteresis is dissipated as heat, as in our archery bow example, but some

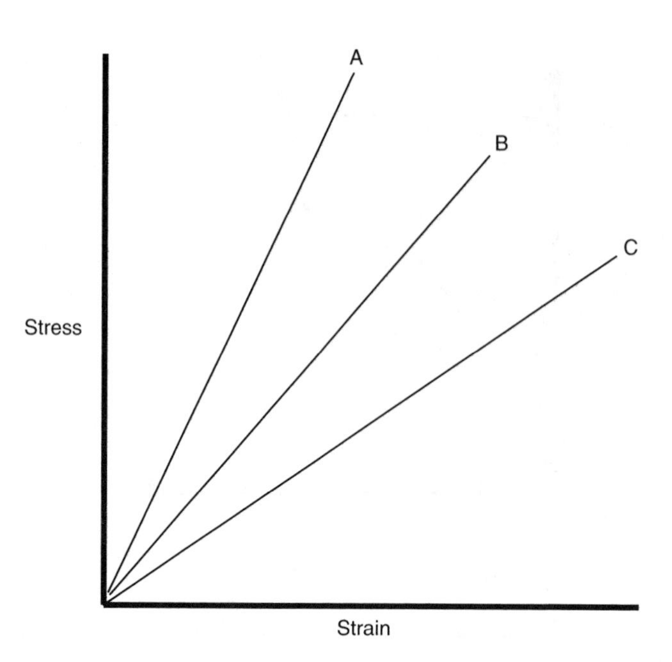

Figure 10.9 The mechanical properties of a rate-dependent biological material vary with the rate of load application. Higher rates of load application (A) cause a stiffer response to an identical load in an intermediate rate (B) and a slow rate of load application (C).

portion of the energy loss is microscopic material breakdown. The accumulation of microscopic damage leads to the failure of the bone in that area, and results in crack initiation.[12–14] Continued fatigue loading can cause the crack to propagate. Eventually, the crack reaches a critical length at which the bone cannot sustain the load and thereby the bone experiences a full-scale fracture.

Finally, the structural configuration of a bone is extremely important in the calculation of mechanical properties. In general, larger diameter bones are stronger and stiffer than smaller diameter bones. They have an increased resistance to bending via an increase in their second moment of area, or moment of inertia, as well as their polar moment. The *second moment of area* represents the quantification of the placement of material in any given cross section with respect to a defined neutral axis of loading, available to resist the load. The second moment will increase as the elements of effective cross-sectional area are placed at a greater distance from the neutral axis of loading. The *polar moment* is related to the amount of material represented in cross section (mean radius) from the neutral axis of torsion, and that is available to resist torsion. Polar moment also increases as the elements of area are placed at a greater distance from the neutral polar axis. Both these geometric characterizations disregard material properties. The second moment of area (I) is characterized by the equation:

$$I = \pi ba^3/4$$

a = major radius of ellipse representing cross section
b = minor radius of ellipse representing cross section

The polar moment is characterized by the equation:

$$J = \mathrm{Tr}/\tau.$$

The second moment of area differs along different axes in an asymmetric beam like a femur or tibia. Thus, the diameter, cross-sectional area, and second moment of area of a bone in any particular plane or planes must be taken into consideration when comparing mechanical properties such as strength and stiffness (Figure 10.10a). The

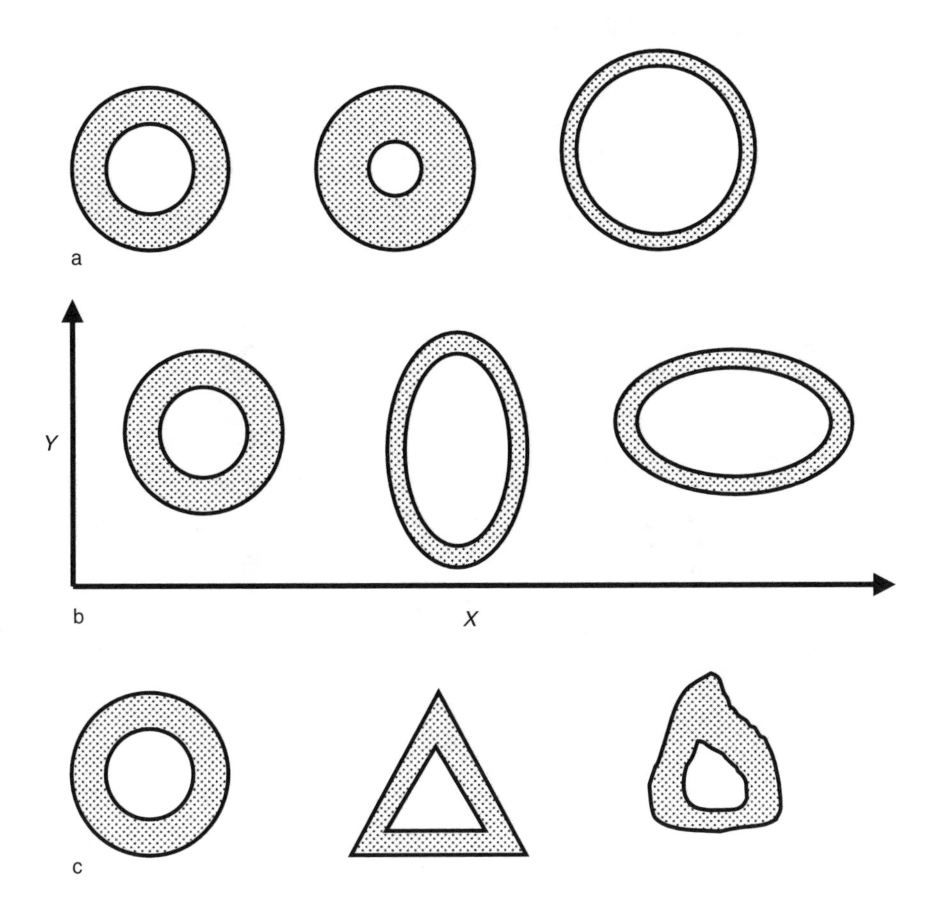

Figure 10.10 (a) A bone (left) can become stiffer and stronger by either increasing the thickness of its cortex (middle), or by increasing its diameter (right). (b) The amount of material distributed in any given axis and its distance from the neutral axis determine the mechanical properties of a structure in bending. A circular cross section (left) provides uniform mechanical properties in all bending directions, whereas an increase in either the *Y*-axis (middle) or the *X*-axis (right) increases resistance to bending in those specific directions. (c) A bone's cross-sectional geometry can reflect the predominant type of loading that the bone is subjected to *in vivo*. Bones with multipurpose loading regimens tend to have circular cross sections (left), while specialized function bones can have more complex cross-sectional shapes like a more triangular form (center), as is found in the proximal tibia (right).

actual cross-sectional configuration of a bone will confer different mechanical properties in different loading directions. A bone that is ovoid in its cross-sectional geometry, and with an evenly distributed cortex thickness, will be more capable of resisting bending when loaded in its greatest cross-sectional diameter versus the secondary diameter (Figure 10.10b). As a result, a bone with complex cross-sectional geometry can have different structural mechanical properties when loaded in different directions, in a triangular shape of the tibia (Figure 10.10c). The cross-sectional geometry of a bone is directly related to the function of a bone *in vivo*. More circular cross sections are found in general function bones (femur) that are loaded in more degrees of freedom, whereas noncircular cross sections tend to be found in more specially loaded bones with reduced degrees of freedom (tibia). Likewise, identical-sized trabecular bone samples from the same bone can have widely disparate mechanical properties in compression, for example, because the orientation of the trabecular

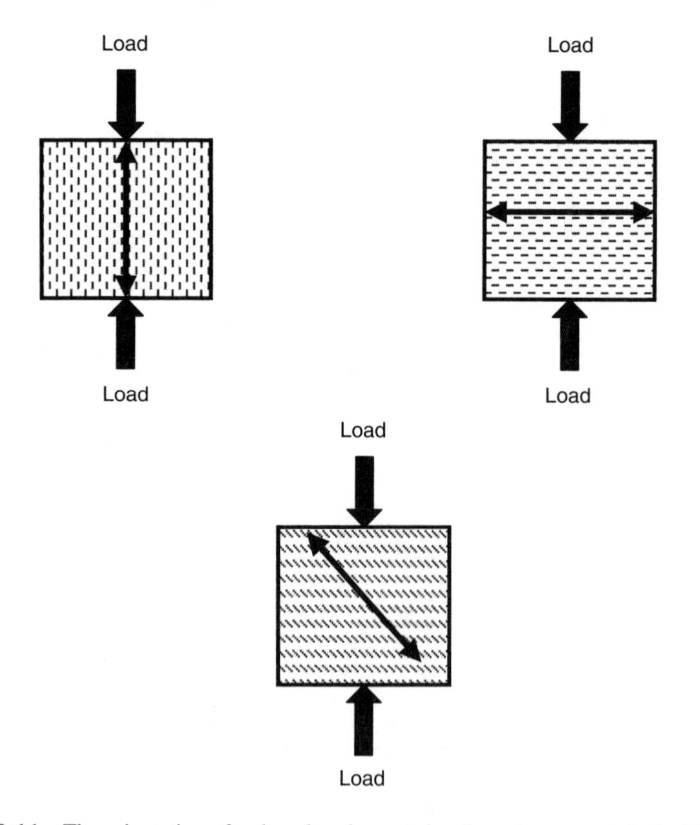

Figure 10.11 The orientation of trabecular elements in a bone has a great deal of influence on the mechanical properties of that bone, and specifically how those properties vary with loading direction. For example, trabeculae whose long axes are directly aligned with a compressive load axis (top left) will generate greatest resistance to compression, while those perpendicular to the load axis (top right) afford the least, and those at 45 (bottom) are intermediate in compression resistance.

elements may differ relative to the loading direction (Figure 10.11). These factors contribute to an intermediate structural-tissue level form of anisotropy in bone.

MECHANICAL PROPERTIES OF BONE

OVERVIEW

Bone can be divided into two basic morphological types, cortical and trabecular, each with its unique morphology, mechanical properties, and function. These properties can vary with any particular specimen or between individuals,[4–9] but also with the particular test method (Table 10.1). Cortical or compact bone comprises the outer shell of bones; it is relatively dense and appears solid in histological sections. Trabecular or cancellous bone is a web-like matrix of individual trabecular rod- or plate-like elements connected to one another in three dimensions. It is concentrated at the epiphyses and metaphyses of long bones and is present in all round bones and irregular-shaped bones such as vertebrae. To give an idea of the relative stiffness of these materials, the compressive moduli of cortical and trabecular bone are approximately 17 GPa and 100–200 MPa, respectively, while that of stainless steel is around 200 GPa. The tensile strength of cortical bone is estimated to be close to 130 MPa, and that of trabecular bone closer to 50 MPa. Thus, the mechanical properties of bone can vary by test, but this is also true for load direction. As an example, the tensile strength of cortical bone in longitudinal loading is approximately 130 MPa, while its longitudinal compressive strength is closer to 200 MPa. In transverse loading, the tensile strength of trabecular bone is significantly lower, approximately 50 MPa. Bone, unlike a uniform material like stainless steel, is extremely anisotropic and thus its material properties vary widely with the direction and orientation of loading.[15–19] Experimental indentation moduli for cortical bone have been shown to vary significantly with loading direction.[15,20] This anisometry in cortical bone has been linked to its constituent layers or lamellae, where each layer has been shown to demonstrate its own unique elastic properties via changes in preferred collagen fibril orientation.[16] It is these multiple layers, piled upon one another, that serve to give cortical bone its generalized material properties.

BONE AS A MINERAL-ORGANIC COMPOSITE

Although cortical and trabecular bone demonstrate similar mechanical properties,[21] cortical bone is a relatively rigid material with a somewhat steeper stress–strain slope, and which yields over a shorter portion of its curve. The mechanical properties of bone are dependent on the specific constituent materials comprising the bone matrix. For example, compressive stress resistance is predominantly a mineral phase property[22,23] while tensile load resistance is predominantly a collagen-related property.[24–26] Depending on the property one is interested in, it may be either the organic or mineral portion of the bone that most influences the test outcome. Moreover, although bone can be broken down to its constituent materials, and while its collective material properties reflect those constituent materials, its conglomerate material properties are unique and not simply a sum of its parts.

Table 10.1
Mechanical Properties of Bone

Cortical Bone

Test Method	Strength	Stiffness	Young's Modulus	Anatomical Location	Author
Three-point bending			5.44 GPa	Proximal tibia	Choi K et al. (1990)
Four-point bending			6.62 ± 1.30 GPa	Tibial diaphyses, fatigue testing	Choi and Goldstein (1992)
		2800 N/mm		Whole femur	Cristofolinin et al. (1996)
Nanoindentation			20.02 ± 0.27 GPa	Femoral midshaft	Turner CH et al. (1999)
			22.4 ± 1.2 GPa	Tibial osteons (longitudinal)	Rho JY et al. (1999)
Tension			162 MPa	Tibia	Vincentelli and Grigorov (1985)
			133 MPa	Femur	Reilly and Burstein (1975)
Microtensile			18.6 ± 3.5 GPa	Middiaphyseal tibia	Rho JY et al. (1993)
Torsion		6.5 Nm/ degree		Whole femur	Cristofolinin et al. (1996)
Pure shear	51.6 MPa				Turner CH et al. (2001)

Trabecular Bone

Test Method	Strength	Stiffness	Young's Modulus	Comments	Author
Cantilevered bending			7.8 ± 5.4 GPa	Tibia	Mente and Lewis (1989)
Three-point bending			4.59 GPa	Proximal tibia	Choi K et al. (1990)
			3.17 ± 1.5 GPa	Tibia	Ku et al. (1987)
			3.81 GPa	Iliac crest	Kuhn et al. (1989)
Four-point bending			5.35 ± 1.36 GPa	Tibia	Choi et al. (1991)
			5.72 ± 1.27 GPa	Proximal tibia	Choi and Goldstein (1992)
Indentation	10.3 MPa		99 MPa	Glenoid trabecular bone	Anglin C et al. (1999)
Nanoindentation			18.14 ± 1.7 GPa	Distal femoral condyle	Turner CH et al. (1999)
			19.4 ± 2.3 GPa	Vertebral	Rho JY et al. (1999)
Unconfined compression	2.35 MPa		247 MPa	Superior femoral head	Brown SJ et al. (2002)

Table 10.1 Continued

Test Method	Strength	Stiffness	Young's Modulus	Comments	Author
	0.56 MPa		51 MPa	Inferior femoral head	Brown SJ et al. (2002)
		58.9 ± 0.6 Nmm	1.41-1.89 GPa	Single trabeculae from human femur head	Bini F et al. (2002)
			126 ± 96.9 MPa	1st and 2nd Lumbar vertebrae	Uchiyana T et al. (1999)
			6.0-81.3 MPa	Calcaneus	Homminga J et al. (1997)
			10.0 ± 2.3 MPa	Femoral had	Homminga J et al. (2002)
	2.22 ± 1.42 MPa		485 MPa	Proximal tibia	Rohl L et al. (1991)
Tension		74.2 ± 0.7 Nmm		Single trabeculae from human femur head	Bini F et al. (2002)
	2.54 ± 1.18 MPa		483 MPa	Proximal tibia	Rohl L et al. (1991)
Microtensile			10.4 ± 3.5 MPa	Proximal tibia	Rho JY et al. (1993)

There exist a great variety of mechanical tests, and the mechanical properties of bone generated by those tests will also vary. One can see, though, that the mechanical properties of cortical and trabecular bone range within certain parameters. Cortical bone is generally at an order of magnitude stiffer and stronger than trabecular bone.

Trabecular or cancellous bone is an extremely anisotropic material and is less stiff than cortical bone, having a stress–strain slope that is less steep, and that yields over a greater portion of its stress–strain curve. This yielding creep behavior is related to the failure of individual trabecular elements over the loading cycle.[27] Trabecular creep initiates after the initial strain is incurred on a loaded specimen and after a transient or tangential equilibrium is established. Cortical bone also exhibits creep behavior, but to a lesser degree than trabecular bone. Trabecular bone material properties vary widely with the principal orientation of the trabecular elements. This variability is perhaps most evident in the vertebral centra of the lumbar spine where the primary orientation of the major trabecular elements is along the superior–inferior axis, which correlates to the primary loading direction in compression.[28,29]

The Mechanobiology of Bone

Bone is a metabolically active tissue capable of adapting to local mechanical conditions, a phenomenon described by Wolff's Law.[30] Simply put, its mass, density, and architecture can change in accordance with prevailing, local mechanical conditions. The skeleton can add or remove bone tissue where appropriate to alter its functional architecture, typically to reduce stress or resulting strain. It has been shown to respond to increased load application by upregulating bone formation *in vivo*[31,32] and *in vitro.*[33] Conversely, in reduced loading scenarios such as in microgravity, bone tissue is removed from the skeleton.[3,34–36]

However, the responsiveness of bone to its local mechanical environment does not stop at the modification of existing tissues. Bone not only responds to local mechanical conditions by increasing and decreasing net formation, but the mechanical signals experienced by osteoprogenitor cells can influence the type of tissue that will form during development and healing,[2,37] as well as the configuration of those tissues.[1] Thus, depending on the local mechanical conditions during fracture healing, for example, the tissues forming within the callus can become either bone, cartilage, or fibrotic.

Noninvasive Estimation of Bone Mechanical Properties

In the clinical setting, noninvasive evaluations of bone mechanical properties are essential for patient evaluation. This is true for assessing healing fractures, osteoporosis, and other skeletal-related pathologies. Noninvasive techniques include quantitative computed tomography (QCT), magnetic resonance imaging (MRI), dual-energy X-ray absorptiometry (DEXA), ultrasound, and plain radiographs. There are two primary groupings for methods by which noninvasive imaging techniques quantify skeletal integrity. The first are densitometric analyses that are performed by QCT,[38] DEXA,[39] ultrasonography,[40] and to some degree, plain radiography. These methods quantify the relative amount of mineral within a bone structure, predicting the material properties of the bone based on known mechanical correlations to the relative mineral content. The second method is reconstructive, predicting bone strength based on the architecture of the bone, and often coupling that information with density estimates. These include MRI, QCT, finite element models (FEM), and plain radiography.[29] These are often used in conjunction with one another, where a QCT series of scans[41] or a DEXA scan[42] may provide data for the construction of an FEM. These models that use a combination of density and architecture may prove to be the most accurate for predicting events like vertebral fractures in osteoporotic women. As the resolution and speed of these scans increase, they will prove more valuable as noninvasive assessors of mechanical integrity.

TESTING CONSIDERATIONS

Overview

What kind of information is needed in order to design a mechanically relevant biomechanical test? First, the individual investigators need to ask themselves two questions: What information do I want from this test? and, How do I interpret the data once I obtain it? Based on the answers to these questions, the investigator will then have to choose the specific test parameters. These include: type of bone (cortical or trabecular), sample location (proximal femur), sample orientation (perpendicular to the diaphyseal axis), sample size (numbers of specimens), sample geometry (block), type of test (indentation), test units (strain/time), test rate (mm/sec), and test type (ramp loading).

What does each test tell you? Test results are meaningless, unless they can be compared between at least two groups, a control and a treatment, or between different treatments. The control group involves identical test parameters but typically utilizes a normal or untreated set of specimens. Depending on the experimental treatment, the controls can be contralateral bones from the same animals, or if a potentially systemic treatment is utilized in the experimental group, controls must be untreated or sham-treated animals of the same species, strain, and mean body size.

How the specimen is fixed in position for a mechanical test is crucial to the outcome of the test. There are two main objectives in fixing a specimen in position: (1) controlling the specimen to avoid unwanted movement and (2) assuring that failure will be within the specimen itself and not due to the fixation device. If a specimen is allowed to slip during the test or if the specimen fails at its junction with the fixation device, then the results of the test must be called into question. This is because stress concentrations resulting from the fixation likely initiated the failure, rather than initiation from within the specimen itself. Because of their unique construction, torsion and tensile tests generally require more elaborate fixation than bending or compression tests. Embedding of whole bones is typically conducted using a low-melting temperature metal alloy or a methylmethacrylate polymer. Both these methods work well to fix specimen position but they each have drawbacks, such as relatively high solidifying temperatures and time to embed, respectively. Other methods of fixation can include multiple locking screws, clamps, or clamps under deep refrigeration, depending on specimen geometry and material properties.

Normalizing for Body Mass

Normalization for body size is an important consideration in biomechanical testing because larger animals tend to have larger bones. If a group of control animals, for instance, randomly have a significantly greater mean body mass than the experimental group, then it is likely their bones are larger. This complicates the results of the test because larger bones will have structural mechanical properties different from smaller ones, and usually higher mechanical values such as stiffness and strength. Thus, normalizations for body mass or direct skeletal measures such as cortex diameter or cross-sectional area must be performed, usually by dividing the outcome values from mechanical tests by normalizing metrics such as mass, diameter, or cross-sectional area. Normalization is crucial because pre-existing mechanical property differences can diminish the subtle differences resulting from experimental treatments, creating false differences, or alternatively, weakening statistically significant differences to nonsignificant levels.

BONE MECHANICS AND FRACTURE RISK

Overview

The relationship between bone mechanics and fracture risk depends on several factors. The organic matrix of bone plays an important role in determining its tensile stiffness,[24–26] whereas the mineral phase largely dictates compressive strength-related properties.[22,23] Further, the spatial distribution of bone tissue influences the structural properties of a whole bone such as bending stiffness and strength. Combining these factors can be useful in forming an accurate picture of fracture potential, as well as in monitoring fracture repair progress. The complimentary processes of bone formation and removal act in tandem to create, maintain, remove, and remodel bone. Any perturbation of this relationship can potentially lead to an increased risk of fracture.

Bone formation or remodeling can occur in response to conditions such as age,[43,44] local mechanical conditions,[31,46,47] metabolic disorder,[48] and even as compensation for tumors.[47] This adaptive response suggests that pathological reduction in mechanical properties due to osteolytic bone defects can be countermeasured by adaptive remodeling of the cortical and/or trabecular bone. Thus, bone is capable of compensating for changes in activity level, age, and disease, and these compensations by the skeleton are an effort to reduce stress and thus minimize the risk of fracture. However, confounding factors such as reduction in cell sensitivity due to advancing age may be difficult to overcome by structural remodeling,[48] and therefore systemic effectors must be explored.

Fracture risk is highly site-specific, depending on the type of bone involved and the loading to which it is subjected. For example, the human spine is composed chiefly of trabecular bone and is predominantly under compressive loading. Therefore, it is the mechanical properties of vertebral trabecular bone that primarily dictate the fracture risk of vertebrae.[49] This is of particular relevance to metabolic bone disease, as turnover rates in trabecular bone are nearly eight times higher than those in cortical bone.[50] As an example, it has been demonstrated that bone loss due to metabolic disorders such as osteoporosis differentially affect the axial skeleton, leaving it more vulnerable to fracture over the course of the disease.[51]

Predicting Fractures

Changes in both the material and structural properties of bone are contingent on numerous factors. Geometric changes at any structural level will significantly influence the mechanical integrity of the structure. For example, increases in the radius of longbone diaphyses will increase resistance to torsion or bending loads by a factor raised to the fourth power. With advanced aging, for example, the outer diameter of a long bone cortex increases as the cortex wall itself thins. The result is an increase in the bending moment and increased resistance to buckling failure, counteracting the increasingly thinner walls of the cortices. Thus, this increase in cortex diameter is thought to be a mechanical adaptation to the thinning cortical walls,[52] and may explain why long bone fractures are relatively uncommon among osteoporotic patients.

Increases in trabecular plate thickness at the microscopic level, however, are more difficult to assess because the continuum properties of trabecular bone volumes are influenced approximately equally by alterations in bone mass and orientation.[53] Similarly, precise estimates of the effects of other changes in trabecular morphology such as plate perforations, trabecular reorientation, and changes in connectivity cannot easily be made. The statistically based empirical relationships described above may provide some insight into fracture prediction; but these relationships have not been verified for diseased bone, or bone undergoing significant adaptation. Future work should continue to address these relationships as they pertain to fracture risk prediction and interdiction.

The long-term objective of the majority of studies designed to characterize the mechanical behavior of bone is to provide the means for accurate fracture risk prediction. Our limited understanding of specific failure mechanisms associated with crack propagation severely reduces our ability to accurately predict fracture risk. By far, the majority of studies to date have tried to relate fracture risk to bone density, but this has only partially succeeded. More recently, attempts to account for architecture by including density distributions, microarchitecture, and finite element models have helped to improve estimates of fracture risk but have not yet been verified in clinical studies.

The use of analytical models such as finite element analyses have the advantage of taking both complex geometric and anisotropic tissue properties into account when attempting to predict the occurrence of fractures. These models incorporate specific geometric measures from individual patients, but are still dependent on appropriate estimates of material properties. They will continue to increase their predictive capacity as higher resolution, noninvasive imaging techniques become available. In addition, accurate failure analysis is dependent on the selection of loading conditions appropriate to the bony region being modeled, as well as appropriate failure criteria within the structure. These parameters can be accurately estimated only through careful locomotion analyses and biomechanical testing. In addition to simple skeletal loading estimates based on force plate data or mass–acceleration estimates, new biomechanical models will have to incorporate additional loading generated by the activity of cross-joint musculature, as currently estimated by electromyography.[54] New noninvasive fracture assessment methods are currently being investigated and refined for future use.

MECHANOBIOLOGY IN TISSUE ENGINEERING

OVERVIEW

The local mechanical environment is a crucial factor in determining cell and tissue differentiation during skeletal development and repair. Beyond the basic architectural response of bone to mechanical load, as described in Wolff's Law, mechanobiology examines the relationship of the local mechanical environment with molecular expression, cell and tissue differentiation, tissue molecular architecture, gross tissue architecture, and tissue mechanical properties.

MECHANOBIOLOGIC MODELS

Early quantitative mechanobiological models were developed *post hoc* by studying tissue differentiation in cases of pseudoarthrosis and oblique fractures.[55] *In vivo* studies have demonstrated that the application of small intermittent dynamic loads to the fracture site can accelerate bony consolidation,[56–58] but a threshold also exists where excessive movement (and thus strain) retards bony healing.[59] The current mechanobiological paradigm, exemplified by Carter et al.,[2] suggests that mechanical load, acting on pluripotential mesenchymal cells recruited to a skeletal defect, can direct their differentiation into bone, cartilage, fibrocartilage, or fibrous tissue. These concepts can be applied to skeletal regeneration in cases of fracture, distraction osteogenesis, joint tissue regeneration, and potentially metabolic diseases like osteoporosis.

Fracture healing involves multiple stages: initial inflammatory, hematoma, cartilage, calcified cartilage, woven bone, and remodeled bone. During the early stages, pluripotential mesenchymal cells, erythrocytes, and platelets migrate to the site of injury. The pluripotential mesenchymal cells have the ability to differentiate into mature mesenchymal tissue-forming cells such as chondroblasts, osteoblasts, or fibroblasts. The cells that arise and the tissue they form are contingent on several factors, including biochemistry, molecular signals, and, perhaps most importantly, the local mechanical environment within the fracture site. A finite element model developed by Carter et al.[2] described the resulting tissue differentiation when a particular mechanical environment was introduced during the first few stages of fracture healing. They found that a relatively high compressive hydrostatic stress history leads to the formation of a cartilaginous matrix, whereas, relatively high tensile strain history leads to the formation of fibrous tissue constituents. A combination of high tensile strain and high compressive stress leads to the formation of fibrocartilage, while bone developed under low to moderate compressive stress and tensile strain conditions.

Carter et al.[2] also created a mechanobiologically based finite element model that accurately predicted tissue differentiation during distraction osteogenesis. In distraction osteogenesis models, an osteotomy (a segmental defect through bone) is created and rigidly fixed using an external fixation device. The defect is initially reduced, and after a latency period the gap size is progressively increased by daily small tensile displacements. As the original segments are drawn apart over the course of distraction, the regenerating tissue in the gap forms new bone. After the defect is distracted to a sufficient length, the bone is rigidly fixed and the newly formed bone undergoes ossification. Finally, when the defect is mechanically sound, the external fixator is removed. The same mechanobiological model that governs tissue formation in fracture healing also appeared to accurately predict tissue differentiation during distraction osteogenesis.

Finally, several studies have looked at the use of mechanobiology for joint tissue repair.[1,60] Cullinane et al.[1] created a defect in a rat femur and mechanically stimulated the healing defect using cyclical bending for 15 min a day, 5 days a week, for a total of 6 weeks. It was found that the mechanical environment transduced into the callus tissues via the bending action generated joint-like structures and tissues in the defect gap. More specifically, there appeared to be articular-like cartilage capping the femur on both sides of the defect. Subchondral bone formed below the cartilage, and fibrous tissue formed along the perimeter of the defect. This pattern of tissue differentiation was also predicted using finite elements similar to the model described by Carter et al.[2]

It has been repeatedly demonstrated that the local mechanical environment can actively direct gene expression, tissue differentiation, and tissue architecture based on controlled stimulation during development or healing. Specific molecular expression,[61] tissue type,[2] and tissue architecture[1] are all influenced by the local mechanical environment. These findings support the importance of mechanical intervention in tissue development, repair, and maintenance. The results of these studies further emphasize the important role that the local mechanical environment plays in the everyday development and repair of the skeleton. For the principles of mechanobiology to be optimized for use in tissue engineering, further studies need to be conducted to determine the precise relationships between the physical environment and gene expression, tissue development, and tissue repair.

TISSUE ENGINEERING MODELS AND BIOMECHANICS

OVERVIEW

Tissue engineering within the context of bone biomechanics relates to the study of tissue development using biomechanical analyses, the study of diseased tissue using biomechanical assays and, as noted above, the use of mechanical intervention to physically engineer tissues. Thus, biomechanics can be utilized as an assay mechanism for evaluating models and therapies, and also as an intervention therapy on its own.

Pathologies that cause significant disability in a patient, including osteoporosis and osteogenesis imperfecta, for example, are difficult to evaluate in humans. As a result, several animal models have been engineered to study not only the disease, but the treatments as well. These models can be as simple as a gene linked to bone density,[62–66] to bone strength,[62,65,66] or to animal activity level, which in turn influences skeletal characters.[67] More specific genetic models identify the influence of constituent portions of the bone matrix such as collagen,[68] which add or detract from overall mechanical properties, simulating diseases like osteogenesis imperfecta. Other genetic models relate the bone quality at specific anatomical sites to genetic factors such as thicker femoral and vertebral cortices, or greater cancellous bone volume in one strain, and fewer trabeculae in the vertebral bodies, femoral neck, and greater trochanter of another.[62,69] Animal models such as these utilize the variability within the host genome in order to create a breed with the desired mechanical traits.

SUMMARY

Biomechanics is the interface between biology and engineering, and is a powerful tool for assessing skeletal pathologies and the modalities used to treat them. Although direct mechanical testing is not always a viable clinical option, mechanical evaluations are useful in direct specimen testing during clinically oriented research, and are the bases of noninvasive clinical analyses. Noninvasive clinical evaluations can take the form of assessing fracture risk, nutritional condition, trauma diagnoses, or general disease state. The more invasive techniques found in biomechanics are useful for directly quantifying, evaluating, and comparing the efficacy of treatment modalities as applied to common skeletal pathologies. These treatment modalities include direct fracture fixation, locally applied growth factors, and newer treatments such as tissue engineering using viral vectors delivering custom growth factor genes to host cells. Also, as the molecular events responsible for the processes of skeletal formation and healing are unraveled, the use of tissue engineering models will become more relevant and more effective. The evaluation of the efficacy of these future models will need to be conducted using the principles of biomechanics.

None of these current or future orthopedic treatment modalities can be evaluated without the use of biomechanical analyses, either utilizing direct testing methods, or image-generated analyses. Biomechanical analyses serve to evaluate, estimate, or simulate the effects of the mechanical environment on the tissues and structures within the body, and it is this process of biomechanical analysis that determines the effect of a disease state, or the effectiveness of a treatment. The primary purpose of skeletal tissues is to provide a mechanical framework upon which the body and environment can interact, and the ultimate goal of a biomechanist is to characterize and quantify that relationship.

References

1. Cullinane, D.M., Fredrick, A., Eisenberg, S.R., Pacicca, D., Elman, M.V., Lee, C., Salisbury, K., Gerstenfeld, L.C. and Einhorn, T.A. (2002). Induction of a neoarthrosis by precisely controlled motion in an experimental mid-femoral defect. *J. Orthop. Res.* 20(3), 579–586.
2. Carter, D.R., Beaupre, G.S., Giori, N.J. and Helms, J.A. (1998). Mechanobiology of skeletal regeneration. *Clin. Orthop.* (355 Suppl), S41–S55.
3. Bikle, D.D., Halloran, B.P. and Morey-Holton. E. (1994). Impact of skeletal unloading on bone formation: role of systemic and local factors. *Acta Astronaut.* 33: 119–129.
4. Eckstein, F., Lochmuller, E.M., Lill, C.A., Kuhn, V., Schneider, E., Delling, G. and Muller, R. (2002). Bone strength at clinically relevant sites displays substantial heterogeneity and is best predicted from site-specific bone densitometry. *J. Bone Miner. Res.* 17(1), 162–171.
5. Gotfredsen, A., Nilas, L., Podenphant, J., Hadberg, A. and Christiansen, C. (1989). Regional bone mineral in healthy and osteoporotic women: a cross-sectional study. *Scand. J. Clin. Lab Invest.* 49(8), 739–749.
6. Keller, T.S., Mao, Z. and Spengler, D.M. (1990). Young's modulus, bending strength, and tissue physical properties of human compact bone. *J. Orthop. Res.* 8(4), 592–603.
7. Morel, J., Combe, B., Francisco, J. and Bernard, J. (2001). Bone mineral density of 704 amateur sportsmen involved in different physical activities. *Osteoporos. Int.* 12(2), 152–157.
8. Ormerod, S., Galea, V., MacDougall, J.D. and Webber, C.E. (1990). Regional bone mineral measurements. *Can. Assoc. Radiol. J.* 41(2), 59–64.
9. Piatkowski, J., Grawe, A., Ehler, E. and Schumacher, G.H. (1985). [Regional differences in the chemical composition of the human tibia]. *Anat. Anz.* 158(4), 315–322.
10. Moore, K.L. and Dalley, A.F. (1999). *Clinically Oriented Anatomy*. 4th ed., P.J. Kelly., Ed. Lippincott Williams & Williams, Philadelphia, p.15.
11. Fyhrie, D.P. and Schaffler, M.B. (1994). Failure mechanisms in human vertebral cancellous bone. *Bone* 15(1), 105–109.
12. Taylor, D. and Prendergast, P.J. (1997). A model for fatigue crack propagation and remodelling in compact bone. *Proc. Inst. Mech. Eng. [H]* 211(5), 369–375.
13. Vashishth, D., Tanner, K.E. and Bonfield, W. (2003). Experimental validation of a microcracking-based toughening mechanism for cortical bone. *J. Biomech.* 36(1), 121–124.

14. Wenzel, T.E., Schaffler, M.B. and Fyhrie, D.P. (1996). *In vivo* trabecular microcracks in human vertebral bone. *Bone* 19(2), 89–95.
15. Fan, Z., Swadener, J.G., Rho, J.Y., Roy, M.E. and Pharr, G.M. (2002). Anisotropic properties of human tibial cortical bone as measured by nanoindentation. *J. Orthop. Res.* 20(4), 806–810.
16. Bensamoun, S., Ho Ba Tho, M.C., Fan, Z. and Rho, J.Y. (2003). Intra and Inter Variation of Elastic Properties of Human Osteon Lamellae. In 49th Annual Meeting of the Orthopaedic Research Society. New Orleans, LA.
17. Hoffmeister, B.K., Smith, S.R., Handley, S.M. and Rho, J.Y. (2000). Anisotropy of Young's modulus of human tibial cortical bone. *Med. Biol. Eng. Comput.* 38(3), 333–338.
18. Kabel, J., van Rietbergen, B., Dalstra, M., Odgaard, A. and Huiskes, R. (1999). The role of an effective isotropic tissue modulus in the elastic properties of cancellous bone. *J. Biomech.* 32(7), 673–680.
19. Liu, D., Weiner, S. and Wagner, H.D. (1999). Anisotropic mechanical properties of lamellar bone using miniature cantilever bending specimens. *J. Biomech.* 32(7), 647–654.
20. Zysset, P.K., Guo, X.E., Hoffler, C.E., Moore, K.E. and Goldstein, S.A. (1999). Elastic modulus and hardness of cortical and trabecular bone lamellae measured by nanoindentation in the human femur. *J. Biomech.* 32(10), 1005–1012.
21. Turner, C.H., Rho, J., Takano, Y., Tsui, T.Y. and Pharr, G.M. (1999). The elastic properties of trabecular and cortical bone tissues are similar: results from two microscopic measurement techniques. *J. Biomech.* 32(4), 437–441.
22. Roy, M.E., Nishimoto, S.K., Rho, J.Y., Bhattacharya, S.K., Lin, J.S. and Pharr, G.M. (2001). Correlations between osteocalcin content, degree of mineralization, and mechanical properties of C. carpio rib bone. *J. Biomed. Mater. Res.* 54(4), 547–553.
23. Wall, J.C., Chatterji, S.K. and Jeffery, J.W. (1978). The influence that bone density and the orientation and particle size of the mineral phase have on the mechanical properties of bone. *J. Bioeng.* 2(6), 517–526.
24. Banse, X., Sims, T.J. and Bailey, A.J. (2002). Mechanical properties of adult vertebral cancellous bone: correlation with collagen intermolecular cross-links. *J. Bone Miner. Res.* 17(9), 1621–1628.
25. Burr, D.B. (2002). The contribution of the organic matrix to bone's material properties. *Bone* 31(1), 8–11.
26. Bowman, S.M., Zeind, J., Gibson, L.J., Hayes, W.C. and McMahon, T.A. (1996). The tensile behavior of demineralized bovine cortical bone. *J. Biomech.* 29(11), 1497–1501.
27. Bowman, S.M., Keaveny, T.M., Gibson, L.J., Hayes, W.C. and McMahon, T.A. (1994). Compressive creep behavior of bovine trabecular bone. *J. Biomech.* 27(3), 301–310.
28. Korstjens, C.M., Mosekilde, L., Spruijt, R.J., Geraets, W.G. and van der Stelt, P.F. (1996). Relations between radiographic trabecular pattern and biomechanical characteristics of human vertebrae. *Acta Radiol.* 37(5), 618–624.
29. Millard, J., Augat, P., Link, T.M., Kothari, M., Newitt, D.C., Genant, H.K. and Majumdar, S. (1998). Power spectral analysis of vertebral trabecular bone structure from radiographs: orientation dependence and correlation with bone mineral density and mechanical properties. *Calcif. Tissue Int.* 63(6), 482–489.
30. Wolff, J. (1892). In *Das gaesetz der transformation der knochen*, A. Hirchwald, Ed. Berlin.
31. Biewener, A.A. and Bertram, J.E. (1994). Structural response of growing bone to exercise and disuse. *J. Appl. Physiol.* 76(2), 946–955.

32. Carter, D.R., Blenman, P.R. and Beaupre, G.S. (1988). Correlations between mechanical stress history and tissue differentiation in initial fracture healing. *J. Orthop. Res.* 6(5), 736–748.
33. Mikuni-Takagaki, Y. (1999). Mechanical responses and signal transduction pathways in stretched osteocytes. *J. Bone Miner. Metab.* 17(1), 57–60.
34. Dehority, W., Halloran, B.P., Bikle, D.D., Curren, T., Kostenuik, P.J., Wronski, T.J., Shen, Y., Rabkin, B., Bouraoui, A. and Morey-Holton, E. (1999). Bone and hormonal changes induced by skeletal unloading in the mature male rat. *Am. J. Physiol.* 276(1 Part 1), E62–E69.
35. LeBlanc, A., Marsh, C., Evans, H., Johnson, P., Schneider, V. and Jhingran, S. (1985). Bone and muscle atrophy with suspension of the rat. *J. Appl. Physiol.* 58(5), 1669–1675.
36. Montufar-Solis, D., Duke, P.J. and Morey-Holton, E. (2001). The Spacelab 3 simulation: basis for a model of growth plate response in microgravity in the rat. *J. Gravit. Physiol.* 8(2), 67–76.
37. Loboa, E.G., Beaupre, G.S. and Carter, D.R. (2001). Mechanobiology of initial pseudarthrosis formation with oblique fractures. *J. Orthop. Res.* 19(6), 1067–1072.
38. Ferretti, J.L., Capozza, R.F. and Zanchetta, J.R. (1996). Mechanical validation of a tomographic (pQCT) index for noninvasive estimation of rat femur bending strength. *Bone* 18(2), 97–102.
39. Beck, T.J., Mourtada, F.A., Ruff, C.B., Scott, W.W., Jr. and Kao, G. (1998). Experimental testing of a DEXA-derived curved beam model of the proximal femur. *J. Orthop. Res.* 16(3), 394–398.
40. Toyras, J., Nieminen, M.T., Kroger, H. and Jurvelin, J.S. (2002). Bone mineral density, ultrasound velocity, and broadband attenuation predict mechanical properties of trabecular bone differently. *Bone* 31(4), 503–507.
41. Fischer, K.J., Jacobs, C.R., Levenston, M.E., Cody, D.D. and Carter, D.R. (1998). Bone load estimation for the proximal femur using single energy quantitative CT data. *Comput. Methods Biomech. Biomed. Eng.* 1(3), 233–245.
42. Sarin, V.K., Loboa Polefka, E.G., Beaupre, G.S., Kiratli, B.J., Carter, D.R. and van der Meulen, M.C. (1999). DXA-derived section modulus and bone mineral content predict long-bone torsional strength. *Acta Orthop. Scand.* 70(1), 71–76.
43. Tanck, E., Blankevoort, L., Haaijman, A., Burger, E.H. and Huiskes, R. (2000). Influence of muscular activity on local mineralization patterns in metatarsals of the embryonic mouse. *J. Orthop. Res.* 18(4), 613–619.
44. Weinans, H. (1998). Is osteoporosis a matter of over-adaptation? *Technol. Health Care* 6(5–6), 299–306.
45. Fyhrie, D.P. and Carter, D.R. (1986). A unifying principle relating stress to trabecular bone morphology. *J. Orthop. Res.* 4(3), 304–317.
46. Hauser, D.L., Kara, M.E. and Snyder, B.D. (2000). Adaptive Remodeling Compensation for the Reduction in Strength Associated with Benign Bone Tumors of the Pediatric Femur. In 46th Annual Meeting, Orthopedic Research Society.
47. Lanyon, L.E. (1996). Using functional loading to influence bone mass and architecture: objectives, mechanisms, and relationship with estrogen of the mechanically adaptive process in bone. *Bone* 18(1 Suppl), 37S–43S.
48. Stanford, C.M., Welsch, F., Kastner, N., Thomas, G., Zaharias, R., Holtman, K. and Brand, R.A. (2000). Primary human bone cultures from older patients do not respond at continuum levels of in vivo strain magnitudes. *J. Biomech.* 33(1), 63–71.
49. Silva, M.J., Keaveny, T.M. and Hayes, W.C. (1997). Load sharing between the shell and centrum in the lumbar vertebral body. *Spine* 22(2), 140–150.

50. Parfitt, A.M. (1987). Bone remodeling and bone loss: understanding the pathophysiology of osteoporosis. *Clin. Obstet. Gynecol.* 30(4), 789–811.
51. Grey, A.B., Stapleton, J.P., Evans, M.C. and Reid, I.R. (1996). Accelerated bone loss in post-menopausal women with mild primary hyperparathyroidism. *Clin. Endocrinol. (Oxf).* 44(6), 697–702.
52. Melton, L.J., Chao, E.Y.S. and Lane, J.M. (1988). Biomechanical aspects of fractures, In Osteoporosis: Etiology, Diagnosis, and Management, B.L. Riggs and Melton, L.J., Ed. Raven Press, New York, p. 111–131.
53. Keaveny, T.M., Niebur, G.L., Yeh, O.C. and Morgan. E.F. (2000). Micromechanics and Trabecular Bone Strength. In 12th Conference of European Society of Biomechanics, Dublin.
54. Biewener, A.A. (2002). Future directions for the analysis of musculoskeletal design and locomotor performance. *J. Morphol.* 252(1), 38–51.
55. Pauwels, F. (1980). Biomechanics of the Locomotor Apparatus. Springer-Verlag, Berlin.
56. Chao, E.Y., Inoue, N., Elias, J.J. and Aro, H. (1998). Enhancement of fracture healing by mechanical and surgical intervention. *Clin. Orthop.* (355 Suppl), S163–S178.
57. Larsson, S., Kim, W., Caja, V.L., Egger, E.L., Inoue, N. and Chao, E.Y. (2001). Effect of early axial dynamization on tibial bone healing: a study in dogs. *Clin. Orthop.* (388), 240–251.
58. Perren, S.M. (2002). Evolution of the internal fixation of long bone fractures. The scientific basis of biological internal fixation: choosing a new balance between stability and biology. *J. Bone Joint Surg. Br.* 84(8), 1093–1110.
59. Claes, L., Eckert-Hubner, K. and Augat, P. (2002). The effect of mechanical stability on local vascularization and tissue differentiation in callus healing. *J. Orthop. Res.* 20(5), 1099–1105.
60. Beaupre, G.S., Stevens, S.S. and Carter, D.R. (2000). Mechanobiology in the development, maintenance, and degeneration of articular cartilage. *J. Rehabil. Res. Dev.* 37(2), 145–151.
61. Cullinane, D.M., Salisbury, K.T., Alkhiary, Y.M., Eisenberg, S.R., Gerstenfeld, L.C. and Einhorn, T.A. (2003). Mechanobiologically directed skeletal regeneration: induced differentiation of cartilage, and collagen type II and GDF-5 expression, and modified matrix architecture. *J. Exp. Biol.*, 206:2459–2471.
62. Akhter, M.P., Iwaniec, U.T., Covey, M.A., Cullen, D.M., Kimmel, D.B. and Recker, R.R. (2000). Genetic variations in bone density, histomorphometry, and strength in mice. *Calcif. Tissue Int.* 67(4), 337–344.
63. Klein, R.F., Shea, M., Gunness, M.E., Pelz, G.B., Belknap, J.K. and Orwoll, E.S. (2001). Phenotypic characterization of mice bred for high and low peak bone mass. *J. Bone Miner. Res.* 16(1), 63–71.
64. Li, X., Mohan, S., Gu, W., Wergedal, J. and Baylink, D.J. (2001). Quantitative assessment of forearm muscle size, forelimb grip strength, forearm bone mineral density, and forearm bone size in determining humerus breaking strength in 10 inbred strains of mice. *Calcif. Tissue Int.* 68(6), 365–369.
65. Li, X., Masinde, G., Gu, W., Wergedal, J., Mohan, S. and Baylink, D.J. (2002). Genetic dissection of femur breaking strength in a large population (MRL/MpJ x SJL/J) of F2 Mice: single QTL effects, epistasis, and pleiotropy. *Genomics* 79(5), 734–740.
66. Turner, C.H., Hsieh, Y.F., Muller, R., Bouxsein, M.L., Rosen, C.J. McCrann, M.E., Donahue, L.R., and Beamer, W.G. (2001). Variation in bone biomechanical properties,

microstructure, and density in BXH recombinant inbred mice. *J. Bone Miner. Res.* 16(2), 206–213.

67. Kaye, M. and Kusy, R.P. (1995). Genetic lineage, bone mass, and physical activity in mice. *Bone* 17(2), 131–135.
68. Misof, K., Landis, W.J., Klaushofer, K. and Fratzl, P. (1997). Collagen from the osteogenesis imperfecta mouse model (oim) shows reduced resistance against tensile stress. *J. Clin. Invest.* 100(1), 40–45.
69. Turner, C.H., Hsieh, Y.F., Muller, R., Bouxsein, M.L., Baylink, D.J., Rosen, C.J., Grynpas, M.D., Donahue, L.R. and Beamer, W.G. (2000). Genetic regulation of cortical and trabecular bone strength and microstructure in inbred strains of mice. *J. Bone Miner. Res.* 15(6), 1126–1131.

SECTION IV

Clincal Opportunities

11 Tissue Engineering of Bone

Sanjeev Kakar and Thomas A. Einhorn

Contents

0-8493-1621-9/05/$0.00+$1.50
© 2005 by CRC Press LLC

INTRODUCTION

Autogenous bone grafting is considered to be the standard of treatment for the management of bone defects. By 2004, over 1.5 million of these procedures will be performed per year to treat various craniofacial disorders such as missing alveolar bone in cleft palates and orthopedic conditions such as fractures and nonunions.[1] Several shortcomings are experienced with this procedure, including difficulties in shaping the graft to fill the defect, the requirement for numerous procedures, lengthy recovery times, and donor site morbidity.[2–4] These limitations have prompted the use of alternate graft materials such as allogenic bone. However, despite its ready availability, the risk of disease transmission, loss of biologic and mechanical properties, and increased cost have limited its use.[5]

With these difficulties, researchers have sought alternatives to current treatment modalities. Tissue engineering represents a field of biological research where promising progress has been made. It is based on the principle of restoring function or replacing damaged or diseased tissues through the application of biological and engineering principles.[6] In terms of its applicability to bone, the aim is to create a bone healing response in a precise anatomic area so that the tissue formed is integrated structurally with the surrounding skeleton and has the biomechanical properties necessary to be durable and effective.[7]

The repair and regeneration of bone by tissue engineering occurs through an ordered sequence of cellular events that are affected by several biological and mechanical factors. In terms of biological factors, the first principle is that all newly formed tissue requires the presence of osteoprogenitor cells capable of forming bone.[7] These cells can be harvested from many sites and if their numbers are low, can be engineered to provide the necessary population of cells.

Once a suitable number of these cells have been obtained, they need to be delivered into the various skeletal defects to ensure that the bone healing response is contiguous and integrates with the surrounding tissues. They are commonly, therefore, grown onto naturally derived or synthetic scaffolds, which act as passive three-dimensional mechanical matrices supporting cell attachment, proliferation, and differentiation. The cells form their own matrix, which is integrated with the host tissue as the implant degrades over time. These properties are considered to be *osteoconductive* as they promote the bone healing response to progress throughout the defects.[7]

The cells that are used are derived from a pluripotent population and are therefore capable of differentiating along several tissue lines. It would be desirable to control cell migration, differentiation, and subsequent tissue formation with the use of growth factors and adhesion molecules contained in or on the surface of the implanted matrix or secreted by cells incorporated in the matrix. The stimuli from these growth factors and adhesion molecules are termed *osteoinductive* as they are capable of determining the osseous nature of the tissue produced at the graft site.[6, 7]

The mechanical environment has also been shown to affect bone formation. Distraction osteogenesis (DO) is the process by which application of a tensile force at an optimal rate and frequency controls new bone formation.[8] Initially popularized by Ilizarov in the 1950s for the management of leg length discrepancies, it has recently become a useful technique for the treatment of numerous traumatic, congenital and acquired conditions of the extremities and craniofacial region.[9]

Using this overall framework of the biological and mechanical principles involved in tissue engineering of bone, this chapter focuses on a detailed description of these factors from their evolution to current-day practice.

BIOLOGICAL COMPONENTS INVOLVED IN TISSUE ENGINEERING OF BONE

CELLS

Bone Marrow Cells

The bone marrow contains osteogenic precursor cells.[10–13] It has been used in the tissue engineering of bone, and in particular the repair of osseous defects. Connolly and Shindell[14] first reported its clinical use in the management of tibial nonunions. Injecting bone marrow into the defects resulted in clinical and radiographic union by six months. Others have described similar successes.[15,16] In a study of delayed unions in eight osteosarcoma patients, Heaney et al. reported bone formation and union in seven cases following percutaneous bone marrow grafting.

The ability of bone marrow cells to heal bony defects can be potentiated by adding osteoinductive agents. Lane et al.[17] demonstrated that combining recombinant human bone morphogenetic protein (BMP)-2 to bone marrow cells resulted in higher union rates with superior mechanical properties compared to treatments using autogenous bone graft or bone marrow.

Despite the success of these procedures, the number of osteoprogenitor stem cells within bone marrow is limited. One stem cell per fifty thousand nucleated cells present in young individuals have osteogenic capacity and this declines dramatically to one per two million in the elderly.[18] As the success of bone marrow grafting depends on the transfer of sufficient osteoprogenitor cells, several investigators have tried to increase the concentrations of these cells. Connolly[19] described a technique involving the injection of marrow concentrate into scaphoid nonunions. Out of the five patients treated, four experienced healing as a result of this application. Bruder et al.[20] expanded the number of osteoprogenitor cells by growing them under special culture conditions. These cells, when added to a composite mixture of ceramics and collagens, were able to bridge critical osseous femoral defects in adult rats.

Mesenchymal Stem Cells

Mesenchymal stem cells (MSCs) comprise a population of resting, undifferentiated cells that have the ability to replicate throughout life. They have a number of advantages over fully differentiated cells including ease of expansion, maintenance of their phenotype and lack of senescence.

MSCs are capable of dividing into clones or differentiating into multiple connective tissues lineages such as muscle, cartilage and bone.[20] Consequently, techniques have been developed to isolate and expand MSC numbers, while ensuring that their phenotype is maintained and there is no loss in the osteogenic or chondrogenic potential.[21] These properties make MSCs a useful source of osteoprogenitor cells for tissue engineering of bone.

Several investigators have reported on the clinical applicability of using MSCs in tissue regeneration. Bruder et al.[20] studied the use of MSCs to heal segmental bone defects in the femora of adult athymic rats. MSCs isolated from human bone marrow were grown in culture, loaded onto a ceramic carrier, and implanted into critical-sized segmental defects. Controls comprised of cell-free ceramics implanted into the contralateral limb. The femurs were harvested and analyzed by high-resolution radiography, immunohistochemistry, quantitative histomorphometry and biomechanical testing. Mesenchymal stem cell-treated defects had evidence of new bone formation by 8 weeks. Biomechanical evaluation confirmed that these femurs were significantly stronger than the controls. These findings demonstrate that human MSCs can regenerate bone in clinically significant osseous defects and may therefore provide an alternative to the use of autogenous bone grafts.

The repair of cranial bone defects is a major challenge for craniofacial surgeons owing to the limited availability of autologous bone graft. Consequently, surgeons have experimented with other materials to find a suitable alternative. Shang et al.[22] studied the use of MSCs to repair these defects. Autologous MSCs were isolated from eight adult sheep, expanded in culture, and added to a calcium alginate composite. Parietal bone defects were created in the animals and repaired by either calcium alginate/MSC composites or calcium alginate alone. New bone formation was observed within the experimental group only, with CT scans revealing almost complete repair of these defects. Importantly, this tissue-engineered bone had the same biomechanical properties as native parietal bone.

In an attempt to enhance the osteogenic potential of MSCs, investigators have studied ways of pretreating these cells.[23] Yoshikawa et al. described a technique whereby hydroxyapatite/MSC composites were cultured in media with or without dexamethasone, a known osteogenic agent, for 2 weeks. After being implanted subcutaneously in rats, the composites were harvested and analyzed for alkaline phosphatase activity and bone Gla protein. The results showed that dexamethasone-treated bone marrow cells exhibited an enhanced osteogenic response immediately after transplantation. In contrast, the untreated composite did not show any bone formation. These results indicate that the inherent osteogenic ability of marrow stromal stem cells can be stimulated using tissue culture technology.

Muscle Cells

Urist, in 1965, first noted that when demineralized bone matrix is implanted into skeletal muscle, a new ossicle of bone is formed. In subsequent reports, this phenomenon of osteoinduction was attributed to the mitogenic effects of BMP on cells in muscle and muscle planes.[24] This was further reported by Katagiri et al.,[25] who incubated myoblasts with BMP2 and demonstrated a downregulation of myogenic markers such as troponin T, but an increase in alkaline phosphatase and osteocalcin expression. This suggests that muscle cells had been transformed into osteoblast-like cells in the presence of this osteoinductive stimulus. Research by Khouri et al.[26] demonstrated that these cells are functionally active by comparing the ability of

BMP3, a muscle flap, and a combination of the two to heal a rat calvarial defect. Results showed that a muscle flap injected with BMP3 was capable of healing the defect completely. In contrast, defects treated with either the muscle flap or BMP3 alone demonstrated only 37% and 64% healing, respectively. These findings support the concept that muscle-based osteoprogenitor cells are functionally effective, and lay the foundation for future investigations regarding the use of skeletal muscle as a source of inducible osteoprogenitor cells for bone healing.

Muscle-derived stem cells have been genetically engineered to express human BMP-4, VEGF, or VEGF specific antagonist (soluble Flt1) to study the interaction between angiogenic and osteogenic factors in bone healing.[27] Using a mouse model, Peng et al. intramuscularly implanted a designated number of transduced cells into the lateral aspect of each femur. Ectopic bone formation was monitored radiographically and histologically for up to 4 weeks postoperatively. Results showed that VEGF acted synergistically with BMP-4 to increase the recruitment of MSCs, enhance cell survival and promote cartilage formation in the early stages of endochondral ossification. In contrast, Flt 1 inhibited this bone healing response elicited by BMP-4. From these studies, the authors concluded that VEGF had an important role in enhancing BMP-4 elicited bone formation and regeneration.

Embryonic Stem Cells

Recent reports have described the use of human embryonic stem cells in tissue engineering. Thomson et al.[28] demonstrated that these cells can be grown *in vitro* from human blastocysts and maintain their developmental potential to form all three germ layers. As a result, cartilage, bone and muscle may be derived from the mesoderm. This development of human embryo technology represents a new therapeutic approach that may be used in the future for regeneration of skeletal tissues. Nevertheless, major challenges exist, not least being the ethical issues, that need to be addressed.

Role of Scaffolds

Scaffolds are commonly used in bone tissue engineering, acting as a conduit for the delivery of cells, genetic material and growth factors to the site of interest.[29] In addition, they support vascular invasion, maintain uniform distribution and retention of cells throughout its three-dimensional lattice, facilitate efficient diffusion of molecules and undergo resorption and replacement by new bone as it is formed.[30]

Several materials have been examined for their use in bone tissue engineering. They can be divided into acellular and cellular systems.[31] The former is comprised of absorbable filler materials that encourage bony ingrowth without any additional cellular component. In contrast, cellular scaffolds are engineered to guide bone development as it is synthesized by cells embedded in the implant.

Acellular Systems

Natural Matrices

Demineralized bone matrix (DBM) is produced from the acid extraction of human cortical bone. Since the earlier observations of Van de Putte and Urist[32,33] on its ability to induce ectopic bone formation, interest has developed within the orthopedic community as to its role in treating bony defects. Tuli and Singh[34] demonstrated its high osteoinductive and osteoconductive properties in the healing of bony defects in a rabbit model. After 12 weeks of treatment with DBM, 13 out of the 16 animal defects had been bridged by new bone formation with no local foreign body or immunogenic reaction to the graft.

Tiedeman et al.[35] reported on the efficacy of DBM used in conjunction with bone marrow in the treatment of patients with bony disorders such as comminuted fractures with associated bone loss. After a follow-up period averaging 19 months, 30 out of 39 patients demonstrated successful bone formation. Overall, the patients grafted with DBM demonstrated healing comparable to those who were treated with autogenous bone graft.

The osteoinductive properties of DBM vary depending on its source. This was highlighted by Rabie and co-workers[36] during a study examining the healing of rabbit parietal bone defects in the presence of DBM extracted from intramembranous bone (imDBM) or endochondral bone (ecDBM). Defects were grafted with endochondral bone mixed with imDBM, endochondral bone with ecDBM, or intramembranous bone with imDBM. Controls consisted of untreated lesions or defects grafted with rabbit skin collagen. Results showed significant new bone formation within the experimental groups compared to the controls, with imDBM possessing higher osteoinductive properties.

Synthetic Polymers

Synthetic matrices have been increasingly used in bone tissue engineering. These scaffolds are bioresorbable, biocompatible, osteoconductive and can be easily molded to fit the individual defect.[37] Meinig et al.[38] demonstrated these properties in treating bone defects with poly-L-lactide (PLLA). PLLA membranes with a pore size of 5–15 μm were implanted into New Zealand White rabbits to cover 1 cm mid-diaphyseal defects of the radii. Untreated defects of a similar size on the contralateral limb served as controls. Results showed that within the experimental animals, cortical bone was seen spanning the defects with no adverse effects. In contrast, controls developed radial-ulnar synostosis with no new bone formation.

Defect size limits the ability of synthetic polymers to promote bone formation. This was highlighted by Gugala and Gogolewski[39] when they attempted to treat 4 cm tibial defects in sheep with PLLA. In the cases where only membranes were used, no osseous repair was noted. In contrast, defects treated with both cancellous bone graft and synthetic membrane demonstrated significant bony repair.

Ceramics

Ceramics comprise a group of biomaterials produced from the heating of natural mineral salts to very high temperatures.[40] Of the various types, calcium phosphate ceramics have been extensively used in the treatment of bony defects owing to their similar architectural properties and surface chemistry in comparison to bone. The most widely used include tricalcium phosphate (TCP) and hydroxyapatite (HA).[41] They display excellent biocompatibility, osteoconductivity and osseointegration.[42–44]

From a functional perspective, calcium phosphate ceramics can be divided into slow and rapid resorbing ceramics.[45] HA is a slow resorbing compound derived from marine coral.[46] A simple hydrothermal treatment process converts it into the more mechanically stable hydroxyapatite form.

Many have investigated the use of HA in bone tissue engineering. Holmes et al.[47, 48] conducted a series of experiments to determine its capabilities of treating osseous defects. In the radial diaphyses of dogs, bilateral cortical windows were created and filled with either HA implants or iliac autografts.[48] Specimens were harvested at 3, 6, 12, 24 and 48 months. Results showed that the HA implants encouraged significant bony ingrowth compared to the grafts.

In light of these findings, HA has been used in the treatment of a number of orthopedic conditions. Thalgott et al.[49] reported its use in spinal reconstruction. Twenty patients underwent circumferential lumbar fusion with HA blocks placed anteriorly and autograft with transpedicular or translaminar facet screw fixation posteriorly. Radiographs demonstrated solid arthrodesis rates of over 90%. Clinical follow-up reflected these positive findings, with over 80% of patients reporting good or excellent pain relief. From these observations, coralline HA is a practical alternative to autograft or allograft in anterior lumbar interbody fusions.

One of the major drawbacks associated with its use is the poor handling properties. To overcome this, Friedman et al.[50] developed HA-based cement to be used in craniofacial reconstruction. With its paste-like consistency, the material can be used to treat defects that were previously not amenable to ceramic fixation. It is rapidly adherent and is directly converted to bone without loss of implant volume. In their study of over 100 patients undergoing craniofacial reconstructive procedures using this cement, success rates were 97% at two years.

Another disadvantage with HA is its slow resorption.[45] Attempts have concentrated on manipulating the thermal conversion process of calcium carbonate to generate a faster resorbing HA. The resulting compound is a composite of calcium carbonate with a thin coating of HA. It initially behaves as a pure HA implant. Once this coating is resorbed after a few months, the remaining calcium carbonate is absorbed much more rapidly.

TCP is an example of a fast resorbing ceramic and has been used in formulations to enhance fracture repair. It undergoes the same remodeling as normal bone and forms tissue with the same structural characteristics to bone.[5] The mechanism by which TCP encourages new bone formation, however, remains unclear. Frost[51] hypothesized that it may be related to the calcium phosphate crystals that are produced as a consequence of TCP dissolution. These particles stimulate osteoclast proliferation, which has an indirect stimulatory effect on osteoblast function.

TCP has been studied as a potential bone filler in traumatic bone injuries. This was reported by Hinz et al.,[52] who used the material to treat calcaneal fractures. Biopsy results demonstrated active bone formation within the scaffold.

Despite these relative successes in achieving osteoinduction, TCP is limited as a bone graft substitute. This is related to its porosity, which is too small to allow complete bony ingrowth before the matrix is resorbed. In order to enhance its osteoinductive properties, Laffargue et al.[53] described a technique of adding rhBMP-2 to TCP cylinders to treat femoral condyle defects in rabbits. They observed increased trabecular bone formation prior to implant resorption.

Cellular Systems

Natural Matrices

Type I collagen is the most abundant protein in the extracellular matrix of bone and promotes mineral deposition by providing binding sites for matrix proteins like osteonectin, which regulate this process. Nevertheless, collagen is a poor bone graft material when used alone. This was demonstrated by Werntz et al.,[54] who noticed that collagen scaffolds were unable to promote healing of diaphyseal defects. In contrast, collagen and bone marrow cell composites induced the repair of these defects. Indeed, the authors commented that this repair was more effective than that seen using cancellous bone graft.

The addition of rhBMP-2 to collagen has similar osteoinductive effects. This was demonstrated by Boyne et al.[55] where they examined combining rhBMP-2 with an absorbable collagen sponge (rhBMP-2/ACS) for use in human maxillary floor reconstruction. Significant bone growth was documented by CT scans in all patients. There were no serious or unexpected immunologic or adverse effects. Histologic examinations of core bone biopsies confirmed the quality of the bone induced by rhBMP-2/ACS. These results suggest that rhBMP-2/ACS may provide an acceptable alternative to traditional bone grafts for maxillary floor reconstruction procedures.

Collagraft®, a composite of fibrillar collagen and porous calcium phosphate ceramic, has been compared in a multicenter prospective randomized trial with iliac crest autografts for the treatment of long bone fractures.[56] Initial results suggest no significant differences between the two groups, thereby lending support to the use of Collagraft as a substitute for autogenous bone grafts. Work by Alvis et al.[57] demonstrated that the osteoconductive properties of Collagraft can be enhanced by adding bone marrow cells. Implants of Collagraft, Collagraft plus bone marrow and bone marrow alone were placed subcutaneously in a rat model. Analysis of the tissue by three weeks revealed new bone formation within the Collagraft–bone marrow composite only.

Synthetic Polymers

Creating osteoconductive scaffolds from nonbiological materials offers many advantages, including excellent biocompatibility, ease of assembly, unlimited supply and no concerns of disease transmission. The ideal material would serve as a scaffold for the growth of new tissue and as a source of growth factors to support cell differentiation. Mechanically, the material should provide the initial strength for a healing process to begin, yet allow load-bearing activity to be gradually shifted from the implant to the developing skeletal tissue.[58]

Polyglycolic acid (PGA) is already used in many areas of modern medicine, for example, suture material. More recently, proponents have investigated its use in bone tissue engineering and, in particular, its effects on fracture healing.[59] Standardized 9 mm defects were created in athymic rat femurs and bridged with titanium miniplates. Half of the defects were treated with PGA constructs containing bovine periosteum-derived cells, whereas the remaining defects were either left untreated or filled with polymer templates alone. After 12 weeks, new bone formation was primarily seen bridging the defects in the experimental groups. Histologic evaluation revealed new bone formation in all experimental animals with islands of cartilage indicative of endochondral bone formation.

In addition to stimulating bone repair, synthetic polymers have been used to create artificial joints. Isogai et al.[60] described a technique in which they were able to create a finger joint from a composite of bovine periosteum, chondrocytes and tenocytes seeded onto a polyglycolic and poly-L-lactic acid copolymer. The composite was implanted subcutaneously into athymic mice. After 20 weeks, new tissues with the shape and dimensions of human phalanges were formed. Histological examination revealed mature articular cartilage and subchondral bone with a tenocapsule that had a structure similar to that of a human finger.

Ceramics

To enhance their osteogenic properties, osteoprogenitor cells have been added to ceramic materials. This was described by Okamura et al.,[61] where they subcutaneously implanted porous HA impregnated with rat marrow cells. Results suggested that HA may facilitate osteogenic differentiation of MSCs as analysis of the composite revealed the appearance of osteoblast-like cells and mineralized bone on the HA surface.

This observation of ceramics supporting MSC differentiation along osteogenic lines led Bruder et al.[62] to study the effects of implants loaded with MSCs on the healing of large segmental femoral defects in a canine model. Animals were treated with either MSC-laden ceramic cylinders (HA and TCP) or empty ceramic cylinders. A third control group comprised of untreated defects. After four months, the femora were harvested and processed for histological examination. Atrophic nonunion was seen within the control group. In contrast, woven and lamellar bony ingrowth was observed in the implants loaded with MSCs.

Quarto et al.[63] reported on a series of three patients in which bone marrow osteoprogenitor cells were placed on macroporous HA scaffolds to treat large bone defects. In all subjects, radiographic analysis revealed abundant callus formation along the implants and good integration at the interfaces with the host bone by 2 months postoperatively.

The problem with using synthetic ceramics for MSC delivery is their brittle nature and lack of interconnecting pores.[29] Natural ceramic composites, on the other hand, combine favorable mechanical properties with an open porous structure. An example is the natural coral exoskeleton, which has been used in several clinical applications.[64] Petite et al.[65] explored its use as a delivery vehicle for MSCs in the treatment of large bony defects in sheep. The authors compared its efficacy to coral scaffold alone to achieve osseous union. Results showed significant increases in clinical union rates in the experimental group compared to controls.

Role of Growth Factors

Growth factors are proteins secreted by cells and function as signaling molecules. They comprise a family of molecules that have autocrine, paracrine or endocrine effects on appropriate target cells. In addition to promoting cell differentiation, they have direct effects on cell adhesion, proliferation and migration by modulating the synthesis of proteins, other growth factors and receptors.[66]

Bone Morphogenetic Proteins

Since the discovery of the osteoinductive properties of DBM,[24] attention has been focused on the role of bone morphogenetic proteins (BMP) in embryological bone formation and bone repair in the postnatal skeleton.[66–68] BMPs are a group of non-collagenous glycoproteins that belong to the transforming growth factor beta (TGFβ) superfamily. They are synthesized locally and predominantly exert their effects by autocrine and paracrine mechanisms. Over 15 different BMPs have been identified and their genes have been cloned.[69] The best-studied examples are BMP-2, BMP-3 and BMP-7 (Osteogenic Protein 1) as these are known to play important roles in bone repair by stimulating MSC differentiation along osteogenic lines.

The importance of BMPs in bone repair has been the subject of several investigations. Cho et al.[67] defined and characterized their temporal expression during murine fracture healing. BMP-2 showed maximal expression on day 1 after fracture, suggesting its role as an early response gene in the cascade of healing events. BMP-3, 4, 7 and 8 exhibited a restricted period of expression from day 14 through day 21, when the resorption of calcified cartilage and osteoblastic recruitment were most active. BMP-5 and 6 were constitutively expressed from day 3 to day 21. These findings suggest that several members of the BMP family are actively involved in fracture healing, with each having a distinct temporal expression pattern and a potentially unique role in this repair process.

Gerhart et al.[70] studied the effects of rhBMP-2 on the healing of segmental bone defects. Fractures were created in the femurs of sheep, stabilized by plate fixation and treated with either bone matrix devoid of bone-inductive proteins (inactive), rhBMP-2 mixed with this inactive bone matrix or autogenous bone graft. A control group with no intervention was also included. Radiographs showed bony union in all defects treated with rhBMP-2 and bone graft. No bone formation was detected in the control and inactive bone matrix groups. Biomechanical testing revealed that the new bone formed in the rhBMP-2 treated group was stronger than that seen in the animals receiving autogenous graft. Long-term analyses of the bone formed by rhBMP-2 revealed that it had undergone a normal sequence of ossification, modeling and remodeling.[71] Results from these and other studies support investigations on the use of BMP-2 in clinical settings as an alternative to autograft for traumatic and reconstructive procedures.

Govender et al.[72] conducted a prospective, randomized, controlled multicenter trial evaluating the effects of rhBMP-2 on the treatment of open tibial fractures. Four hundred and fifty patients were randomized to receive either intramedullary (IM) nail fixation alone or IM fixation plus an implant containing either 0.75 or 1.5 mg/ml of rhBMP-2 at the time of definitive treatment. The implant was placed over the fracture site at the time of closure. Routine soft-tissue management was used in all patients. Results showed that the 1.5 mg/ml rhBMP-2 group had accelerated times to union, improved wound healing, reduced infection rates and fewer secondary invasive interventions.

In addition to its use in fracture healing, BMPs have also been utilized in spine surgery. Sandhu et al.[73] conducted a study analyzing the efficacy of rhBMP-2 in comparison to autograft in sheep anterior spinal fusion. Comparisons were made using radiographic, mechanical and histological analyses after 6 months of treatment. Radiographs revealed complete fusion in all of the rhBMP-2-treated animals compared to only 40% in the autograft group. Biomechanical testing revealed that segments treated with rhBMP-2 were 20% stiffer in flexion than were those treated with autograft. The authors concluded that treatment with rhBMP-2 improves fusion rates and strength of repair compared to using autograft in anterior spinal fusions.

With similar positive results seen in a rhesus monkey model,[74] Boden et al.[75] conducted a prospective, randomized controlled clinical study evaluating the use of rhBMP2 to achieve posterolateral lumbar spine fusion in patients who had single-level disc degeneration, grade 1 or less spondylolisthesis, mechanical low back pain with or without leg pain and at least 6 months failure of nonoperative treatment. Patients were randomized into groups receiving autograft/Texas Scottish Rite Hospital (TSRH) pedicle screw instrumentation (controls), rhBMP-2/TSRH, or rhBMP-2 alone without internal fixation. Results showed a significantly improved radiographic fusion rate between those receiving rhBMP-2 (100%) compared to those in which an autograft had been used (40%). Clinical symptoms improved at a faster rate in the rhBMP-2 group with successful posterolateral spine fusion being present after 1 year follow-up.

As with BMP-2, BMP-7 (OP-1) has proved to be efficacious in animal models. Cook et al.[76] studied the effects of rhBMP-7 on the healing of ulnar and tibial fractures in a monkey model. Ulnar and tibial defects were treated with a composite of rhBMP-7 and bovine type 1 collagen. Controls comprised of defects treated with collagen carrier alone or autogenous cancellous bone graft. Radiographs demonstrated that five of the six ulnae and four of the five tibiae treated with rhBMP-7 had completely healed by 6–8 weeks postoperatively. In contrast, none of the defects treated with the collagen carrier or bone graft had demonstrated bony union. Histological evaluation revealed the formation of new cortices with areas of woven and lamellar bone and normal-appearing bone marrow elements. Mechanical testing of the defects treated with BMP showed higher torsional strengths to failure than those treated with autogenous bone graft.

The osteogenic effect of BMP-7 has been used in a prospective, randomized double-blind trial to evaluate its ability to heal fibular defects in patients who have undergone high tibial osteotomy.[77] Results showed significant new bone formation in the patients treated with BMP compared to controls. In a larger study, Friedlaender et al.[78] assessed the efficacy of rhBMP-7 over iliac crest bone graft to treat patients with tibial nonunions. Nine months after surgery, 81% of the BMP-7 treated nonunions and 85% of those treated with bone graft had achieved clinical union. The authors concluded that while no statistical difference was noted between the two groups, BMP-7 was a safe and effective alternative to bone graft in the treatment of tibial nonunions.

Transforming Growth Factor β

Transforming Growth Factor β (TGFβ) influences a number of cell processes. These include stimulating MSC growth and differentiation, acting as a chemotactic factor for fibroblast and macrophage recruitment, and enhancing collagen and other ECM product secretion.[79]

TGFβ has been used to stimulate bone regeneration. This was demonstrated by Lind et al.,[80] where they continuously administered TGFβ for 6 weeks to rabbits in which unilateral plated tibial defects had been created. The control group comprised of defects treated with solvent without the growth factor. Results showed that TGFβ had a positive effect on fracture repair, with increased bending strengths and callus formation observed in the experimental group.

Critchlow et al.[81] performed a study to test the hypothesis that the anabolic effects of TGFβ on bony repair are dependent on the mechanical stability at the fracture site. Unilateral tibial fractures were produced in a rabbit model and held in either an unstable or stable configuration using plastic or steel plates, respectively. TGFβ-2 was injected into the calluses 4 days after fracture. In animals with unstable mechanical fixation, TGFβ-2 did not have an anabolic effect on callus formation. In contrast, those with stable mechanical constructs developed enlarged calluses. Histological analyses revealed that the calluses comprised almost entirely of bone compared to those seen within the unstable group, which predominantly comprised of cartilage. These findings demonstrate that stable fracture fixation is important for TGFβ-2-mediated skeletal repair.

From the above studies, it can be seen that TGFβ augments fracture healing in the experimental models. It is difficult, however, to draw definitive conclusions regarding these effects as reported studies testing several isoforms of TGFβ at different dosages in numerous animal models yield inconsistent results. Consequently, some believe that the anabolic effects of TGFβ may be due to its potentiation of BMPs.[82] Ripamonti et al.[83] reported enhanced BMP-7 effects on bone differentiation when low doses of TGFβ-1 were added. Combinations of BMP-7 and TGFβ-1 yielded a 2 to 3 times increase in the cross-sectional area of newly generated ossicles compared to BMP-7 alone. The tissue had distinct morphological differences, with larger amounts of endochondral bone formation compared to BMP-7-generated specimens.

Fibroblast Growth Factors

Fibroblast growth factors (FGF) are a group of structurally related compounds that share between 30% and 50% sequence homology. Acidic FGF (aFGF, FGF1) and basic FGF (bFGF, FGF2) are the most well-studied members of this family, with bFGF considered to be more potent. It stimulates angiogenesis, endothelial cell migration and is mitogenic for fibroblasts, chondrocytes, and osteoblasts.[84, 85]

During fracture repair, FGFs differ in their temporal and spatial expression.[86] In the early stages, FGF 1 and 2 are localized to the proliferating periosteum. This expression is then limited to osteoblasts during intramembranous bone formation and in the chondrocytes and osteoblasts during endochondral bone formation.

In light of their active involvement during fracture repair, investigators have studied the potential therapeutic roles of FGF in bone formation. Nakamura et al.[87]

studied these effects by injecting bFGF into mid-diaphyseal transverse tibial fractures in dogs. Controls were injected with carrier molecules. Specimens were harvested at 2, 4, 8, 16 and 32 weeks and were assessed in terms of callus formation, morphology and strength. Results showed that bFGF had positive effects on callus formation, remodeling rates, maximum load, bending stress and energy absorption.

Radomsky et al.[88] demonstrated that, by combining FGF with hyaluronic acid, they were able to achieve enhanced bone-forming potential. Bilateral fibula fractures were created in a primate model. The experimental site was randomly chosen and injected with a bFGF/hyaluronic acid composite gel. The contralateral fibula was left untreated and acted as a control. Increased callus formation and mechanical strength were noted in the treated defects. Radiographic and histologic analyses demonstrated that the callus size, periosteal reaction, vascularity and cellularity were consistently greater in the treated osteotomies than in the controls. Similar findings were reported by Lisignoli et al.[89] Segmental radial fractures were produced in rats and treated with either a biodegradable hyaluronic acid scaffold or a hyaluronic acid polymer–MSC composite that had been grown in a medium with or without supplemental bFGF. Enhanced mineralization of the bone defects was noted in the presence of the composites grown in bFGF. These results suggest that FGF has the potential of being used as an adjunct to promote skeletal repair.

Insulin-like Growth Factor

Insulin-like growth factors (IGF) exert an anabolic effect on bone metabolism. Two types have been described, IGF-1 and IGF-2, which stimulate osteoblast and osteoclast cell proliferation and matrix synthesis.[79] Reductions in IGF levels have been linked to age-related declines in bone mineral density.[90] Jehle et al.[91] conducted a cross-sectional study of the relationship between serum IGF levels and bone metabolism in patients with osteoporosis. Serum parameters including IGF-1 and IGF binding proteins (IGFBP) 1 through 6 were measured. Dual-energy X-ray absorptiometry was used to determine lumbar spine bone mineral density. Compared to age- and sex-matched controls, patients with osteoporosis showed a 73% decrease in free IGF-1, a 29% decrease in total IGF-1, a 10% decrease in IGFBP-3 and a 52% decrease in IGFBP-5 levels. These reductions were most evident in patients who had sustained vertebral fractures. The authors concluded that derangements in IGF system components reflect alterations in bone metabolism and a subsequent increase in susceptibility to fractures in these patients.

For tissue engineering purposes, researchers have tended to favor IGF-1 over IGF-2 due to its greater stimulatory effects on osteoblast function and its expression during fracture healing.[92,93] Thaller et al.[94] first examined its ability to promote healing of critical-sized calvarial defects in rats. Animals received either subcutaneous administration of IGF-1 or were left untreated to act as controls. Within the experimental group, repair commenced within 1 week of treatment, with complete bone formation seen by 6 weeks. Delayed osseous repair was detected in the control animals by 8 weeks.

Shen et al.[95] demonstrated similar positive findings using MSCs transfected with IGF-1 gene. The cells were systemically injected into mice that had sustained closed, mid-diaphyseal femoral fractures. Their findings demonstrated that the cells preferentially localized to the fracture site and exerted a positive effect on the repair process. This resulted in enhanced callus formation and ossification in the experimental group compared to controls.

This positive effect on bone formation has been used to promote spinal fusion. Kandziora et al.[96] compared the efficacy of IGF-1 with TGFβ-1 to autologous bone grafts in cervical fusion. After C3-4 discectomy, stabilization was achieved using either a titanium cage, a titanium cage with autologous bone graft, or a titanium cage with IGF-1 and TGFβ-1. After 12 weeks, animals treated with IGF-1/TGFβ-1 had significantly higher fusion rates than the bone-grafted animals. This could be attributed to the increased callus mineral density seen in the IFG-1 group.

Platelet-Derived Growth Factor

Platelet-derived growth factor (PDGF) is synthesized by numerous cell types including platelets, macrophages and endothelial cells. It consists of two polypeptide A and B chains, which share 60% amino acid sequence homology.[97] PDGFs possess strong mitogenic properties and stimulate the proliferation of osteoblasts.[98, 99] This is particularly important in fracture healing where they exhibit differential spatial and temporal expression.[100] Nash et al.[101] examined the efficacy of PDGF in bone formation using a rabbit tibial osteotomy model. Each osteotomy was injected with either collagen or collagen containing PDGF. An increase in callus formation and a more advanced stage of endosteal and periosteal osteogenic differentiation were seen in the experimental group compared to the controls after 28 days. Osteotomies treated with PDGF were not statistically different in strength from the nonoperated contralateral bones. In the control group, however, the osteotomies were statistically weaker than their intact contralateral bones. From these observations, it appears that exogenous PDGF has a stimulatory effect on fracture healing.

PDGF has been used clinically to stimulate periodontal regeneration.[102] Patients with periodontal osseous defects underwent reconstructive flap surgery and either received PDGF-BB and IGF-1 (50 or 150 μg/ml) or underwent no further intervention. Results showed that those receiving low-dose PDGF showed no increase in bone regeneration compared to controls. In contrast, patients treated with the higher doses developed statistically significant increases in alveolar bone formation.

Based on these promising findings, Giannobile et al.[103] examined the use of adenoviral vectors encoding for PDGF-A gene on root-lining cells. Results showed that this genetic delivery vehicle stimulated root lining cell proliferation and may provide beneficial results in periodontal tissue engineering.

Gene Therapy

Gene therapy is an emerging technology in the field of bone tissue engineering. It involves the transfer of genetic material into a cell's genome, thereby altering its synthetic function. For this process, the selected gene's messenger ribonucleic acid (mRNA) is reversely transcribed into complementary deoxyribonucleic acid (cDNA). It is then inserted into a plasmid and placed into a vector (viral or nonviral) carrier that facilitates gene transfer into the targeted cell lines. Successful gene transfer using nonviral vectors is termed *transfection*, whereas with viral carriers it is known as *transduction.*

Viruses are efficient vectors owing to their increased ability to infect host cells. They can be divided into *integrating* or *nonintegrating* subtypes, depending on their effects on the host cell's genome (Table 11.1). The former group are designed to integrate their genetic material into the cells DNA without causing the replication of the virus or inducing an immunological response.[104–106] Examples include adeno-associated viruses and retroviruses. Adeno-associated viruses are small DNA viruses

Table 11.1
Properties of Present Vectors

Vector[a]	Advantages	Disadvantages
Integrating viral		
Retrovirus		
MMLV-based	Straightforward production	Require target cell division
	No viral proteins made	Possible insertional mutagenesis
	Extensive use in human trials	
Lentivirus-based	Transduce nondividing cells	More development required
AAV	Site-specific integration[b]	Difficult to produce
	Nonpathogenic	Small packaging capacity (4 kb)
	Transduce nondividing cells	
	No viral proteins made	
Viral nonintegrating	Straightforward production	Inflammatory
Adenovirus	High titers	Immunogenecity of transduced cells
	Transduce nondividing cells	
HSV	Large packing capacity	Difficult to produce
	High titers	Cytotoxicity
	Transduce nondividing cells	
Nonviral		
Naked DNA	Simple	Few cells transfect well
	Nonimmunogenic	
	Inexpensive	
	Safe	
Liposomes	As above	Gene expression usually transient and low
Particle bombardment (gene gun)	Used in conjunction with plasmid DNA	Cumbersome; requires specialized equipment
DNA–ligand complexes	May be targetable	Possible antigenicity
	Receptor-mediated uptake often efficient	Low expression

[a] All types of vectors are the subject of considerable research. This table summarizes the present state of development. MMLV = Moloney murine leukemia virus; AAV = adeno-associated virus; HSV = herpes simplex virus.

[b] Wild-type AAV integrates in a site-specific manner. Recombinant virus appears as if it does not.

Reprinted with permission from Evans, C.H. et al. (1999). Gene therapy for rheumatic diseases. *Arthritis Rheum.*, 42, 1.

originating from the Parvovirus family. They have the advantages of being able to infect nondividing cells and of stable integration of its DNA into the host cell's genome at a precise location on chromosome 19.[107] Disadvantages include difficulty in generating high titers of the recombinant viruses, their small size limiting the amounts of exogenous DNA and their inability to insert DNA at specific sites in the host cell's genome.

Retroviruses are the best developed viral vectors for gene therapy and are being used in many clinical trials.[108] They are small RNA viruses that, once inside the cell, have their RNA transcribed into double-stranded DNA by the cell's reverse transcriptase. The DNA is integrated into the cell's genome and is expressed throughout the duration of the cell's lifecycle. The main drawbacks of these vectors are that they only infect and transduce actively replicating cells and they randomly integrate into the host cell's DNA. Concerns about possible mutagenesis resulting from placement of retroviral sequences could result in the activation of an oncogene leading to the development of a malignant tumor.

Nonintegrating viruses do not insert their genetic material into the host cell's DNA; but instead maintain it within the nucleus as an unintegrated, episomal form.[104] The adenovirus (Table 1) are the more studied virus of this type and has the advantage of being able to infect both dividing and some nondividing cells, thereby achieving a high level of transient gene expression. They, however, induce the formation of adenoviral antigens, which results in an immune response against infected cells, thereby resulting in a loss of gene expression after a short period of time.

Herpes simplex viruses (HSV) are capable of infecting both dividing and nondividing cells, carry large amounts of DNA, and infect many different cell types. The demonstration of HSV vector-associated toxicity has led to the development of a second generation of vector to minimize these effects.

Nonviral vectors possess limited immunogenecity and are safer than viral vectors. They are much cheaper and easier to produce in large quantities than viral carriers. The most commonly used nonviral vectors are liposomes, which are phospholipid vesicles that fuse with the cell membrane and deliver its contents into the cell.[109] Their main disadvantage, however, is their poor rate of genetic transfer.[105] Consequently, modern day gene therapy applications employ transduction methods owing to their greater efficiency over transfection techniques.[106]

The two main approaches to gene therapy involve *in vivo* and *ex vivo* gene transfer.[106] The *in vivo* technique involves the direct transfer of genetic material into the host. It is a technically easier method to perform but is limited by an inability to perform *in vitro* safety testing on transfected cells.

In vivo gene therapy has been used to promote fracture repair through the expression of BMP-2.[110] Segmental defects were created in the femora of New Zealand White rabbits and animals were divided into three groups: a positive control with no intervention, a negative control receiving adenoviral vectors infected with a luciferase gene and an experimental group receiving viral vectors encoding for BMP-2. Results demonstrated that BMP-2 exerted an anabolic effect with evidence of fracture healing in the treated animals. In contrast, the repair tissue in the control animals consisted of a fibrotic response with minimal osseous union. These observations lend support to the use of local adenoviral delivery of an osteoinductive gene to promote fracture repair in defects that would have otherwise progressed to nonunion.

Lumbar interbody fusions have been used to treat a number of conditions including spinal instability, tumors and disc disruptions.[111,112] Reports vary, but between 4% and 40% of these procedures are unsuccessful.[113,114] Gene therapy has been investigated as a potential mechanism to enhance spinal fusion in these cases. Alden et al.[115] reported on the use of an adenoviral construct containing the BMP-2 gene to achieve spinal fusion. Athymic rats were divided into three groups and the spinous process–lamina junction injected with either BMP-2 gene viral constructs, β galactosidase gene viral vectors, or both. Results showed enhanced bone formation in the BMP-2 group compared to the controls. Additionally, well-developed vasculature, cartilage and cancellous bone were found within the paraspinal muscles where BMP-2 vectors were injected. There was no evidence of neural compromise in the BMP-2 treated animals suggesting that this direct *in vivo* model may be a safe method to achieve spinal fusion.

Gene therapy has been described for treating craniofacial disorders.[116] Using an athymic rat model, Lindsey investigated whether nasal bone reconstruction could be enhanced using recombinant adenoviral vectors encoding for BMP-2 gene. Results showed significant osseous repair in the BMP-2-treated animals compared to the controls.

The use of direct *in vivo* techniques is not without problems. The generation of adenoviral vectors induces a florid immune response. Not only is this deleterious to the immunocompetent host[117] but it also limits the effectiveness of gene expression. This was demonstrated by Alden et al.,[118] where they investigated the endochondral response to BMP-2 adenoviral vector injections in immunocom petent and deficient animals. The former group showed evidence of acute inflammation without ectopic bone formation at the injection sites. In the athymic nude rats, BMP-2 gene therapy induced mesenchymal stem cell chemotaxis and proliferation, with subsequent differentiation into chondrocytes. The chondrocytes secreted a cartilaginous matrix, which underwent mineralization and subsequent replacement by bone. The study demonstrated that within immunocompetent animals, the endochondral response is limited by the immune response to adenoviral constructs.

The indirect *ex vivo* approach to gene therapy is technically more demanding than the *in vivo* method. It involves the removal of cells from a tissue biopsy and genetically modifying them *in vitro* before transfer back into the host. This offers the advantage of selecting the cells with the greatest gene expression and testing them for any abnormal behavior before reimplantation.[119]

Using the principles of *ex vivo* gene transfer, Lieberman et al.[120] generated BMP-2 producing bone marrow cells and investigated their ability to heal segmental femoral fractures in syngeneic rats. A group treated with rhBMP-2 served as a positive control. Negative controls included uninfected rat bone marrow cells, DBM and β-galactosidase transduced cells. Results showed that the BMP-2 transduced cells induced greater trabecular bone formation in the fracture site compared to any of the control groups. Others have reported similar results using the *ex vivo* method of gene therapy to promote this repair process.[121]

MECHANICAL FACTORS

Limb lengthening was first described by Codivilla in 1905[122] for the treatment of limb length discrepancies. It was not until the work of Ilizarov[123, 124] 50 years later that the technique of distraction osteogenesis (DO) gained popularity as a method for enhancing bone regeneration. Currently, the concept of DO is applied for the correction of a variety of orthopedic deformities and malformations with predictable results.[125] This was highlighted by Rozbruch,[126] where patients with leg length discrepancy, malalignment and nonunion following high tibial osteotomy were treated with distraction. Bone union was achieved with correction of the deformities and limb length inequalities.

DO has been used as a method for correcting craniofacial defects.[127] McCarthy et al.[128] reported its use in mandibular reconstruction and achieved bone lengthening from 18 to 24 mm. Successful follow-up of the patients demonstrated that this technique may be used for early reconstruction of maxillofacial deformities without the need for bone grafts, blood transfusion, or intermaxillary fixation.

BIOLOGY OF DISTRACTION OSTEOGENESIS

Distraction osteogenesis generates new tissue through the application of tensile forces to developing callus in a controlled osteotomy.[129, 130] It is characterized by three separate stages: (1) the latency phase that immediately follows osteotomy; (2) the active or distraction phase that permits active separation of bony segments; and (3) the consolidation phase where active distraction has ended and healing of the callus begins.[131–133] The period of time for each stage varies, depending on the anatomic site and the size of the osseous defect needing repair.

Bone formation results primarily from intramembranous ossification.[134] The early events after the latency period closely resemble fracture healing with a localized inflammatory response and hematoma formation.[135] The callus within the distraction gap comprises a number of different cell types with mesenchymal-like cells at the center, surrounded by fibroblast-like cells secreting a collagen-rich matrix.[130] Chondrocyte-like cells can also be seen at the interface between trabecular bone and osteoblasts, secreting a mineralized matrix.[131]

PHYSIOLOGICAL FACTORS GOVERNING DISTRACTION OSTEOGENESIS

Blood Supply

Studies of distraction have demonstrated the importance of blood vessel formation during this process. Rowe et al.[136] analyzed this angiogenic process during mandibular DO. Osteotomies were created in the right hemimandible of rats and a distraction device was applied. Distraction commenced after a 3-day latency period for 6 days. Results demonstrated that mandibular DO was associated with an intense vascular response during the early stages of distraction. Similar findings were reported by Choi et al.[137] Using scanning electron microscopy, they studied the spatial and temporal expressions of new blood vessels during DO. They showed that proliferations of periosteal and medullary blood vessels occurred primarily during the latency and distraction periods.

Li et al.[138] demonstrated that the angiogenic response is directly related to the distraction rate. Intense proliferations of capillary precursor cells were noticed within the fibrous interzone of the distracted gap, with maximum values occurring between 0.7 and 1.3 mm per day.

Latency Period

The latency period relates to the time between an osteotomy and the distraction of bone ends. Fluctuations in its length affect the tissue formed within the regenerate. This was demonstrated by the work of White and Kenwright.[139] Tibial osteotomies were created in New Zealand White rabbits and divided into groups receiving immediate distraction or after a latency of 7 days. Results showed that a delay in distraction influenced the osteogenic response. Immediate distraction resulted in the formation of fibrous tissue within the osteotomy defect. In contrast, those within the delayed distraction group produced a large volume of callus with areas of proliferating cartilage. This difference may be related to the well-developed capillary network found within the latter animals. The authors hypothesized that during the latency period, damaged blood vessels have the necessary time to repair and can better withstand the tensile forces of distraction.

Warren et al.[140] further investigated the effects of gradual distraction versus acute lengthening in rat mandibular DO. Animals either underwent immediate lengthening of 3mm or gradual distraction of 0.25mm twice a day after a three-day latency period. Results showed a marked elevation of critical extracellular matrix molecules (osteocalcin and collagen I) during the consolidation phase of the gradually distracted groups compared with acute lengthening. These findings suggest that gradual distraction osteogenesis promotes successful osseous bone repair by regulating the expression of bone-specific extracellular matrix molecules. In contrast, decreased production or increased turnover of proteins like collagen and osteocalcin may lead to fibrous union during acute lengthening.

Distraction Rates

The rate and rhythm of distraction appear to determine the success of osseous repair. This was highlighted by the work of Ilizarov,[141] in which he investigated the effects of different distraction rates (0.5, 1.0, or 2.0mm per day) and frequencies (1 step per day, 4 steps per day, 60 steps per day) on limb lengthening in canine tibiae. Results showed that a distraction rate of 1.0mm per day led to the best results, with optimum preservation of periosseous tissues, bone marrow and blood. Distraction of 0.5mm per day led to premature consolidation of the regenerate, whereas rates of 2.0mm adversely affected the surrounding tissues.

Farhadieh and co-workers[142] examined the effects of several distraction rates on the biomechanical, mineralization and histologic properties of regenerate tissue. A uniaxial distractor was applied to the angle of the mandible and varying rates of distraction were applied (1, 2, 3, and 4mm/day). After 5 weeks of distraction, results showed that the biomechanical, mineralization and histologic properties were significantly superior in the 1 mm group compared to the 4mm group.

From these studies, it can be seen that the biological nature of the regenerate is related to the stress–strain forces generated within the distraction gap. Meyer et al.[129] demonstrated that the magnitude of these forces directly influences the phenotypic differentiation of the cells within the distraction gap. Li et al.[143] reported similar findings using an experimental model of tibial leg lengthening. The authors' aims were to determine the morphology of the collagenous proteins present and their genes expressed within the regenerate at four different rates. At distraction rates of 0.3 mm per day, remodeled bone ends were separated by central areas of intramembranous bone and fibrocartilage. In the osteotomies distracted at 0.7 mm or 1.3 mm per day, new bone formation was seen at the corticotomy ends and were separated by central areas of fibrous tissue and cartilage. Fibrous tissue with sparse bone formation was seen within the animals distracted at 2.7 mm per day. Type I collagen was mainly expressed by the fibroblasts in the fibrous tissue, the bone surface cells, and to a limited extent by the osteocytes. Type II collagen was produced by the chondrocytes. These results suggest that osteoblasts and chondrocytes within the regenerate originate from the same pool of osteoprogenitor cells, and their differentiation and expression of types I and II collagen genes are affected by the different rates of distraction.

CONCLUSION

The field of tissue engineering to repair or regenerate the musculoskeletal system is developing rapidly and expanding in its applications. To date, strategies have met with limited success within the clinical setting. With ongoing research to enhance the osteogenic potential of cell concentrates, to develop better delivery systems and gene therapy applications for growth factors and osteoinductive substances, this technology will add to current treatment modalities and greatly enhance the management of musculoskeletal injuries and diseases in the future.

References

1. Deutche Banc. (2001). Alex Brown. Estimates and Company Information.
2. Banwart, J.C., Asher, M.A. and Hassanein. R.S. (1995). Iliac crest bone graft harvest donor site morbidity: a statistical evaluation. *Spine* 20, 1055.
3. Fowler, B.L., Dall, B.E. and Rowe. D.E. (1995). Complications associated with harvesting autogenous iliac bone graft. *Am. J. Orthop.* 24, 895.
4. Goulet, J.A. et al. (1997). Autogenous iliac crest bone graft: complications and functional assessment. *Clin. Orthop.* 339, 76.
5. Parikh, S.N. (2002). Bone graft substitutes: past, present and future. *J. Postgrad. Med.* 48, 142.
6. Alsberg, E., Hill, E.E. and Mooney, D.J. (2001). Craniofacial tissue engineering. *Crit. Rev. Oral. Biol. Med.* 12, 64.
7. Fleming, J.E., Cornell, C.N. and Muschler, G.F. (2000). Bone cells and matrices in orthopedic tissue engineering. *Orthop. Clin.*, 31, 357.
8. McCarthy, J.G. et al. (2001). Distraction osteogenesis of the craniofacial skeleton. *Plast. Reconstr. Surg.* 107, 1812.

9. Swennen, G. et al. (2001). Craniofacial distraction osteogenesis: a review of the literature. Part 1: clinical studies. *Int. J. Oral. Maxillofac. Surg*. 30, 89.
10. Beresford, J.N. (1989). Osteogenic stem cells and the stromal system of bone and marrow. *Clin. Orthop*. 240, 270.
11. Connolly, J.F. et al. (1989). Development of an osteogenic bone marrow preparation. *J. Bone Joint Surg*. 71A, 684.
12. Connolly, J.F. et al. (1991). Autologous marrow injection as a substitute for operative grafting of tibial nonunions. *Clin. Orthop*. 266, 259.
13. Grundel, R.E. et al. (1991). Autogeneic bone marrow and porous biphasic calcium phosphate ceramic for segmental bone defects in canine ulna. *Clin. Orthop*. 266, 244.
14. Connolly, J.F. and Shindell, R. (1986). Percutaneous marrow injection for an ununited tibia. *Nebr. Med. J*. 71, 105.
15. Healey, J.H. et al. (1990). Percutaneous bone marrow grafting of delayed union and nonunion in cancer patients. *Clin. Orthop*., 256, 280.
16. Garg, N.J., Gaur, S. and Sharma, S. (1993). Percutaneous autogenous bone marrow grafting in 20 cases of ununited fracture. *Acta Orthop. Scand*. 64, 671.
17. Lane, J.M. et al. (1999). Bone marrow and recombinant human bone morphogenetic protein-2 in osseous repair. *Clin. Orthop*. 361, 216.
18. Werntz, J.R. et al. (1996). Qualitative and quantitative analysis of orthotopic bone regeneration by marrow. *J. Orthop. Res*. 14, 85.
19. Connolly, J. (1995). Injectable bone marrow preparations to stimulate osteogenic repair. *Clin. Orthop*. 313, 8.
20. Bruder, S.P. et al. (1998). Bone regeneration by implantation of purified, culture-expanded human mesenchymal stem cells. *J. Orthop. Res*. 16, 155.
21. Bruder, S.P., Jaiswal, N. and Haynesworth, S.E. (1997). Growth kinetics, self renewal and the osteogenic potential of purified human mesenchymal stem cells during extensive subcultivation and following cryopreservation. *J. Cell Biochem*. 64, 278.
22. Shang, Q. et al. (2001). Tissue engineered bone repair of sheep cranial defects with autologous bone marrow stromal cells. *J. Craniofac. Surg*. 12, 586.
23. Yoshikawa, T., Ohgushi, H. and Tamai, S. (1996). Immediate bone forming capability of prefabricated osteogenic hydroxyapatite. *J. Biomed. Mater. Res*. 32, 481.
24. Urist, M.R. (1965). Bone: formation by autoinduction. *Science* 150, 893.
25. Katagiri, T. et al. (1994). Bone morphogenetic protein-2 converts the differentiation pathway of C2C12 myoblasts into the osteoblast lineage. *J. Cell Biol*. 127, 1755.
26. Khouri, R.K. et al. (1996). Repair of calvarial defects with flap tissue: role of bone morphogenetic proteins and competent responding tissues. *Plast. Reconstr. Surg*. 98, 103.
27. Peng, H. et al. (2002). Synergistic enhancement of bone formation and healing by stem cell-expressed VEGF and bone morphogenetic protein-4. *J. Clin. Invest*. 110, 751.
28. Thomson, J.A. et al. (1998). Embryonic stem cell lines derived from human blastocysts. *Science* 282, 1145.
29. Doll, B. et al. (2001). Critical aspects of tissue engineered therapy for bone regeneration. *Crit. Rev. Eur. Gene Exp*. 11, 173.
30. Bruder, S.P. and Fox, B.S. (1999). Tissue engineering of bone. *Clin. Orthop*. S367, S68.
31. Burg, K.J.L., Porter, S. and Kellam, J.F. (2000). Biomaterial developments for bone tissue engineering. *Biomaterials* 21, 2347.
32. Van de Putte, K.A. and Urist, M.R. (1965). Experimental mineralization of collagen sponge and decalcified bone. *Clin. Orthop*. 40, 48.
33. Van de Putte, K.A. and Urist, M.R. (1966). Osteogenesis in the interior of intramuscular implants of decalcified bone matrix. *Clin. Orthop*. 43, 257.

34. Tuli, S.M. and Singh, A.D. (1978). The osteoinductive property of decalcified bone matrix. An experimental study. *J. Bone Joint Surg*. 60, 116.
35. Tiedeman, J.J. et al. (1995). The role of a composite, demineralized bone matrix and bone marrow in the treatment of osseous defects. *Orthopaedics* 18, 1153.
36. Rabie, A.B., Wong, R.W. and Hagg, U. (2000). Composite autogenous bone and demineralized bone matrices used to repair defects in the parietal bone of rabbits. *Br. J. Oral Maxillofac. Surg*. 38, 565.
37. Oreffo, O.C. and Triffitt, J.T. (1999). Future potentials for using osteogenic stem cells and biomaterials in orthopedics. *Bone* 25, 5S.
38. Meinig, R.P. et al. (1996). Bone regeneration with resorbable polymeric membranes: treatment of diaphyseal bone defects in the rabbit radius with poly(L-lactide) membrane. A pilot study. *J. Orthop. Trauma* 10, 178.
39. Gugala, Z. and Gogolewski, S. (1999). Regeneration of segmental diaphyseal defects in sheep tibiae using resorbable polymeric membranes: a preliminary study. *J. Orthop. Trauma*, 13, 187.
40. Ladd, A.L. and Pliam, N.B. (1999). Use of Bone Graft Substitutes in Distal Radius Fractures. *J. Am. Acad. Orthop. Surg*., 7, 279.
41. Hollinger, J.O. and Battistone, G.C. (1986). Biodegradable bone repair materials. *Clin. Orthop*. 207, 290.
42. Ohgushi, H., Goldberg, V.M. and Caplan, A.I. (1989). Heterotopic osteogenesis in porous ceramics induced by marrow cells. *J. Orthop. Res*. 7, 568.
43. Goshima, J., Goldberg, V.M. and Caplan, A.I. (1991). The osteogenic potential of culture expanded rat marrow mesenchymal cells assayed *in vivo* in calcium phosphate ceramic blocks. *Clin. Orthop*. 262, 298.
44. Ripamonti, U. (1996). Osteoinduction in porous hydroxyapatite implanted in heterotopic sites of different animal models. *Biomaterials* 17, 31.
45. Fleming, J.E., Cornell, C.N. and Muschler, G.F. (2000). Bone cells and matrices in orthopaedic tissue engineering. *Orthop. Clin*. 31, 357.
46. Chiroff, R.T. et al. (1975). Tissue ingrowth of replamineform implants. *J. Biomed. Mater. Res*. 6, 29.
47. Holmes, R.E. et al. (1984). A coralline hydroxyapatite bone graft substitute. Preliminary report. *Clin. Orthop*. 188, 252.
48. Holmes, R.E., Bucholz, R.W. and Mooney, V. (1987). Porous hydroxyapatite as a bone graft substitute in diaphyseal defects: a histometric study. *J. Orthop. Res*. 5, 114.
49. Thalgott, J.S. et al. (2002). Anterior lumbar interbody fusion with processed sea coral (coralline hydroxyapatite) as part of a circumferential fusion. *Spine* 27, E518.
50. Friedman, C.D. et al. (1998). BoneSource hydroxyapatite cement: a novel biomaterial for craniofacial skeletal tissue engineering and reconstruction. *J. Biomed. Mater. Res*. 43, 428.
51. Frost, H. (1991). A new direction for osteoporosis research. *Bone* 12, 249.
52. Hinz, P. et al. (2002). A new resorbable bone void filler in trauma: early clinical experience and histologic evaluation. *Orthopedics* 25, S597.
53. Laffargue, P. et al. (1999). Evaluation of human recombinant bone morphogenetic protein-2-loaded tricalcium phosphate implants in rabbits' bone defects. *Bone* 25, S55.
54. Werntz, J.R., Lane, J.M. and Piez, C. (1986). The repair of segmental bone defects with collagen and marrow. *Orthop. Trans*. 10, 262.
55. Boyne, P.J. et al. (1997). A feasibility study evaluating rhBMP-2/absorbable collagen sponge for maxillary sinus floor augmentation. *Int. J. Periodontics Restorative Dent*. 17, 11.

56. Cornell, C.N. et al. (1991). Multicenter trial of Collagraft as bone graft substitute. *J. Orthop. Trauma* 5, 1.
57. Alvis, M. et al. (2003). Osteoinduction by a collagen mineral composite combined with autologous bone marrow in a subcutaneous rat model. *Orthopaedics* 26, 77.
58. Laurencin CT et al. (1996). Tissue engineered bone-regeneration using degradable polymers: the formation of mineralized matrices. *Bone* 19, 93S.
59. Puelacher, W.C. et al. (1996). Femoral shaft reconstruction using tissue-engineered growth of bone. *Int. J. Oral Maxillofac. Surg*. 25, 223.
60. Isogai, N. et al. (1999). Formation of phalanges and small joints by tissue-engineering. *J. Bone J. Surg. Am.* 81, 306.
61. Okamura, M. et al. (1997). Osteoblastic phenotype expression on the surface of hydroxyapatite ceramics. *J. Biomed. Mater. Res*. 37, 122.
62. Bruder, S.P. et al. (1998). The effect of implants loaded with autologous mesenchymal stem cells on the healing of canine segmental bone defects. *J. Bone Joint Surg. Am.* 80, 985.
63. Quarto, R., Mastrogiacomo, M. and Cancedda, R. (2001). Repair of large bone defects with the use of autologous bone marrow stromal cells. *N. Engl. J. Med*. 344.
64. Roux, F.X. et al. (1988). Madreporic coral: a new bone graft substitute for cranial surgery. *J. Neuro. Surg*. 69, 510.
65. Petite, H. et al. (2000). Tissue engineered bone regeneration. *Nat. Biotechnol.* 18, 959.
66. Johnson, E.E., Urist, M.R. and Finerman, G.A. (1988). Repair of segmental defects of the tibia with cancellous bone grafts augmented with human bone morphogenetic protein. A preliminary report. *Clin. Orthop*. 236, 249.
67. Cho, T.J., Gerstenfeld, L.C. and Einhorn, T.A. (2002). Differential temporal expression of members of the transforming growth factor β superfamily during murine fracture healing. *J. Bone. Miner. Res*. 17, 513.
68. Ripamonti, U. and Duneas, N. (1998). Tissue morphogenesis and regeneration by bone morphogenetic proteins. *Plast. Reconstr. Surg*. 101, 227.
69. Croteau, S. et al. (1999). Bone morphogenetic proteins in orthopaedics: from basic science to clinical practice. *Orthopaedics* 22, 686.
70. Gerhart, T.N. et al. (1993). Healing segmental femoral defects in sheep using recombinant human bone morphogenetic protein. *Clin. Orthop*. 293, 317.
71. Kirker-Head, C.A. et al. (1995). Long-term healing of bone using recombinant human bone morphogenetic protein 2. *Clin. Orthop*. 318, 222.
72. Govender, S. et al. (2002). Recombinant human bone morphogenetic protein 2 for treatment of open tibial fractures. A prospective, controlled, randomized study of four hundred and fifty patients. *J. Bone J. Surg. Am.* 84, 2123.
73. Sandhu, H.S. et al. (1995). Histologic evaluation of the efficacy of rhBMP-2 compared with autograft bone in sheep spinal anterior interbody fusion. *Spine* 20, 2669.
74. Boden, S.D. et al. (1998). Laparoscopic anterior spinal arthrodesis with rhBMP2 in a titanium interbody threaded cage. *J. Spinal Disord.* 11, 95.
75. Boden, S.D. et al. (2002). Use of recombinant human bone morphogenetic protein-2 to achieve posterolateral lumbar spine fusion in humans: a prospective, randomized clinical pilot trial: 2002 Volvo Award in clinical studies. *Spine* 27, 2662.
76. Cook, S.D. et al. (1995). Effect of recombinant human osteogenic protein-1 on healing of segmental defects in non-human primates. *J. Bone Joint Surg. Am.* 77, 734.
77. Geesink, R.G.T., Hoefnagels, N.H.M. and Bulstra, S.K. (1999). Osteogenic activity of OP1 bone morphogenetic protein (BMP7) in a human fibular defect. *J. Bone. J. Surg. Br.* 81, 710.

78. Friedlaender, G.E. et al. (2001). Osteogenic protein 1 (bone morphogenetic protein 7) in the treatment of tibial nonunions. *J. Bone Joint Surg. Am.* 83, S1-151.
79. Khan, S.N. et al. (2000) Bone growth factors. *Orthop. Clin.* 31, 375.
80. Lind, M. et al. (1993). Transforming growth factor β enhances fracture healing in rabbit tibiae. *Acta Orthop. Scand.* 64, 553.
81. Critchlow, M.A., Bland, Y.S. and Ashhurst, D.E. (1995). The effect of exogenous transforming growth factor β2 on healing fractures in the rabbit. *Bone* 16, 521.
82. Centrella, M. et al. (1994). Transforming growth factor beta gene family members and bone. *Endocr. Rev.*, 15, 27.
83. Ripamonti, U. et al. (1997). Recombinant transforming growth factor-beta1 induces endochondral bone in the baboon and synergizes with recombinant osteogenic protein-1 (bone morphogenetic protein-7) to initiate rapid bone formation. *J. Bone Miner. Res.* 12, 1584.
84. Ingber, D.E. and Folkman, J. (1989). Mechanochemical switching between growth and differentiation during fibroblast growth factor stimulated angiogenesis in vitro: role of extracellular matrix. *J. Cell Biol.* 109, 317.
85. Hurley, M.M. et al. (1993). Basic fibroblast growth factor inhibits type 1 collagen gene expression in osteoblastic MC3T3E1 cells. *J. Biol. Chem.* 268, 5588.
86. Rundle, C.H. et al. (2002). Expression of the fibroblast growth factor receptor genes in fracture repair. *Clin. Orthop.* 403, 253.
87. Nakamura, T. et al. (1998). Recombinant human basic fibroblast growth factor accelerates fracture healing by enhancing callus remodeling in experimental dog tibial fracture. *J. Bone Miner. Res.* 13, 942.
88. Radomsky, M.L. et al. (1999). Novel formulation of fibroblast growth factor 2 in a hyaluronan gel accelerates fracture healing in nonhuman primates. *J. Orthop. Res.* 17, 607.
89. Lisignoli, G. et al. (2002). Osteogenesis of large segmental radius defects enhanced by basic fibroblast growth factor activated bone marrow stromal cells grown on non-woven hyaluronic acid-based polymer scaffold. *Biomaterials* 23, 1043.
90. Bennett, A.E. et al. (1984). Insulin-like growth factors I and II: aging and bone density in women. *J. Clin. Endocrinol. Metab.* 59, 701.
91. Jehle, P.M. et al. (2003). Serum levels of insulin-like growth factor (IGF)-I and IGF binding protein (IGFBP)-1 to -6 and their relationship to bone metabolism in osteoporosis patients. *Eur. J. Intern. Med.* 14, 32.
92. Bak, B., Jorgensen, P.H. and Andreassen, T.T. (1990). Dose response of growth hormone on fracture healing in the rat. *Acta Orthop. Scand.* 61, 54.
93. Andrew, J.G. et al. (1993). Insulin like growth factor gene expression in human fracture callus. *Calcif. Tissue Int.* 53, 97.
94. Thaller, S.R., Dart, A. and Tesluk, H. (1993). The effects of insulin like growth factor 1 on critical size calvarial defects in Sprague–Dawley rats. *Ann. Plast. Surg.* 31, 430.
95. Shen, F.H. et al. (2002). Systemically administered mesenchymal stromal cells transduced with insulin-like growth factor-I localize to a fracture site and potentiate healing. *J. Orthop. Trauma* 16, 651.
96. Kandziora, F. et al. (2002). Comparison of BMP-2 and combined IGF-I/TGFβ1 application in a sheep cervical spine fusion model. *Eur. Spine J.* 11, 482.
97. Solheim, E. (1998). Growth factors in bone. *Int. Orthop.* 22, 410.
98. Canalis, E., McCarthy, T.L. and Centrella, M. (1989). Effects of platelet derived growth factor on bone formation *in vitro*. *J. Cell. Physiol.* 140, 530.
99. Canalis, E. (1981). Effect of platelet-derived growth factor on DNA and protein synthesis in cultured rat calvaria. *Metabolism* 30, 970.

100. Andrew, J.G. et al. (1995). Platelet-derived growth factor expression in normally healing human fractures. *Bone* 16, 455.
101. Nash, T.J. et al. (1994). Effect of platelet-derived growth factor on tibial osteotomies in rabbits. *Bone* 15, 203.
102. Howell, T.H. et al. (1997). A phase I/II clinical trial to evaluate a combination of recombinant human platelet-derived growth factor-BB and recombinant human insulin-like growth factor-I in patients with periodontal disease. *J. Periodontol.* 68, 1186.
103. Giannobile, W.V. et al. (2001). Platelet-derived growth factor (PDGF) gene delivery for application in periodontal tissue engineering. *J Periodontol.* 72, 815.
104. Evans, C.H. and Robbins, P.D. (1995). Possible orthopaedic applications of gene therapy. *J. Bone J. Surg.* 77A, 1103.
105. Salypongse, A.N., Billiar, T.R. and Edington, H. (1999). Gene therapy and tissue engineering. *Clin. Plast. Surg.* 26, 663.
106. Robbins, P.D. and Ghivizzani, S.C. (1998). Viral vectors for gene therapy. *Pharmacol. Ther.* 80, 35.
107. Samulski, R.J. et al. (1991). Targeted integration of adeno-associated virus (AAV) into human chromosome 19. *EMBO J.* 10, 3941.
108. Miller, A.D. (1992). Human gene therapy comes of age. *Nature* 357, 455.
109. Musgrave, D.S., Fu, F.H. and Huard, J. (2002). Gene therapy and tissue engineering in orthopaedic surgery. *J. Am. Acad. Orthop. Surg.* 10, 6.
110. Baltzer, A.W. et al. (2000). Genetic enhancement of fracture repair: healing of an experimental segmental defect by adenoviral transfer of the BMP-2 gene. *Gene Ther.* 7, 734.
111. Patil, P.V. et al. (2000). Interbody fusion augmentation using localized gene delivery. *Trans. Orthop. Res. Soc.* 25, 360.
112. Goldstein, S.A. (2000). *In vivo* nonviral delivery factors to enhance bone repair. *Clin. Orthop. S*379, S113.
113. Steinmann, J.C. and Herkowitz, H.N. (1992). Pseudoarthrosis of the spine. *Clin. Orthop.* 284, 80.
114. Zoma, A. et al. (1987). Surgical stabilization of the rheumatoid cervical spine: a review of indications and results. *J. Bone J. Surg.* 49B, 8.
115. Alden, T.D. et al. (1999). Percutaneous spinal fusion using bone morphogenetic protein-2 gene therapy. *J. Neurosurg.* 90, 109.
116. Lindsey, W.H. (2001). Osseous tissue engineering with gene therapy for facial bone reconstruction. *Laryngoscope* 111, 1128.
117. Brody, S.L. et al. (1994). Acute responses of non human primates to airway delivery of an adenovirus vector containing the human cystic fibrosis transmembrane conductance regulator cDNA. *Hum. Gene Ther.* 5, 821.
118. Alden, T.D. et al. (1999). *In vivo* endochondral bone formation using a bone morphogenetic protein 2 adenoviral vector. *Hum. Gene Ther.* 10, 2245.
119. Chen, Y. (2001). Orthopaedic applications of gene therapy. *J. Orthop. Sci.* 6, 199.
120. Lieberman, J.R. et al. (1999). The effect of regional gene therapy with bone morphogenetic protein-2-producing bone-marrow cells on the repair of segmental femoral defects in rats. *J. Bone Joint Surg.* A 81, 905.
121. Breitbart, A.S. et al. (1999). Gene enhanced tissue engineering: applications for bone healing using cultured periosteal cells transduced retrovirally with the BMP7 gene. *Ann. Plast. Surg.* 42, 488.
122. Codivilla, A. (1905). On the means of lengthening in the lower limbs, the muscles and tissues which are shortened through deformity. *Am. J. Orthop. Surg.* 2, 353.

123. Ilizarov, G.A. et al. (1978). Characteristics of systemic growth regulation of the limbs under the effects of various factors influencing their growth and length [in Russian]. *Ortop. Travmatol. Protez.* 8, 37.
124. Ilizarov, G.A., Pereslitskikh, P.F., and Barabash, A.P. (1978). Closed directed longitudino-oblique or spinal osteoclasia of the long tubular bones (experimental study) [in Russian]. *Ortop. Travmatol. Protez.* 11, 20.
125. Ilizarov, G.A. (1990). Clinical applications for a tension stress effect for limb lengthening. *Clin. Orthop.* 250, 34.
126. Rozbruch, S.R. et al. (2002). Distraction osteogenesis for nonunion after high tibial osteotomy. *Clin. Orthop.* 394, 227.
127. Davies, J., Turner, S. and Sandy J.R. (1998). Distraction osteogenesis— a review. *Br. Dent. J.* 185, 462.
128. McCarthy, J.G. et al. (1992). Lengthening the human mandible by gradual distraction. *Plast. Reconstr. Surg.* 89, 1.
129. Meyer, U. et al. (2001). Mechanical tension in distraction osteogenesis regulates chondrocyte differentiation. *Int. J. Oral Maxillofac. Surg.* 30, 522.
130. Lewinson, D. et al. (2001). Expression of vascular antigens by bone cells during bone regeneration in a membranous bone distraction system. *Histochem. Cell Biol.* 116, 381.
131. Sato, M. et al. (1999). Mechanical tension stress induces expression of bone morphogenetic protein(BMP) 2 and BMP 4, but not BMP6, BMP7 and GDF5 mRNA during distraction osteogenesis. *J. Bone Miner. Res.* 14, 1084.
132. Tay, BK-B. et al. (1998). Histochemical and molecular analysis of distraction osteogenesis in a mouse model. *J. Orthop. Res.* 16, 636.
133. Isefuko, S., Joyner, C.J. and Simpson, H.R.W. (2000). A murine model of distraction osteogenesis. *Bone* 27, 661.
134. Einhorn, T.A. and Lee, C.A. (2001). Bone regeneration. New findings and potential clinical applications. *J. Am. Acad. Orthop. Surg.* 9, 157.
135. Tajana, G.F., Morandi, M. and Zembo, M. (1989). The structure and development of osteogenic repair tissue according to Ilizarov technique in man: characterization of the extracellular matrix. *Orthopedics* 12, 515.
136. Rowe, N.M. et al. (1999). Angiogenesis during mandibular distraction osteogenesis. *Ann. Plast. Surg.* 42, 470.
137. Choi, I.H. et al. (2000). Vascular proliferation and blood supply during distraction osteogenesis: a scanning electron microscopic observation. *J. Orthop. Res.* 18, 698.
138. Li, G. et al. (1999). Effect of lengthening rate on angiogenesis during distraction osteogenesis. *J. Orthop. Res.* 17, 362.
139. White, S.H. and Kenwright, J. (1990). The timing of distraction of an osteotomy. *J. Bone J. Surg. (Br.)* 72, 356.
140. Warren, S.M. et al. (2000). Rat mandibular distraction osteogenesis: part III. Gradual distraction versus acute lengthening. *Plast. Reconstr. Surg.* 107, 441.
141. Ilizarov, G.A. (1989). The tension–stress effect on the genesis and growth of tissues: Part II. The influence of the rate and frequency of distraction. *Clin. Orthop.* 239, 263.
142. Farhadieh, R.D. et al. (2000). Effect of distraction rate on biomechanical, mineralization, and histologic properties of an ovine mandible model. *Plast. Reconstr. Surg.* 105, 889.
143. Li, G et al. (2000). Tissues formed during distraction osteogenesis in the rabbit are determined by the distraction rate: localization of the cells that express the mRNAs and the distribution of types I and II collagens. *Cell Biol. Int.* 24, 25.

12 Clinical Challenges and Contemporary Solutions: Craniofacial and Dental

James P. Bradley

Contents

INTRODUCTION

Joseph Murray, who performed the first successful human organ transplant in 1968 and received the Nobel Peace Prize in 1990, listed the chronologic evolution of surgery as follows:

(1) Ablative surgery
(2) Reconstructive surgery
(3) Transplantation surgery
(4) Tissue engineering or inductive surgery.[1]

Currently, exciting progress is being made in this fourth phase of surgery (tissue engineering and inductive surgery). Based on this progress, the field of craniofacial surgery is changing. Better options are becoming available to correct facial and skull deformities and to replace defects from birth, injury, or disease.[2]

0-8493-1621-9/05/$0.00+$1.50
© 2005 by CRC Press LLC

Today, functional and cosmetic corrections rely on the transfer of tissues or on the placement of implantable prostheses. The forehead flap used for total and subtotal nasal reconstruction was developed thousands of years ago in India and is still used today with only minor modifications.[3] Microsurgery used for reconstruction after oropharyngeal cancer ablation became widespread in the 1970s as a way to provide vascularized tissue from distant sites.[4–6] Limitations to these techniques include: (1) donor site morbidity and (2) suboptimal results secondary to relying on tissue that is inherently different from the tissue being replaced.[2,5] Manufactured implants for both hard- and soft-tissue substitution have also been adopted as a solution.[7,8] Since the description of the repair of a calvarial defect with a gold plate was first given in 1600, improvements have been made to synthetic biomaterials.[9] However, limitations with the use of foreign materials exist. These include infection, immunologic reaction, and breakage.[9] In addition, for the pediatric patient, neither approach allows for the appropriate growth.[10]

A recent trend in plastic surgery is the use of more tissue-inductive treatments. Skin expansion has been used for breast reconstruction,[11,12] hair transplantation,[13,14] resurfacing after skin cancer ablation,[15,16] and closure of large donor sites.[17] Wound care with vacuum-assisted closure (VAC) suction dressing has been used to promote healing in chronic wounds,[18] deep sternal wounds,[19] enterocutaneous fistula,[20] and diabetic foot wounds.[21] In craniofacial reconstruction, distraction osteogenesis has been used in the correction of micrognathia,[22] facial asymmetry,[23] midface hypoplasia,[24] and temporomandibular joint reconstruction.[25]

Clinical challenges in craniofacial and dental reconstruction have evolving solutions. Ideally, missing parts and birth deformities should be replaced with "like" tissue that is durable and will grow with the patient. In addition, better solutions to current problems should minimize donor tissue problems.

REPLACEMENT OF MISSING PARTS

Central to congenital craniofacial problems is a structural defect. Often when a defect occurs, the void is filled by surrounding tissues in an abnormal way.[26] For instance, hypertelorism (abnormal separation or widening of the orbital cavities) is strongly associated with an encephalocele (herniation of the forebrain and/or its lining) (Figure 12.1). The earlier the correction may be performed (reduction and closure of encephaolcele and translocation of the orbits), the sooner the functional matrix may be restored. Once the functional matrix is restored, normal growth may occur.[27] Therefore, correction of congenital defects in the pediatric patient is often a functional necessity.

SKULL DEFECTS

Aplasia cutis, congenital absence of tissue on the vertex of the head, in its mild form involves the skin. In more severe forms, absence of skin, cranial bone, and dura lining may occur (Figure 12.2).[28] Staged treatment involves skin grafting to the open area of the head, followed by tissue expansion and then large flap rotation and

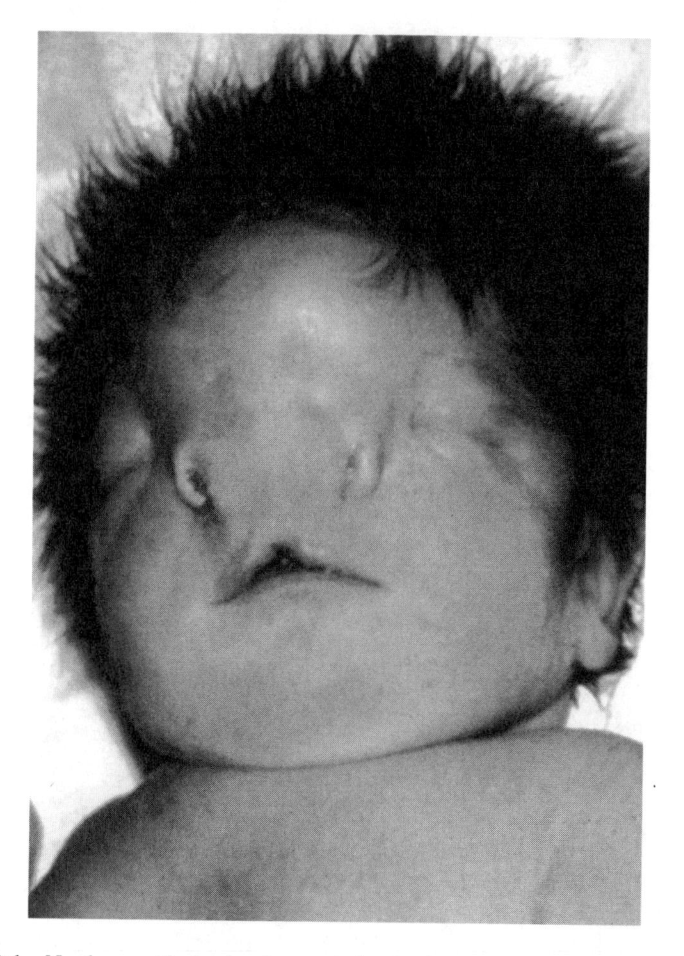

Figure 12.1 Newborn with forehead encephalocele (herniation of forebrain) and hypertelorism (wide separation of bony orbits).

eventual bone grafting.[28] However, these techniques may fail, particularly in cases of Adams–Oliver syndrome, in which the vascularity of the scalp is abnormal.[29] In such cases, the condition is life-threatening because of the exposed neural tissue and sagittal sinus.[30]

Young children, particularly children below 2 years of age, have the ability to heal calvarial defects. It is suggested that the biomolecular bone- and matrix-inducing phenotype of immature dura mater regulates this process.[31] However, even in young patients, critical size defects of the calvaria exist, and are challenging for the clinician. Often such defects remain unrepaired until years after birth when adequate donor sites become available. Protective helmets are necessary during this waiting period. Animal models of critical size skull defect will be important in developing better methods of repair.[32]

In pediatric cranial reconstruction, autologous bone grafting is most often used. Split thickness cranial bone grafts may not be harvested until approximately 7 years

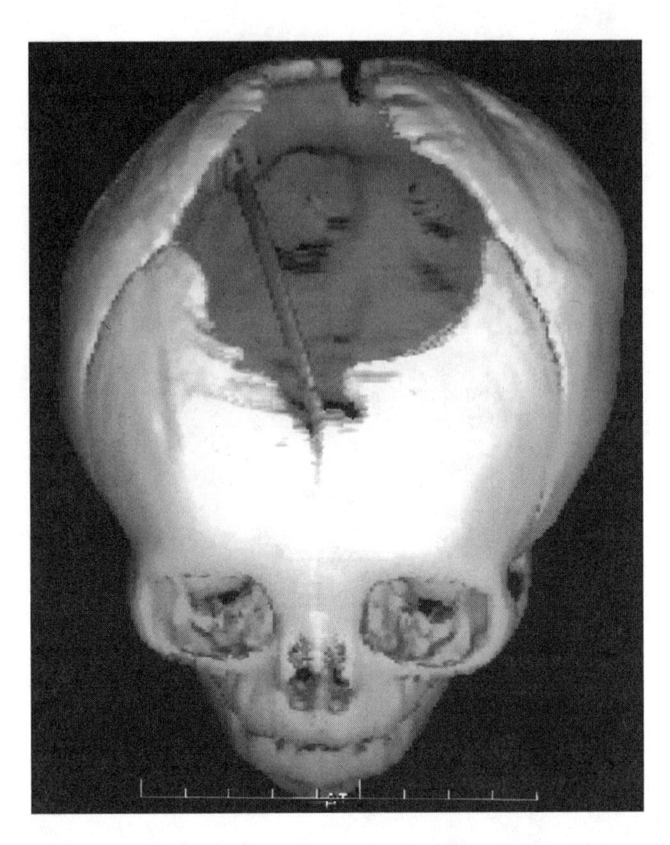

Figure 12.2 3D CT scan of a newborn diagnosed with cutis aplasia and a large defect of the apical skull. A ventriculoperitoneal shunt was placed to control CSF leakage through the defect.

of age when a diploic space has developed. Split rib grafts and iliac crest grafts are other bone graft options. The use of calcium phosphate bone substitutes, like bone source and norion, has become more widespread.[33,34] However, usage in children is not universally accepted because of long-term concerns, particularly in the areas of appositional growth like in the upper face and skull.

Adult cranial defects may be from trauma, oncogenic resection, infection, or an emergent decompressive craniotomy. Common reconstructive options include autogenous bone grafts (calvarial, rib, iliac), cryopreserved autogenous bone (saved from initial decompressive craniotomy), methylmethacrylate cranioplasty,[33] and resorbable or titanium plates with or without a bone substitute.[35,36] Use of plates alone, above, and below the cranial defect relies on protective bone regeneration.[37] More often, a calcium phosphate bone substitute is used in conjunction with plating. It has been suggested that bone substitutes should not be used with large autogenous bone graft because the substitutes act as a barrier to osteoconduction within the bone graft.[38] For biosynthetic closure of cranial defects, medical modeling allows for the manufacture of a precise size and shape allograft from a 3D CT scan. Despite numerous options, the failure rate may be high for both autogenic grafts (13–30%)[39] and allogenic (20–35%) alternatives.[40]

EAR DEFECT: MICROTIA

Craniofacial microsomia occurs in 1 in 5600 newborns. One of the striking features is microtia or near absence of an ear.[41] Total ear reconstruction remains one of the most difficult problems in craniofacial surgery. An autogenous cartilage framework is preferred by most surgeons over alloplastic implants (silicone or polypropylene).[42] Alloplastic implants for ear reconstruction are susceptible to infection and extrusion.[43] To reduce this risk, a temporoparietal vascularized flap must be used for coverage of the implant. Although the rib cartilage framework has better long-term durability, consistently obtaining an optimal shape may be difficult because of limited donor tissue or lack of experience. Total ear reconstruction is staged in two procedures (Nagata technique[44]) or four procedures (Brent technique[45]) and may take years to complete.

When scarring exists in the site of the ear defect, current techniques of total ear reconstruction are not acceptable. Such scarring may result from trauma or previous failed attempts at reconstruction. Currently, these patients are treated with osteointegrated implants and ear prosthesis.[43]

Tissue engineering offers the potential to grow autogenous cartilage in a precise auricular form based on polymer constructs.[46] This has been successfully done when implanted subcutaneously in the nude mouse. External molding stents were necessary until cartilage formation began. But once the neocartilage matured, the overlying skin adhered to the framework to replicate the precise complex convolutions of the auricle. However, long-term maintenance in an immunocompetent host has not been observed.

NASAL DEFECT

Arhinia, congenital absence of the nose, is rare.[47,48] It may be associated with a Tessier midline number 0 or number 14 cleft (Figure 12.3). Heminasal birth defects may also occur. Reconstruction of the nose is challenging because it is a prominent three-dimensional structure with three layers to consider: skin, cartilage, and nasal mucosal lining. In children, problems that occur with facial growth make nasal reconstruction even more difficult. In addition, the reconstructed nose will not grow and will need serial revisions. More commonly, adult nasal defects occur from skin cancer ablations or trauma. Acceptable cosmetic results may be achieved, but require multiple stages.[49] Donor tissue is needed from ear or septal cartilage, surrounding skin or mucosa and/or forehead skin. For larger defects, particularly with older patients, osteointegrated implants and nasal prosthesis may be used.[50]

DEFORMITIES

CLEFT LIP AND PALATE

One of the more common congenital deformities, the cleft lip and palate, occurs in 1 in 200 to 1 in 600 newborns.[51] It includes deficiencies and improper orientation of lip skin, orbicularis oris muscle, mucosa, alveolar and palatal bone, palatal oral and

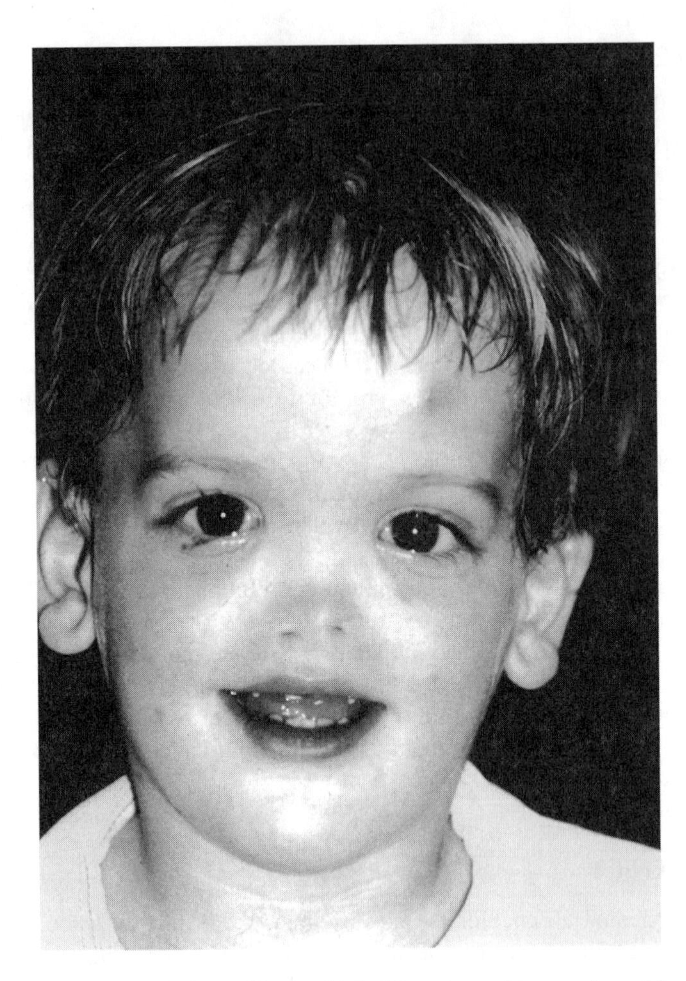

Figure 12.3 Four-year-old boy with an arhinia. There is an absence of nasal bone and cartilage. The nostrils are small blind pouches.

nasal mucosa, and levator veli palatine muscle.The timing of surgical correction is aimed at minimizing the detrimental effect of surgical scarring on facial growth. Despite best treatment plans, 18–45% of patients require orthognathic jaw advancement surgery at maturity.[52]

Advances in neonatal presurgical nasoalveolar molding (PNAM) have allowed for orthodontic preoperative correction of alveolar ridge misalignment, nasal columellar lengthening, and nasal cartilage repositioning.[53] Rotation advancement for unilateral cleft lip and straight line repair for bilateral cleft lip repairs allow for anatomic placement of misaligned tissues. The technique of closure of alveloar defects with iliac bone grafting has not changed considerably. Yet, there is a fairly high rate of bone exposure and persistent fistulization.[54] The healing of alveolar defects in dogs and primates with bone morphogenetic protein-2 (BMP-2) in carriers offers a potentially better clinical alternative.[55,56] BMP-2 has also been used for periodontal ridge augmentation[57] and in endosseous dental implants.[58]

More severe facial clefting deformities are less frequent but are often more challenging (Figure 12.4). These facial defects may be larger and involve more vital tissues. For these patients, rotation of local tissues and subsequent bone grafting offer closure of the clefts but lead to facial growth disturbances. A better solution with like tissue substitution and the potential for normal facial growth is needed.

Treacher–Collins Syndrome

Oro-mandibular dysostosis or Treacher–Collins syndrome is autosomal dominant and manifests as bilateral malar defects (Figure 12.5).[59] Penetrance of this deformity is variable. With hypoplastic or absent zygomas, lateral eye support is missing. Autogenous bone is used for reconstruction of zygomatic arches, lateral orbital walls, and orbital floors. If procedures are performed before 6–8 years of age,

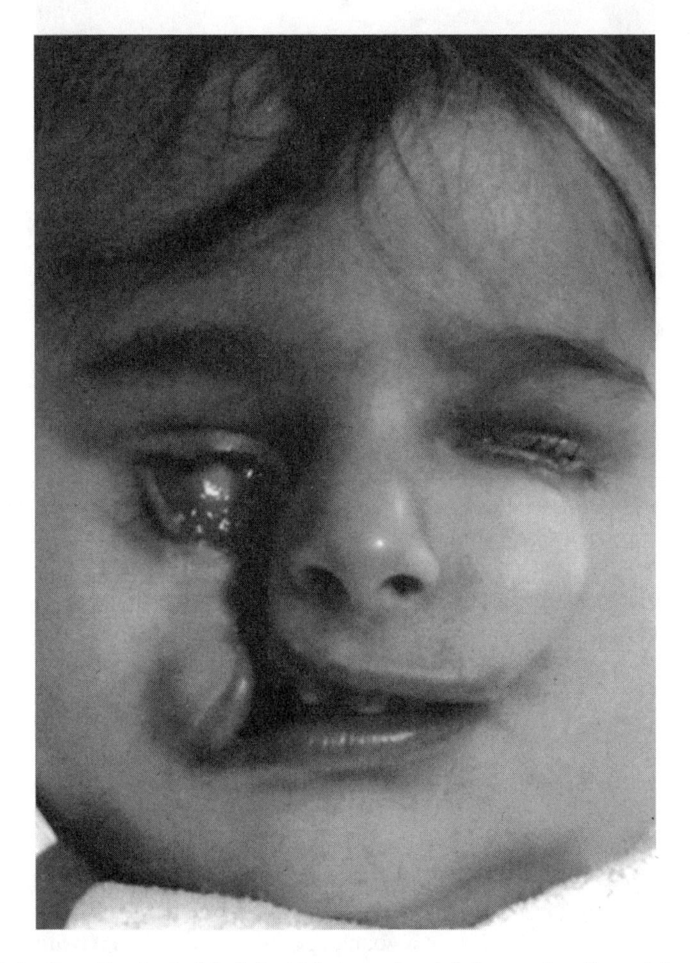

Figure 12.4 Rare Tessier facial clefts (right=number 4, left=number 5) result in deficits of the eyelid, zygomatic-maxillary bone, and oral mucosa.

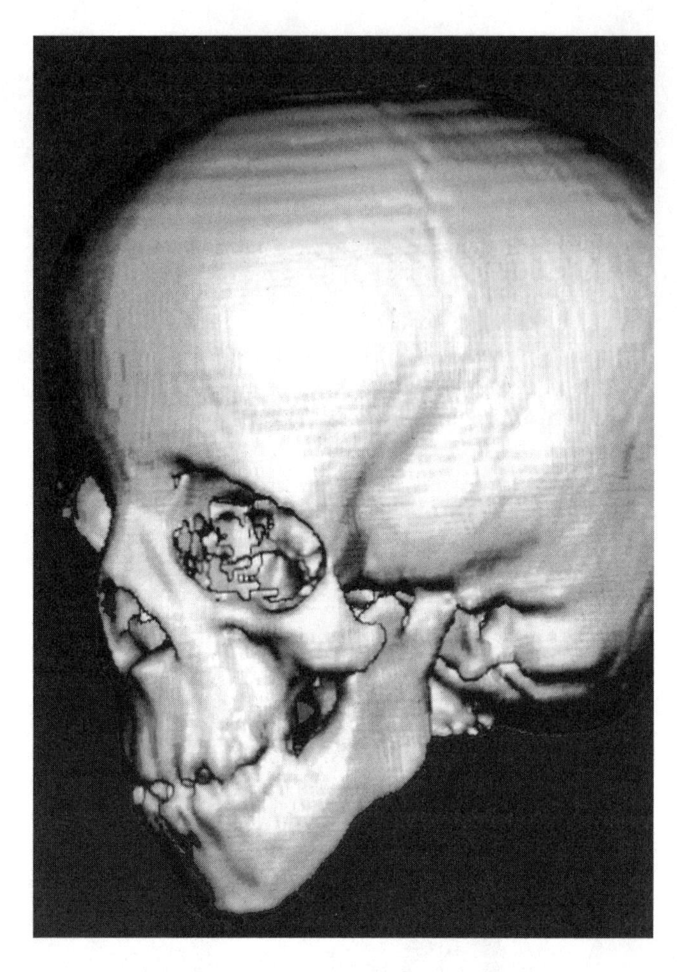

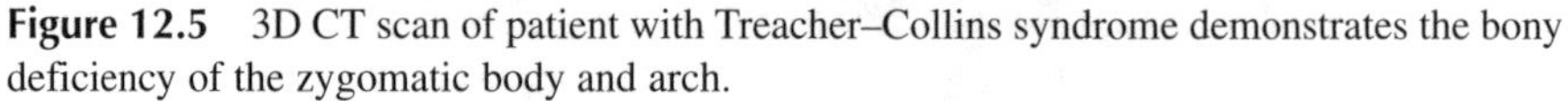

Figure 12.5 3D CT scan of patient with Treacher–Collins syndrome demonstrates the bony deficiency of the zygomatic body and arch.

the bone grafts usually dissolve over time. Attempts at using cranial bone on a temporalis muscle leash did not prove to retard this resorptive bone problem. In mature patients, silastic implants may be used to camouflage defects.

Craniofacial Dysostosis Syndromes

Patients with craniofacial dysostosis syndromes have craniosynostosis (premature cranial suture fusion) at birth, midface hypoplasia (and sometimes orbital deformities) in early childhood, and orthognathic malocclusions at maturity.[59] Correction for craniosynostosis involves the release of the affected suture and reshaping of the cranial vault. Healing of cranial defects is usually not a problem because of the strong osteogenic influence of the dura at a young age.[31] However, with an infection, entire bone grafts may be lost, leaving a problematic large cranial defect.[60] Often, secondary correction of such a defect must wait until the patient is older.

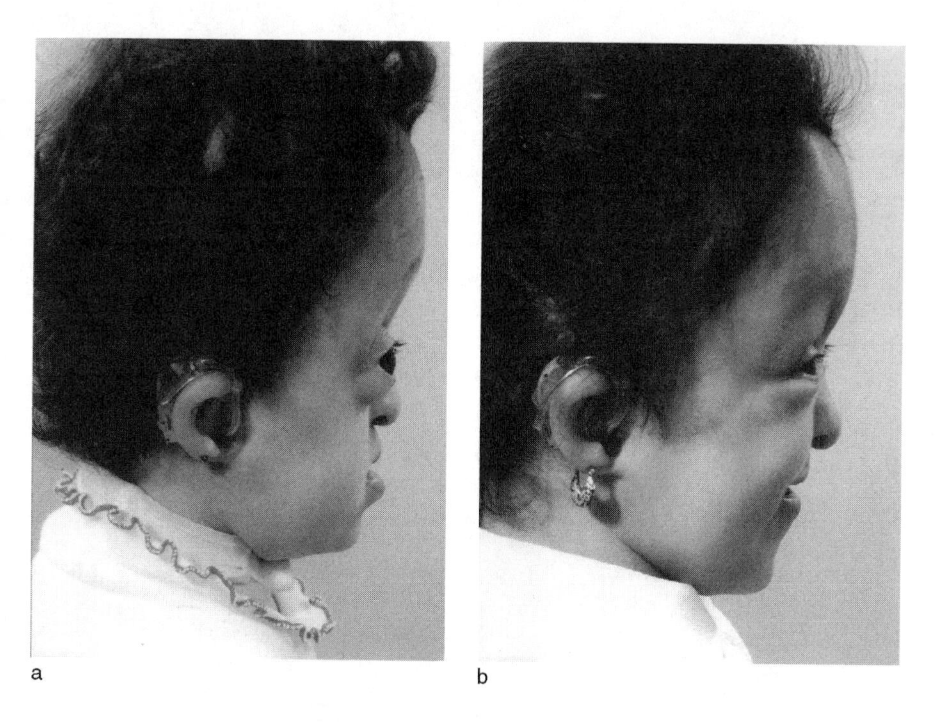

Figure 12.6 Girl with congenital facial deformity and midface hypoplasia who had undergone forehead and midface distraction osteogenesis (monobloc procedure). (a) Preoperative: ocular exposure is temporarily controlled by tarsorrhaphy (lateral eyelid closure). (b) Postoperative: the advancement allowed for improvement in her airway and removal of the tracheostomy.

For correction of midface hypoplasia in the growing patient, distraction osteogenesis is used.[24] With distraction osteogenesis, no bone grafts are needed. Also, gradual expansion of the soft tissues may allow for further advancement and less relapse. Disadvantages include additional procedures for device removal. A shorter consolidation time (faster bone repair) after completion of distraction would be an advantage. Introduction of appropriate mitogenic factors may lead to improvement (see Figure 12.6a, b).

Mandibular Anomalies

Correction of the abnormal mandible is necessary for airway obstruction, facial asymmetries, and malocclusion.[22,23,61] Distraction osteogenesis, with internal or external devices, is used for mandibular advancement in the growing patient. Distraction has been used to avoid a tracheostomy, remove a tracheostomy, or alleviate sleep apnea.[61] In addition, transport distraction osteogenesis has been used in temporomandibular joint reconstruction.[25] Unless a severe class II malocclusion (micrognathia) exists, traditional orthodontics followed by an orthognathic procedure is preferred in the mature patient.

Patients with severe micrognathia (Nager's syndrome) or ankylosis of the TMJ present a clinical challenge. Rib grafts for mandibular reconstruction in the patient with severe micrognathia are used but have unpredictable growth. With excessive rib graft growth, mandibular setbacks are necessary at maturity. Transport gap arthroplasty attemps to provide a cartilaginous cap for the neocondyle.[25] Yet, even with intensive physical therapy, ankylosis may recur.

SUMMARY

In conclusion, clinical challenges in craniofacial and dental surgery center around filling tissue deficits to allow for functional recovery. Tissue engineering may have distinct advantages over native tissue or prosthetic substitution, including growth with the patient, avoiding multiple surgeries from infant to maturity.[2,62]

References

1. McCarthy, J.G. (1999i). *Distraction of the Craniofacial Skeleton.* Springer-Verlag, Berlin.
2. Grikscheit, T.C. and Vacanti, J., P. (2002). The history and current status of tissue engineering: The future of pediatric surgery. *J. Pediatr. Surg.* 37, 277.
3. McCarthy, J.G. (1990). *Plastic Surgery*, W.B. Saunders, Philadelphia, 132.
4. Baker, D.C., Shaw, W.W. and Conley, J. (1980). Microvascular free dermis — fat flaps for reconstruction after ablative head and neck surgery. *Arch. Otolaryngol.* 106, 449.
5. Sharzer, L.A. et al. (1976). Intraoral reconstruction in head and neck cancer surgery. *Clin. Plast. Surg.* 3(3), 495–509.
6. Morais-Besteiro, J. et al. (1990). Microvascular flaps in head and neck reconstruction. *Head Neck.* 12, 21.
7. Constantino, P.D. (1994). Synthetic biomaterials for soft-tissue augmenation and replacement in the head and neck. *Otolaryngol. Clin. North Am.* 27, 223.
8. Schultz, R.C. (1981). Reconstruction of facial deformities with alloplastic material. *Ann. Plast. Surg.* 7, 434.
9. Fallopius, G. (1600), in *Opera omnia Francofurti*, Wecheli, A., Ed. 1.
10. Rubin, J.P. and Yaremchuk, M.J. (1997). Complications and toxicities of implantable biomaterials used in facial reconstructive and aesthetic surgery: a comprehensive review of the literature. *Plast. Reconstr. Surg.* 100, 1336.
11. Argenta, L.C. (1984). Reconstruction of the breast by tissue expansion. *Clin. Plast. Surg.* 11, 257.
12. Gibney, J. (1987). The long-term results of tissue expansion for breast reconstruction. *Clin. Plast. Surg.* 14, 509.
13. Kabaker, S.S. et al. (1986). Tissue expansion in the treatment of alopecia. *Arch. Otolaryngol. Head Neck Surg.* 112, 720.
14. Guzel, M.Z. et al. (2000). Aesthetic results of treatment of large alopecia with total scalp expansion. *Aesthet. Plast. Surg.* 24, 130.
15. Bauer, B.S. et al. (2001). The role of tissue expansion in the management of large congenital pigmented nevi of the forehead in the pediatric patient. *Plast. Reconstr. Surg.* 107, 668.
16. Antonyshyn, O. et al. (1988). Tissue expansion in head and neck reconstruction. *Plast. Reconstr. Surg.* 82, 58.
17. Radovan, C. (1984). Tissue expansion in soft-tissue reconstruction. *Plast. Reconstr. Surg.* 74, 482.

18. Evans, D. (2001). Topical negative pressure for treating chronic wounds: a systematic review. *Br. J. Plast. Surg.* 54, 238.
19. Fleck, T.M. et al. (2002). The vacuum-assisted closure system for the treatment of deep sternal wound infections after cardiac surgery. *Ann. Thorac. Surg.* 74, 1596.
20. Cro, C., et al. (2002). Vacuum assisted closure system in the management of entercutaneous fistulae. *Postgrad. Med. J.* 78, 364.
21. Clare, M.P. (2002). Experience with the vacuum assisted closure negative pressure technique in the treatment of non-healing diabetic and dysvascular wounds. *Foot Ankle Int.* 23, 896.
22. McCarthy, J.G. et al. (1992). Lengthening the human mandible by gradual distraction. *Plast. Reconstr. Surg.* 89, 1.
23. Polley, J.W. and Figueroa, A.A. (1997). Distraction osteogenesis: its application in severe mandibular deformities in hemifacial microsomia. *J. Craniofac. Surg.* 8, 422.
24. Chin, M. and Toth, B.A. (1997). Le Fort III advancement with gradual distraction using internal devices. *Plast. Reconstr. Surg.* 100, 819.
25. Stucki-McCormick, S.U. (1997). Reconstruction of the mandibular condyle using transport distraction osteogenesis. *J. Craniofac. Surg.* 8, 48.
26. Kawamoto, H.K. (1976). The kaleidoscopic world of rare craniofacial clefts: order out of chaos (Tessier classification). *Clin. Plast. Surg.* 3, 529.
27. Moss-Salentin, L. (1997). Melvin L Moss and the functional matrix. *J. Dent. Res.* 76, 181.
28. Vinocr, C.D. et al., Surgical management of aplasia cutis congenital. *Arch. Surg.* 111, 1160, 1976.
29. Beekmans, S.J.A. and Wiebe, M.J. (2001). Surgical treatment of aplasia cutis in the Adams–Oliver syndrome. *J. Craniofac. Surg. 0* 6, 569.
30. Abbott, R., et al. (1991). Aplasia cutis congenital of the scalp: Issues in its management. *Pediatr. Neurosurg.* 17, 182.
31. Greenwald et al. (2000). Biomolecular mechanisms of calvarial bone induction: immature versus mature dura mater. *Plast. Reconstr. Surg.* 105, 1382.
32. Hollinger, J.O. and Kleinschmidt, J.C. (1990). The critical size defect as an experimental model to test bone repair materials. *J. Craniofac. Surg.* 1, 60.
33. Gosain, A.K. (1999). Biomaterials in the face: benefits and risks. *J. Craniofac. Surg.* 10, 404.
34. Gosain, A.K. (1997). Hydroxyapatite cement paste cranioplasty for the treatment of temporal hollowing after cranial vault remodeling in a growing child. *J. Craniofac. Surg.* 8, 506.
35. Reddi, S.P. et al. (1999). Hydroxyapatite cement in craniofacial trauma surgery: indications and early experience. *J. Craniomaxillofac. Trauma* 5, 7.
36. Frodel Jr., J.L. (1999). The use of bone grafts in the upper craniomaxillofacial skeleton. *Facial Plast. Surg.* 15, 25.
37. Lemperle, S.M. (1998). Bony healing of large cranial and mandibular defects protected from soft-tissue interposition: a comparative study of spontaneous bone regeneration, osteoconduction, and cancellous autografting in dogs. *Plast. Reconstr. Surg.* 10, 660.
38. Jarcho, M. (1981). Calcium phosphate ceramic as hard tissue prosthetics. *Clin. Orthop.* 157, 259.
39. Gregory, C.F. (1972). *Clin. Orthop. Rel. Res.* 87, 156.
40. Enneking, W.F. and Mindell, E.R. (1991). *J. Bone Joint Surg.* 73, 1123.
41. McCarthy, J.G. (1997). Craniofacial microsomia. In *Grabb and Smith's Plastic Surgery*, 5 ed. Aston, S., Beasely, W. and Thorne, C.W., Eds. Lippincott, Williams and Wilkins, New York.

42. Firmin, F. (1998). Ear reconstruction in cases of typical microtia. Personal experience based on 352 mictoic ear corrections. *Scand. J. Plast. Reconstr. Surg. Hand Surg*. 32, 35.
43. Thorne, C.H. (2001). Auricualar reconstruction: indications for autogenous and prosthetic techniques. *Plast. Reconstr. Surg.* 107, 1241.
44. Nagata, S. (1993). A new method of total reconstruction of the auricle for microtia. *Plast. Reconstr. Surg.*, 92, 187.
45. Brent, B. (1999). *Plast. Reconstr. Surg*. 104, 319.
46. Cao, Y. M.D. et al. (1997). Transplantation of chondrocytes utilizing a polymer-cell construct to produce tissue-engineered cartilage in the shape of a human ear. *Plast. Reconstr. Surg.*, 100, 297.
47. Cohen, D. and Goitein, K. (1986). Arhinia. *Rhinology* 24, 287.
48. Hollinger, J.O. and Winn, S.R. (1999). Tissue engineering of bone in the craniofacial complex. *Ann. NY Acad. Sci.* 18, 379.
49. Burget, G.C. and Menick, F.J. (1986). Nasal reconstruction: seeking a fourth dimension. *Plast. Reconstr. Surg.* 78, 145.
50. Roumanas, E.D. et al. (2002). Implant-retained prostheses for facial defects: an up to 14-year follow-up report on the survival rates of implants at UCLA. *Int. J. Prosthodont*. 15, 325.
51. Millard, R. (1976). *Cleft Craft,* 1. Lippincott, Williams and Wilkins, New York, p. 102.
52. Linton, J.L. (1998). Comparative study of diagnostic measures in borderline surgical cases of unilateral cleft lip and palate and noncleft Class III malocclusions. *Am. J. Orthod. Dentofacial Orthop.* 113, 526.
53. Maull, D.J. (1999). Long-term effects of nasoalveolar molding on three-dimensional nasal shape in unilateral clefts. *Cleft Pal. Craniofac. J.* 36, 391.
54. Lehman, J.A. (1978). Closure of anterior palate fistulae. *Cleft Palate J*. 15, 33.
55. Nagao, H. et al. (2002). Effect of recombinant human bone morphogenetic protein-2 on bone formation in alveolar ridge defects in dogs. *Int. J. Oral Maxillofac. Surg.* 31, 66.
56. Rigamonti, U. et al. (2001). Periodontal tissue regeneration by combined applications of recombinant human osteogenic protein-1 and bone morphogenetic protein-2. A pilot study in Chacma baboons (*Papio ursinus*). *Eur. J. Oral Sci.* 109, 241.
57. Barboza, E.P. et al. (2000). Ridge augmentation following implantation of recombinant human bone morphogenetic protein-2 in the dog. *J. Peridontol.* 71, 488.
58. Cochran, D.L. et al. (1999). Recomninant human bone morphogenetic protein-2 stimulation of bone formation around endosseous dental implants. *J. Periodontol.* 70, 139.
59. Hunt, J.A. and Hobar, P.C. (2002). Common craniofacial anomalies: the facial dysostoses. *Plast. Reconstr. Surg*., 110, 1714.
60. Fialkov, J.A. et al. (2001). Postoperative infections in craniofacial reconstructive procedures. *J. Craniofac. Surg*. 12, 362.
61. Cohen, S.R. (1999). Alternatives to tracheostomy in infants and children with obstructive sleep apnea. *J. Pediatr. Surg*. 34, 182.
62. Vacanti, C.A. et al. (1993). Tissue-engineered growth of bone and cartilage. *Transplant Proc*. 25, 1019.

13 Opportunities in Spinal Applications

John M. Rhee and Scott D. Boden

Contents

INTRODUCTION

The normal human spine is made up of multiple, independent motion segments (vertebrae), which articulate via a three-joint complex (the disc and two facet joints). Although preservation of motion is physiologic and ideal, certain disease states of the spine require immobilization of these independent motion segments through a process known as spinal fusion. Approximately 200,000 spinal fusions are performed each year in the United States. Typically, indications for fusion include instability (e.g., spondylolisthesis, trauma), iatrogenically created instability (e.g., after laminectomy), correction of spinal deformity (e.g., scoliosis or kyphosis), reconstruction of spinal segments after the removal of tumors or infections, and pain due to arthritic conditions of the spine.

0-8493-1621-9/05/$0.00+$1.50
© 2005 by CRC Press LLC

Spinal fusion is a complex process in which the goal is to bridge bone across, and thus stop motion at, a normally mobile motion segment. As such, it presents a formidable environment, biomechanically and biologically, for healing. Traditionally, the biomechanical challenges have been addressed through the use of spinal instrumentation, which can impart rigid fixation and thus stability to the segments being fused. The biological challenges have been addressed primarily through the use of autologous bone graft, which is harvested from the iliac crest and transplanted to the intended levels of fusion to provide a biologic stimulus to healing. Autograft bone is the only currently available material that can provide the cells, growth and differentiation factors, and osteoconductive scaffold needed for fusion. Nevertheless, even though iliac crest autograft bone remains the gold standard grafting material, there remain two significant problems related to its use. First, there may be up to a 30% prevalence of complications[1] such as chronic donor site pain, infection, fracture, and hematomas—related to the harvest of autologous bone graft for spinal fusion. Second, a high percentage of patients still fail to achieve bony healing across the intended fusion levels despite the use of autograft bone. This condition, known as a nonunion or pseudarthrosis, can leave patients with persistent pain, instability, or deformity. Anywhere from 5% to 44% of patients who undergo posterolateral lumbar spinal fusions (the most common type of spinal fusion) may develop this problem.[2,3]

These inadequacies in outcomes associated with autologous bone grafting represent opportunities for improvement. Although in the past the focus was directed toward the biomechanical issues related to achieving fusion, recent advances in molecular biology have placed the impetus on solving the problems associated with spinal fusion from biological and bone tissue engineering approaches. The ultimate goals are to eliminate the need for autologous donor bone graft and thus its attendant morbidity while at the same time achieving 100% fusion rates. In this chapter, we review emerging technologies in bone tissue engineering applied to solving these critical problems associated with spinal fusion. In particular, the potential roles of bone morphogenetic proteins (BMPs) and gene therapy will be discussed.

THE FUNCTIONS OF BONE GRAFT

Bone graft and bone graft substitutes must be able to potentially serve three distinct functions during the process of fusion. First, the graft material should ideally be osteoinductive, defined as the ability to induce *de novo* bone formation at a nonbony site. The classic test for osteoinduction is the ability of the substance to produce *de novo* bone when implanted into a rat subcutaneous pouch model. Of the multiple cytokines involved with the many aspects of bone formation, only one family of proteins, the bone morphogenetic proteins (BMPs), is osteoinductive. Although desirable, the use of an osteoinductive graft is not always necessary or sufficient for fusion to occur. It is not necessary in all cases, because successful fusions can be achieved in the anterior cervical spine with freeze-dried allografts, which possess essentially no osteoinductivity, or even with coralline hydroxyapatite, which is a purely osteoconductive substance. In this setting, bone formation occurs by creeping substitution from cells provided by the surrounding decorticated host bone. Not all osteoinductive graft materials are sufficient for fusion. For example, demineralized bone matrix products have varying degrees of osteoinductivity but because they cannot predictably achieve fusion when used alone in the posterolateral spine, they are technically graft extenders rather than substitutes. Autograft, which includes live osteoblasts, is osteogenic, meaning it contains cells that can directly synthesize new bone matrix.

Second, the graft material should be osteoconductive, defined as the ability to provide a scaffold for new bone formation in a bony environment, even in the absence of implanted osteoinductive factors. Purely osteoconductive materials facilitate bone formation by acting as a matrix for bone ingrowth, but they lack osteogenic factors capable of inducing osteoblast proliferation and differentiation. Examples of osteoconductive materials include collagen, polymers, and various ceramics (defined as inorganic, nonmetallic materials that consist of metallic and nonmetallic elements bonded together primarily by ionic and/or covalent bonds; examples include hydroxyapatite, tricalcium phosphate, and calcium sulfate). Graft materials vary in their relative osteoconductive vs. osteoinductive potentials. Freeze-dried structural allografts are relatively osteoconductive but only weakly osteoinductive, whereas BMPs are highly inductive but by themselves not osteoconductive, because they are proteins and require a scaffold to support bone formation. Spinal fusion is thus the result of an intricate interplay between osteoinductive and osteoconductive factors.

Third, in certain fusion beds, it may be beneficial for the graft to possess structural mechanical properties. For example, when reconstructing the anterior column of the spine, a strut graft capable of replacing excised vertebral bodies or discs may be necessary. This last function does not, in contrast to the first two functions, require a biologically active bone graft material. Nonbiological materials, such as metal cages, can serve this purpose and be filled with bone graft or substitutes to fulfill the remaining two functions.

The ideal bone graft material should meet the following criteria. It should be plentiful, cost-effective, come in various shapes and sizes to accommodate a variety of fusion beds, have no risk of disease transmission, have no immunogenicity, avoid any donor site morbidity, and achieve 100% fusion success rates. None of the currently available bone grafting materials meets all of these criteria. As noted, autograft is associated with donor site morbidity and cannot provide 100% fusion rates. Allograft has the potential for disease transmission and is relatively lacking in osteoinductivity. BMPs are highly osteoinductive but are currently expensive and possess no structural properties. Furthermore, even though the conditions for successful fusion with BMP-2 have been studied for anterior lumbar interbody fusions using tapered cylindrical cages,[4] the same is not true for all fusion beds (e.g., posterolateral intertransverse fusion or anterior corpectomy constructs). Because no single graft material fulfills all of these criteria, combinations of different graft materials are commonly used to make up for deficiencies in each individual graft type—for example, structural allograft with cancellous iliac autograft, or threaded cages with recombinant BMP. Tissue-engineered biologic agents may be very useful in this regard as osteogenic factors in conjunction with osteoconductive and structural graft materials.

BONE MORPHOGENETIC PROTEINS

In 1965, an orthopedic surgeon, Dr. Marshall Urist, made a seminal discovery upon identifying what he called an "inducible bone forming substance" when he noted *de novo* bone formation after the implantation of purified bovine demineralized bone matrix into a rat subcutaneous pouch.[5] This active bone-forming substance within the bone matrix was subsequently purified and named "bone morphogenetic protein". BMPs are now known to comprise a number of related proteins belonging to the transforming growth factor beta (TGF-β) superfamily of molecules,[6] which in turn consist of a variety of growth and differentiation factors related by primary amino acid sequence. BMPs are highly conserved, with a high degree of amino acid sequence homology among species as diverse as fruitflies and humans. For example, the decapentaplegic protein of *Drosophila melanogaster*, which is important for proper limb formation and establishment of the dorsal–ventral axis, shares many sequence similarities with human BMP-2 and BMP-4.[7]

BMPs play a critical role in human embryonic development, from pattern formation to tissue development. In animal studies, the expression of various BMPs has been demonstrated in skeletal repair states such as fracture healing[8] and spinal fusion.[9] Although BMPs are colloquially referred to as being "growth factors," they are in fact differentiation factors rather than mitogens (cf., platelet-derived growth factor (PDGF) or TGF-β itself). BMPs have been demonstrated to cause differentiation of stem cells along the osteoblastic lineage *in vitro*. BMPs can also induce cartilage formation. The proportion of intramembranous versus endochondral bone formation achieved experimentally depends on numerous factors, including the concentration of BMP, the carrier, species, and the site of implantation (Table 13.1).

EFFECT OF BMP CONCENTRATION

The concentration of BMP applied to the fusion bed has a major effect on its osteoinduction. The most important predictor for the efficacy of BMP in forming bone is the concentration of BMP expressed in milligrams of protein per volume of matrix implanted. Below a threshold dose, no bone will be formed. Above that threshold, consistent bone formation is observed. However, the threshold differs depending on the BMP type, the carrier used, the anatomic location (i.e., anterior interbody vs.

Table 13.1
Factors Affecting BMP Activity *In Vivo*

Concentration	A threshold dose exists that is specific to species, site, carrier, and host healing potential
Carrier	Should be biocompatible, maintain a space for fusion to occur, provide for controlled release and retention of BMP, and be resorbed as fusion progresses
Species	The osteogenic efficacy of BMP in one species cannot be assumed in another
Site	Posterior fusion beds are more challenging environments for healing than anterior fusion beds

posterolateral), species, and factors that affect host osteogenesis such as smoking, diabetes, use of steroids, or chemotherapy.

There is a large discrepancy between the physiologic concentration of BMP found in bone and that required to achieve spinal fusion with commercially available products. The physiologic concentration in bone is approximately 0.002 mg/kg of bone,[10] but that required to achieve spinal fusion is many orders of magnitude higher. Several potential explanations for this discrepancy exist. First, locally applied BMP diffuses away from the site of application and is rapidly degraded. Thus, high implantation doses of BMP may be necessary to provide sufficient local concentrations for a long enough period of time to stimulate osteogenesis in responsive cells. Second, inhibitors of BMP, such as *chordin* and *noggin*, exist *in vivo* that limit the activity of BMP to regions that are physiologically useful, such as sites of fracture healing.[11–13] Since spinal fusion is not a physiologic process, high doses of BMP may be needed to overcome these natural inhibitors. Third, in the normal process of bone growth and repair, a cascade of BMPs and growth factors are likely involved in a specific temporal and spatial sequence that is regulated. Although a single BMP can have the same osteogenic activity as the combined effect of all of the factors present in the bone matrix, larger concentrations may be necessary under these non-physiologic circumstances.

Effect of Carrier

Although a number of growth and differentiation factors are involved in the normal process of osteogenesis, BMPs are the only known factors capable of independently initiating the entire cascade of *de novo* bone formation.[14–18] Despite this unique ability of the BMPs when tested in model systems such as the rat subcutaneous pouch, the actual clinical efficacy of BMP in promoting spinal fusion is critically dependent on the carrier that is used. The overwhelming importance of the carrier has been demonstrated in numerous studies, including in a nonhuman primate model of posterolateral intertransverse process fusion. When rhBMP-2 was delivered on a collagen sponge in this model, fusion rates were low, presumably because the paravertebral muscles compressed the sponge, overtaking the space available for fusion, and the sponge resorbed too quickly.[19] When a polyethylene barrier was placed dorsal to the sponge to prevent muscle-related compression, fusion rates improved. Furthermore, when bulking agents, such as allograft chips or ceramic granules, were added to the collagen sponge carrier, a 100% fusion rate was achieved.[20] Studies such as these have highlighted one important role of the carrier: maintaining a space for bone formation to occur.

In addition to creating a space for osteoconduction, the ideal carrier of BMP should also possess the following characteristics. First, it should be biocompatible and nonimmunogenic to avoid an excessive immune response that might hinder fusion. Second, the carrier should provide for the controlled release of BMP into the fusion bed to create sufficient concentrations of BMP to initiate osteoinduction when the responsive cells arrive. Third, the carrier should allow for the controlled retention of BMP so that it can affect responsive cells for a long enough time frame to ensure fusion, rather than allowing all of the BMP to be released at once. Fourth, the carrier should ideally be resorbed at a rate equal to that of bone formation. If resorption occurs too quickly, pseudoarthrosis might occur. On the other hand, if resorption occurs too slowly, bone ingrowth may be hindered, leading again to pseudoarthrosis or, alternatively, to weakening of the newly formed bone. Finally, the ideal carrier would localize the BMP to the site of fusion to prevent unwanted ossification of nearby structures.

Just as the ideal graft material is not currently available, identifying the ideal carrier remains elusive. There are four major types of carriers: allograft bone, natural polymers (e.g., hyalurans, fibrin, chitosan), synthetic polymers (e.g., polylactic acid, polyglycolide), and inorganic materials (e.g., hydroxyapatite, tricalcium phosphate, calcium phosphate cements). Each carrier type has its own set of advantages and disadvantages. The inorganic materials like hydroxyapatite, for example, retain BMPs well and thus provide for prolonged delivery of osteoinductive factors to the fusion bed. However, their sustained existence at the site of osteogenesis may exert a negative effect, namely, weakening of the newly formed bone.[21] Collagen-based natural matrices, such as demineralized bone matrix, are advantageous in that they are naturally present in bone. However, these carriers can be immunogenic and may potentially transmit disease.[22] Allografts are made of human bone but have the potential for disease transmission and incorporate very slowly. Finally, synthetic polymers avoid the potential problem of disease transmission, but their breakdown products can be associated with inflammatory responses.[23] One way to overcome the inherent limitations in any one carrier is to use composite carriers composed of different carrier types. Ceramic and collagen composites, for example, have been used in animal models of posterolateral fusion.[24]

Effect of Species and Site

In addition to the effects of carrier and concentration, the species and the anatomic site in which bone formation occurs are major determinants of BMP efficacy. It is generally accepted that rodents provide more permissive environments for bone formation. This may be related to qualitative or quantitative differences in stem cells or rates of bone formation among species. For example, a given osteoinductive agent at a certain dose may be sufficient to induce spinal fusion in a rabbit at one dose, but that same dose may be unable to consistently do the same in a nonhuman primate or human model. Therefore, it cannot be assumed that the osteogenic efficacy of BMP at any specific dose in a lower animal model will directly translate to a higher animal model.

Even within a single species, different healing environments are associated with progressive levels of difficulty in forming bone. A metaphyseal defect, for example, is more permissive than a long bone fracture, which is in turn more permissive than an anterior interbody spine fusion model, which is in turn more permissive than a posterolateral spine fusion model. There are several potential explanations for the observed differences between the anterior and posterior spine. The anterior spine is typically under compression and consists of relatively large, cancellous surface areas for healing. The posterior spine, in contrast, may be either loaded in neutral or under tension. In addition, especially after laminectomy, smaller surface areas remain for healing posteriorly (i.e., the transverse processes and lateral pars), and these surfaces are separated by relatively larger distances. It therefore cannot be assumed that the success of BMP in one environment can be guaranteed in another.

Sources of BMP

BMPs are currently available from two sources: extracts and recombinant. Extracts, such as NeOsteo (Centerpulse Biologics, Austin, TX), are extracted from bovine or other bone sources. Extracts contain a mixture of differentiation and mitogenic factors. NeOsteo, for example, contains a mixture of BMP-2, -3, -4, -5, -6, and -7. In addition, it contains TGF-β1, TGF-β2, TGF-β3, fibroblast growth factor-1 (FGF-1), and type I collagen. One advantage of BMP extracts is that a variety of osteogenic factors are present that may more closely mimic the physiologic situation involved in bone formation. Another advantage of extracts may be the presence of BMP heterodimers. BMPs naturally exist as dimers: both homodimers (composed of two identical subunits) and heterodimers (composed of two different subunits). Evidence suggests that some heterodimers are more biologically active than homodimers.[25] On the other hand, BMP extracts also possess some disadvantages. These include the potential for disease transmission, presence of inhibitors of osteogenesis normally present within the bone matrix, and relatively low concentrations of BMP. Variations in BMP content of the starting raw material can also lead to variability in the BMP concentration of the final product.

The other source of commercially available BMP comes through the use of recombinant DNA technology. Recombinant human BMP-2 (rhBMP-2; Medtronic Sofamor Danek, Memphis, TN) and recombinant human BMP-7 (rhBMP-7 — also known as OP-1; Stryker Biotech, Hopkinton, MA) are manufactured through this process. The cDNA of the BMP is placed in a vector, which is then transfected into a mammalian cell line, such as the Chinese hamster ovary (CHO). The BMP sequence is amplified through the use of selectable markers in order to produce multiple copies within the cell. The cells are then expanded in culture and are allowed to produce BMP protein. Because the cells are mammalian, they prepare and process the BMP as a final product in the form of fully functional, glycosylated dimers, just as is normally done *in vivo*. The recombinant BMP protein is then purified from the medium. A major advantage of recombinant BMP is the ability to reproducibly generate large quantities of a single, highly purified BMP with consistent biological activity. One disadvantage is that heterodimers, which may be more biologically potent, are currently difficult to manufacture through recombinant technology because cells must be made to produce equal amounts of each subunit, and it is difficult to isolate heterodimers because they share chemical properties similar to their homodimer counterparts.

Current Uses of BMP

BMP-2

BMP-2 was isolated and cloned in 1988.[6] It is currently commercially available as rhBMP-2 (Infuse; Medtronic Sofamor Danek, Memphis, TN) and is approved by the

Food and Drug Administration (FDA) as a bone graft substitute for use in anterior lumbar interbody fusions with a tapered cylindrical cage for the surgical management of degenerative disc disease. A prospective, randomized, clinical trial comparing rhBMP-2 with autograft in this setting has demonstrated equivalence or better with autogenous iliac crest bone graft in terms of patient outcomes and radiographic fusion.[4] One hunderd and forty-three patients received rhBMP-2 and 136 received autograft bone from the iliac crest. The clinical fusion success at 24 months was 94.5% for the rhBMP-2 groupvs. 88.7% for the autograft group. The radiographic fusion success as measured by computed tomography (CT) scan was over 99% in both groups. The average back pain scores were slightly lower at 24 months with rhBMP-2 vs. autograft. Thirty-two percent of the autograft group reported some degree of donor site pain at 24 months; 66.1% of the rhBMP-2 group were back to work at 24 months, compared to 56.1% in the autograft group. Blood loss and operative times were also less with rhBMP-2.

Currently, rhBMP-2 is not yet FDA approved or proven efficacious for use in posterolateral lumbar spinal fusions, although a pilot study has been performed.[26] Numerous, promising preclinical animal studies of posterolateral lumbar fusion have demonstrated improved fusion rates with rhBMP-2 vs. autograft bone in species ranging from rabbits to monkeys,[20,27–29] but these results cannot be extrapolated to humans for the reasons noted previously. BMP-2 usage in cervical spine fusions has not been extensively studied. In one prospective clinical trial assessing the use of rhBMP-2 in one- and two-level anterior cervical discetomies and fusions, 33 patients were randomized to undergo operative treatment and fusion with either fibular allograft bone filled with rhBMP-2 or filled with cancellous autograft bone obtained from the iliac crest. The results showed no major advantage of using rhBMP-2 in these patients.[30] At 6 months, fusion was achieved in all patients. At 12 months, neck pain scores were similar in all groups (except those who underwent two-level arthrodesis with autograft; these patients had higher pain scores), and donor site pain was negligible in the autograft group. Although it is possible that BMP-2 will be demonstrated to have efficacy in cervical and posterolateral lumbar fusions, further study is necessary.

BMP-7 (OP-1)

OP-1 has similar osteoinductive effects as BMP-2 *in vitro.*[10] Also, as with rhBMP-2, preclinical studies have shown the efficacy of rhBMP-7 (OP-1) as a bone graft substitute in animal models of anterior and posterior spinal fusion.[31–35] Early human trials have demonstrated promising but in conclusive results with the use of OP-1 for posterolateral spine fusion. In a study of 16 patients with spinal stenosis and degenerative spondylolisthesis randomized to treatment by laminectomy and uninstrumented fusion with either OP-1 plus autograft or autograft alone, fusion rates were 75% in the OP-1 group compared to 50% in the autograft alone group at 6 months.[36] A 20% improvement in the Oswestry score (a measure of disability related to lumbar spine disorders) was achieved in 83% of the OP-1 group compared to 50% of the controls. Because of the relatively small numbers of patients, however, statistical significance was not achieved for either the clinical or radiographic outcomes.

Although larger numbers and further follow-up may demonstrate otherwise, the inability of the study thus far to demonstrate fusion rates in the high 90–100% range has tempered enthusiasm for OP-1. Furthermore, this study examined the more limited role of OP-1 as an autograft enhancer rather than as a substitute.

In a separate pilot study, OP-1 has been examined as a bone graft substitute in posterolateral lumbar fusions. Thirty-six patients with lumbar spinal stenosis and degenerative spondylolisthesis were treated with decompression and uninstrumented fusion with either OP-1 or iliac autograft bone.[37] Clinical success, defined as a 20% reduction in Oswestry score and a solid fusion, was 32% higher in the OP-1 group compared to the autograft control at 6 months. However, this difference did not achieve statistical significance.

Bovine BMP Extract (bBMPx; NeOsteo)

Bovine BMP extract (NeOsteo; Centerpulse Biologics, Austin, TX) contains a mixture of BMPs and other factors extracted from bovine cortical bone through a chemical purification process. The exact composition of the finished product depends on that of the original raw materials used, and thus may vary somewhat. A dose–response effect of bBMPx has been shown in a preclinical nonhuman primate model of posterior intertransverse fusion.[38] Animals received either monkey DBM plus iliac autograft or DBM plus varying concentrations of bBMPx. At 18–24 weeks postoperatively in this challenging fusion model, 54% of animals receiving the highest dose of bBMPx achieved fusion compared to 21% of those receiving autograft. A threshold dose of bBMPx between 1.5 and 3.0 mg per side was required to obtain fusion in any animals, and a dose of 25 mg per side was required for 100% successful fusions. A human posterolateral lumbar fusion pilot trial has been performed in Europe and demonstrated successful fusions at the 25mg per side dose.[38]

Safety of BMP

A number of theoretical safety issues arise with the use of BMP in spinal fusion. These include uncontrolled bone formation, causing neural impingement or unwanted ossification of surrounding tissues or adjacent segments, carcinogenicity, systemic toxicity, and immune responses. Animal studies, including those in nonhuman primates, have failed to demonstrate recurrent stenosis secondary to bony regrowth over laminectomy sites when BMP is used to achieve concomitant posterolateral arthrodesis.[19,39,40] The safety of using BMP in the presence of a dural tear is unclear, but may depend on the nature of the durotomy or the type of BMP that is used. In a canine model, the direct application of rhBMP-2 with a collagen sponge to a pinhole durotomy did not cause intradural bone formation or neurologic compromise.[41] In a more extreme experimental situation, however, OP-1 with a type I collagen carrier did cause intradural ossification when implanted into the subdural space with subsequent closure of the durotomy over the implanted OP-1.[42] Local adverse effects on surrounding tissues have not been noted in human trials using either rhBMP-2 or OP-1.

BMPs are known to be expressed in certain tumors, such as malignant bone tumors,[43] pancreatic cancer cells,[44] gastric cancer cell lines,[45] and prostate cancer cells.[46] However, there is no evidence to suggest that BMPs are carcinogenic. As mentioned previously, BMPs are differentiation factors rather than mitogens. In addition, human and animal studies have not demonstrated any systemic toxicity, such as disseminated bone formation. BMP-2 has no known systemic effects. BMP-7 (OP-1) does have systemic effects, including a possible therapeutic role in the treatment of acute renal failure.[47] In the case of rhBMP-2, the lack of carcinogenic effect and systemic toxicity may be due to its rapid systemic clearance.[48] Immune responses have not been problematic in human trials thus far. In a randomized clinical trial of anterior lumbar interbody fusion, 0.7% of the patients receiving rhBMP-2 demonstrated anti-rhBMP-2 antibodies at 3 months postoperatively, compared to 0.8% of those receiving autologous iliac crest bone.[4] Overall, the rate of transient antibody response is about 5%, but has not affected bone induction results.

It appears, therefore, that although the potential for iatrogenic complications exists with the use of BMP, these have not been clinically evident so far. Nevertheless, because long-term (5 years or greater) follow-up is not available in human trials, the specter of safety issues must be kept in mind and respected. Care must be taken to ensure that BMP is placed accurately and contained in the area of fusion. Also, the surgeon should be aware that using BMP in the face of a durotomy may have deleterious consequences. Finally, a safety profile must be established for each BMP/carrier configuration prior to clinical use.

GENE THERAPY

An alternative strategy for delivering osteogenic proteins to a fusion mass involves the use of local gene therapy. Rather than directly placing the osteogenic protein into the fusion bed, local gene therapy involves delivery of the gene to the osteogenic protein. This is achieved by inserting the gene of interest (i.e., the transgene) into a capable cell, which then transcribes and translates the transgene, and secretes its product into the local environment to promote bone formation. Currently, work on gene therapy for spinal fusion remains preliminary. Preclinical animal studies have provided inconsistent results.[49,50] Host immune responses associated with adenoviral vector usage remain one major limiting factor. Nevertheless, gene therapy approaches remain theoretically attractive and may become successful in the future with further advances in bone tissue engineering.

The main advantage over direct protein delivery is that gene therapy can allow for a more sustained, high-level expression of the osteogenic protein in the fusion bed. With the direct application approach, the carrier must be relied upon to provide controlled retention and release of the factor. The factor must also be placed in large amounts so that concentrations sufficient for inducing osteogenesis are present, despite diffusion away from the fusion bed or local breakdown of the factor. However, with gene therapy, cells containing the gene of interest can provide sustained local expression. In addition, the level and timing of transgene expression can potentially be manipulated through the use of inducible vectors. As more is learned about the biology of spinal fusion, the ability to time the delivery of osteogenic factors like BMPs to match a critical state of readiness among the cells in the fusion bed may lead to higher, faster fusion rates and provide a more elegant alternative to the current approach of simply overloading the area with BMP.

Viral vs. Nonviral Vectors

The transgene can be delivered to target cells by either viral or nonviral vectors (Table 13.2). Viral vectors can be highly efficient in their ability to transduce, or deliver the transgene, to the target cell. They are also advantageous as vectors for genes requiring a fairly high level of expression for a prolonged timeframe. Viral vectors include adenoviruses, retroviruses, adeno-associated viruses, and herpes viruses. In order to prevent uncontrolled infection by the virus, the vector is usually rendered replication deficient by removing specific viral genes necessary for replication.

Each type of viral vector possesses advantages and disadvantages. Adenoviruses are double-stranded DNA viruses that bind to cell surface receptors, are endocytosed into the cell, then release their DNA into the cytoplasm. Adenoviruses have an extrachromosomal life cycle, meaning that they do not incorporate their genes into the genome of the transfected cell. As a result, insertional mutagenesis of the transfected cell is unlikely. However, the overall transgene expression may be diminished because of the immune response directed against the transfected cell due to the immunogenicity associated with the expression of viral genes.[49] Furthermore, because they do not integrate into the cellular genome, the duration of gene expression is shorter than with the use of retroviruses and occurs only for the life cycle of the transfected cell, as their DNA is not passed on to the progeny cells.

In contrast to adenoviruses, integrating viruses include retroviruses, which are single-stranded RNA viruses. After being taken up by the target cell, retroviral reverse transcriptase produces a double-stranded DNA copy, which then integrates into the target cellular genome during cell division. The risk of insertional mutagenesis in the target cell is higher than with the use of nonintegrating vectors. Because of its life cycle, retroviruses must be transduced into dividing cells in order to express the desired transgene. Adeno-associated viruses are another example of integrating viruses. Unlike other conditions for which gene therapy is being investigated—such as muscular dystrophy or cystic fibrosis—where the transgene must be

Table 13.2
Vectors Used for Gene Therapy

Type	DNA vs. RNA	Pros	Cons
Extrachromosomal viral (e.g., adenovirus)	DNA	Minimal risk of insertional mutagenesis	Host immune response Shorter expression times
Integrating viral (e.g., retrovirus)	RNA	Longer expression times	Risk of insertional mutagenesis Must be transduced into dividing cells
Plasmids	DNA	Safe, nonviral	Less efficient transduction Shorter expression times

expressed in all of the affected cells for the lifetime of the patient, requirements for BMP expression during spinal fusion are temporary. Therefore, the risk associated with the use of integrating viruses as vectors may outweigh the benefits for spinal fusion applications.

Transgenes can also be inserted into target cells through nonviral means. Plasmids containing the transgene can be transduced into target cells, but the efficiency of transduction is less than with viral vectors and expression times are shorter than with integrating viral vectors. A major advantage of nonviral vectors is their safety. If the transgene encodes an intranuclear factor such as LIM mineralization protein-1 (LMP-1),[51] plasmid vectors may be sufficient. LMP-1 is an intracellular protein that acts via secretion of soluble osteoinductive factors, which subsequently induce BMPs and BMP receptor expression. Thus, LMP-1 requires a relatively short expression time because it possesses an amplifying effect, in that its primary function is to start a cascade of bone forming events in *other* cells rather than directly inducing osteoblast differentiation itself. This mechanism is in contrast to that of BMP, which needs to be expressed when cells capable of responding are present. A protocol using low-dose adenovirus to deliver the LMP-1 *ex vivo* to autologous leukocytes was successful for spine fusion in immune competent adult rabbits.[52] The desired gene can be transfected directly *in vivo* or *ex vivo*. With *ex vivo* techniques, the transgene is transduced into cells while outside the body, which are subsequently implanted into the host. Both approaches have been attempted in preclinical studies. Direct (*in vivo*) gene therapy with injection of Ad-BMP-2 into mouse triceps muscle has been shown to induce osteogenesis.[53] In nude mice, the same approach also induced bone formation but failed to achieve spinal fusion when injected into paraspinal musculature.[54] *Ex vivo* attempts using marrow-derived stem cells in rabbits transfected with Ad-BMP-2 demonstrated bone formation, but this occurred in only one out of five animals, and no animal successfully fused.[50] These and other studies have demonstrated the proof of concept that local gene therapy with BMP is capable of inducing osteogenesis. However, they have yet to demonstrate the feasibility of consistently achieving spinal fusion with gene therapy.

CONCLUSIONS

Spinal fusion is a complex biomechanical and biological process that remains challenging to consistently achieve. Despite significant advances in surgical technique and spinal instrumentation, nonunions still occur. Furthermore, donor site morbidity associated with the procurement of autograft remains problematic. Bone tissue engineering advances have led to the successful clinical use of BMPs. However, many questions and problems remain even with the advent of BMP, such as the need for large concentrations, the search for the ideal carrier, and the appropriate use of BMP in the challenging healing environment of the posterolateral spine. Gene therapy approaches may be beneficial in the future in addressing some of these issues, but this technology currently remains to be perfected. Much work in the field of bone tissue engineering needs to be done before the ultimate goals of spinal fusion—that is, achieving 100% fusion rates with no donor site morbidity—are reached.

References

1. Arrington, E.D. et al. (1996). Complications of iliac crest bone graft harvesting. *Clin. Orthop.* 329: 300.
2. DePalma, A.F. and Rothman, R.H. (1968). The nature of pseudarthrosis. *Clin. Orthop.* 59: 113.
3. Steinmann, J.C. and Herkowitz, H.N. (1992). Pseudarthrosis of the spine. *Clin. Orthop.* 284: 80.
4. Burkus, J.K. et al. (2002). Anterior lumbar interbody fusion using rhBMP-2 with tapered interbody cages. *J. Spinal Disord. Tech.* 15:337.
5. Urist, M.R. (1965). Bone: formation by autoinduction. *Science* 150:893.
6. Wozney, J.M. et al. (1988). Novel regulators of bone formation: molecular clones and activities. *Science* 242:1528.
7. Wozney, J.M. (2002). Overview of bone morphogenetic proteins. *Spine* 27:S2.
8. Bostrom, M.P. et al. (1995). Immunolocalization and expression of bone morphogenetic proteins 2 and 4 in fracture healing. *J. Orthop. Res.* 13:357.
9. Morone, M.A. et al. (1998). The Marshall R. Urist Young Investigator Award. Gene expression during autograft lumbar spine fusion and the effect of bone morphogenetic protein 2. *Clin. Orthop.* 351:252.
10. Wang, E.A. et al. (1990). Recombinant human bone morphogenetic protein induces bone formation. *Proc. Natl. Acad. Sci. USA* 87:2220.
11. Abe, E. et al. (2000). Essential requirement of BMPs-2/4 for both osteoblast and osteoclast formation in murine bone marrow cultures from adult mice: antagonism by noggin. *J. Bone Miner. Res.* 15:663.
12. Aspenberg, P., Jeppsson, C. and Economides, A.N. (2001). The bone morphogenetic proteins antagonist Noggin inhibits membranous ossification. *J. Bone Miner. Res.* 16:497.
13. Reddi, A.H. (2001). Interplay between bone morphogenetic proteins and cognate binding proteins in bone and cartilage development: noggin, chordin and DAN. *Arthritis Res.* 3:15.
14. Boden, S.D. (1999). Bioactive factors for bone tissue engineering. *Clin. Orthop.* S84.
15. Boden, S.D. (2001). Clinical application of the BMPs. *J. Bone Joint Surg. Am.* 83-A (Suppl 1):S161.
16. Boden, S.D. Schimandle, J.H. (1995). Biologic enhancement of spinal fusion. *Spine* 20:113S.
17. Boden, S.D. and Sumner, D.R. (1995). Biologic factors affecting spinal fusion and bone regeneration. *Spine* 20:102S.
18. France, J.C. et al. (1999). A randomized prospective study of posterolateral lumbar fusion. Outcomes with and without pedicle screw instrumentation. *Spine* 24:553.
19. Martin et al., G.J., Jr. (1999). Posterolateral intertransverse process spinal arthrodesis with rhBMP-2 in a nonhuman primate: important lessons learned regarding dose, carrier, and safety. *J. Spinal Disord.* 12:179.
20. Akamaru, T. et al. (2003). Simple carrier matrix modifications can enhance delivery of recombinant human bone morphogenetic protein-2 for posterolateral spine fusion. *Spine,* 28:429.
21. Bruder, S.P. et al. (1998). The effect of implants loaded with autologous mesenchymal stem cells on the healing of canine segmental bone defects. *J. Bone Joint Surg. Am.* 80:985.
22. Sampath, T.K. and Reddi, A.H. (1983). Homology of bone-inductive proteins from human, monkey, bovine, and rat extracellular matrix. *Proc. Natl. Acad. Sci. USA* 80:6591.

23. Brekke, J.H. (1996). A rationale for delivery of osteoinductive proteins. *Tissue Eng.* 2:97.
24. Minamide, A. et al. (2001). Evaluation of carriers of bone morphogenetic protein for spinal fusion. *Spine* 26:933.
25. Israel, D.I. et al. (1996). Heterodimeric bone morphogenetic proteins show enhanced activity *in vitro* and *in vivo*. *Growth Factors* 13:291.
26. Luque, E. (2000). Latest clinical results using demineralized bone materials and rhBMP-2: The Mexican experience. In *Total Spine: Advanced Concepts and Constructs*. North American Spine Society, Cancun, Mexico, p. 114.
27. Sandhu, H.S. et al. (1995). Evaluation of rhBMP-2 with an OPLA carrier in a canine posterolateral (transverse process) spinal fusion model. *Spine* 20:2669.
28. Boden, S.D. et al. (1996). Video-assisted lateral intertransverse process arthrodesis. Validation of a new minimally invasive lumbar spinal fusion technique in the rabbit and nonhuman primate (rhesus) models. *Spine* 21:2689.
29. Schimandle, J.H., Boden, S.D., and Hutton, W.C. (1995). Experimental spinal fusion with recombinant human bone morphogenetic protein-2. *Spine* 20:1326.
30. Baskin, D.S. et al. (2002). ACDFP with cornerstone-SR (allograft and plate: rhBMP-2 vs. autograft). Presented at the 18th Annual Meeting of the American Association of Neurological Surgeons and Congress of Neurological Surgeons Section on Disorders of the Spine and Peripheral Nerves. February 27 to March 2, Orlando, FL, p. 26.
31. Patel, T.C. Erulkar, J.S. and Grauer, J.S. (2000). OP-1 Overcomes the Inhibitory Effects of Nicotine on Lumbar Fusion. New Orleans, LA, NASS.
32. Cunningham, B.W. et al. (2000). Posterolateral Spinal Arthrodesis Using Osteogenic Protein-1: An in vivo time-course study using a canine model. New Orleans, LA, NASS.
33. Grauer, J.N. et al. 2000. Young Investigator Research Award winner. Evaluation of OP-1 as a graft substitute for intertransverse process lumbar fusion. *Spine* 26:127.
34. Cook, S.D. et al. (1994). *In vivo* evaluation of recombinant human osteogenic protein (rhOP-1) implants as a bone graft substitute for spinal fusions. *Spine* 19:1655.
35. Cunningham, B.W. et al. (1999). Osteogenic protein versus autologous interbody arthrodesis in the sheep thoracic spine. A comparative endoscopic study using the Bagby and Kuslich interbody fusion device. *Spine* 24:509.
36. Patel, T.C. et al. (2001). A safety and efficacy study of OP-1 (rhBMP-7) as an adjunct to posterolateral lumbar fusion. North American Spine Society 16th annual meeting, Seattle, WA.
37. Patel, T.C. et al. (2001). A pilot safety and efficacy study of OP-1 (rhBMP-7) in posterolateral lumbar fusion as a replacement for the iliac crest autograft. The North American Spine Society 16th Annual Meeting, Seattle, WA.
38. Damien, C.J. et al. (2002). Purified bovine BMP extract and collagen for spine arthrodesis: preclinical safety and efficacy. *Spine* 27:S50.
39. Boden, S.D. et al. (1999). Posterolateral lumbar intertransverse process spine arthrodesis with recombinant human bone morphogenetic protein 2/hydroxyapatite-tricalcium phosphate after laminectomy in the nonhuman primate. *Spine* 24:1179.
40. David, S.M. et al. (1995). Lumbar spinal fusion using recombinant human bone morphogenetic protein-2(rhBmP-2): a randomized, blinded and controlled study. *Trans Int Soc Study Lumbar Spine*. International Society for the Study of the Lumbar Spine.
41. Meyer Jr., R.A., et al. (1999). Safety of recombinant human bone morphogenetic protein-2 after spinal laminectomy in the dog. *Spine* 24:747.
42. Paramore, C.G. et al. (1999). The safety of OP-1 for lumbar fusion with decompression—a canine study. *Neurosurgery* 44:1151.
43. Laitinen, M. et al. (1997). Measurement of total and local bone morphogenetic protein concentration in bone tumours. *Int. Orthop.* 21:188.

44. Kleeff, J. et al. (1999). Bone morphogenetic protein 2 exerts diverse effects on cell growth in vitro and is expressed in human pancreatic cancer *in vivo*. *Gastroenterology* 116:1202.
45. Katoh, M. and Terada, M. (1996). Overexpression of bone morphogenic protein (BMP)-4 mRNA in gastric cancer cell lines of poorly differentiated type. *J. Gastroenterol.* 31:137.
46. Barnes, J. et al. (1995). Bone morphogenetic protein-6 expression in normal and malignant prostate. *World J. Urol.* 13:337.
47. Vukicevic, S. et al. (1998). Osteogenic protein-1 (bone morphogenetic protein-7) reduces severity of injury after ischemic acute renal failure in rat. *J. Clin. Invest.* 102:202.
48. Poynton, A.R. and Lane, J.M.: (2002). Safety profile for the clinical use of bone morphogenetic proteins in the spine. *Spine* 27:S40.
49. Boden, S.D. et al. (2000). Gene therapy for spine fusion. *Clin. Orthop.* 379 Suppl:S225.
50. Riew, K.D. et al. (1998). Induction of bone formation using a recombinant adenoviral vector carrying the human BMP-2 gene in a rabbit spinal fusion model. *Calcif. Tissue Int.* 63:357.
51. Boden, S.D. et al. (1998). LMP-1, a LIM-domain protein, mediates BMP-6 effects on bone formation. *Endocrinology* 139:5125.
52. Viggeswarapu, M. et al. (2001). Adenoviral delivery of LIM mineralization protein-1 induces new-bone formation *in vitro* and *in vivo*. *J. Bone Joint Surg. Am.* 83-A:364.
53. Musgrave, D.S. et al. (1999). Adenovirus-mediated direct gene therapy with bone morphogenetic protein-2 produces bone. *Bone* 24:541.
54. Alden, T.D. et al. (1999). Percutaneous spinal fusion using bone morphogenetic protein-2 gene therapy. *J. Neurosurg.* 90:109.

Index